THEORY OF ELECTRIC POLARIZATION

VOLUME I

Dielectrics in static fields

THEORY OF ELECTRIC POLARIZATION

C. J. F. BÖTTCHER

Department of Physical Chemistry,
University of Leiden, The Netherlands

Second edition completely revised by

O. C. VAN BELLE, P. BORDEWIJK AND A. RIP

Department of Physical Chemistry,
University of Leiden, The Netherlands

VOLUME I

Dielectrics in static fields

ELSEVIER

AMSTERDAM - LONDON - NEW YORK - TOKYO

ELSEVIER SCIENCE PUBLISHERS B.V.
Sara Burgerhartstraat 25
P.O. Box 211, 1000 AE Amsterdam, The Netherlands

First Edition: 1973
Second Impression: 1993

ISBN: 0-444-41019-8

Printed in The Netherlands

FOREWORD

As it is customary that an author of my age revises his own book, I should begin this foreword with an apologetic explanation. Having accepted the presidency of the Science Policy Council of The Netherlands, when the government founded this advisory body in 1966, I soon realized that this function did not leave sufficient spare time for the thorough revision of a scientific book. Fortunately the three main co-workers in my research unit at Leiden University kindly offered to take on this task. I then decided that I should not try to interfere, so that the considerable adaptation which the text needed after twenty years has been entirely their work and their responsibility.

I would like to express my gratitude for what they have achieved and my hope that the revised edition will be accorded a reception similar to that given to the first edition and its unchanged second printing.

C. J. F. Böttcher

PREFACE TO THE SECOND EDITION

A period of twenty years is sufficiently long for a book to become outdated. In the case of dielectrics, the years between 1952 and the present have seen a great number of new developments, so that many subjects in the theory of electric polarization that are of great interest now, especially the dynamic behaviour of dielectrics, have not or insufficiently been dealt with in the 1952 edition of this book. Although a number of these subjects have been treated in monographs or books published since then, none of these books started from the same purposes as the first edition of "Theory of Electric Polarization", *i.e.*, a comprehensive treatment that can be used as a handbook as well as a textbook. In presenting a revised edition of this work, it was our aim to maintain this basic purpose, and to incorporate new material, thus giving an up-to-date and systematic treatment of the classical theory of dielectrics, in which much attention is paid to the fundamental concepts, and to the connection between theory and experimental results.

To prevent the increase of the book to an unmanageable size due to the large amount of new material, and to speed up the publication, we decided to split the book into two volumes. Volume I, the present volume, contains the theory of the static behaviour of dielectrics, and Volume II will contain the theory of the dynamic behaviour of dielectrics, as well as chapters on special subjects (determination of permanent dipole and multipole moments, polarization of solids).

In preparing the revised edition, not only additional material had to be incorporated. We also used the opportunity to make some changes in the order of presentation and to introduce a number of other improvements. Furthermore a historical introduction was added.

In the first volume of the revised edition the fundamental theory of electrostatics has been maintained as a starting-point. As in the first edition the Legendre polynomials and their value in solving potential problems have been treated extensively in Chapters I and II. The chapter on multipoles was reduced to one section in Chapter I, while the relation to the associated

Legendre polynomials was considered too much of a side track to keep it in this revised edition. Instead more attention has been paid to the problem of comparing multipole moments of molecules standardized according to different conventions. In the derivation of the relation between the electric field, the dielectric displacement, and the polarization, use has now been made of the concept of a dipole density.

The chapter on "Polarization and Energy" was placed directly behind the chapters on fundamental electrostatics, and more attention has been paid to the multipole terms in the electrostatic interaction energy between two molecules. Some material contained in this chapter in the first edition was postponed to Chapters IV, V and VII. The chapter "The Reaction Field" now contains a section on "Reaction Field and Energy"; more attention has been paid to the case of an ellipsoidal cavity.

The chapters on polar and non-polar dielectrics in static fields of the first edition were completely reorganized. The material has been split into a chapter "The Dielectric Constant in the Continuum Approach to the Environment of the Molecule" (Ch. V), and a chapter "Statistical-mechanical Theories of the Dielectric Constant" (Ch. VI). In both chapters polar as well as non-polar dielectrics are treated. In Chapter V the generalization of the continuum theory to the case of ellipsoidal cavities has been incorporated. Chapter VI contains new sections on dielectric virial coefficients and on the Kirkwood–Fröhlich equation; the application of this equation to polymers and to associating liquids is discussed in detail.

The developments in the field of non-linear dielectrics (*e.g.*, the so-called anomalous saturation effect) made it necessary to devote a new chapter to this subject (Ch. VII). In this chapter also the enlarged section on electrostriction has been incorporated.

As in the first edition the mathematics used in the book are reviewed shortly in a number of appendices; due to the revisions in the treatment of the multipoles the appendix on the associated Legendre polynomials could be abandoned. Because of these appendices the book can also be used by students with knowledge of elementary calculus and vector algebra only.

When used as a textbook for an advanced course in the theory of dielectrics it is convenient to leave out the sections dealing with ellipsoidal shapes (sections 9d, e, 20, 22, 30, 33) and to skip the small print everywhere. Furthermore section 5 (general multipoles) can be broken off after the definition of ideal multipoles. The Chapters IV and VII can be included at will, but it should be noted that in Ch. VII only sections 41 (introduction), 42 (normal saturation), and 45 (electrostriction), are independent of the statistical-mechanical theory developed in Ch. VI.

For introductory courses in dielectrics we suggest the following sections and parts of sections: 1, 2, 3 up to application b, 6, 7, 8, 9a, b, c, 10, 11 up to and including eqn. (3.9), 17, 18, 19, 25, 26, 27, 28, 29, 31, 40 up to a, 41 up to eqn. (7.7).

Leiden, November 1972

O. C. van Belle
P. Bordewijk
A. Rip

CONTENTS

IMPORTANT SYMBOLS

(The numbers indicate the page where the symbol is introduced)

a	radius of sphere or spherical cavity, half of principal axis of ellipsoid
A	first virial coefficient, 231
A	affinity, 306
A_n	abbreviation of $(1 - n\alpha_1\alpha_2/s^6)$ or $(1 - n\alpha^2/r^6)$, 121, 236
A_α	$(\alpha = a, b, c)$ shape factor for ellipsoid, 79
$\boldsymbol{\mathsf{A}}$	depolarizing tensor for ellipsoid *in vacuo*, 316
$\mathbf{\mathsf{A}}$	3N-dimensional tensor connected with the polarizability of a system, 209
$\mathscr{A}$	first dielectric virial coefficient, 232
Å	Ångstrom unit (10^{-8} cm)
b	half of principal axis of ellipsoid
B	second virial coefficient, 231
$\mathscr{B}$	second dielectric virial coefficient, 232
c	half of principal axis of ellipsoid
C	third virial coefficient, 231
C	proportionality constant of the Van der Waals attractive potential, 227
C	orthogonal transformation matrix, 263, 340
$\mathscr{C}$	third dielectric virial coefficient, 232
d	density
D	Debye unit (10^{-18} e.s.u. of electric moment), 11
$\boldsymbol{D}$	dielectric displacement, 59
e	electric point charge, total positive charge, total charge, 9, 40

e_0	charge of electron, 11
$\boldsymbol{e}$	unit vector in the direction of the external field, 206
$\mathbf{e}$	3N-dimensional vector connected with the unit vector in the direction of the external field, 213
$\boldsymbol{E}$	electric field, 11, 59
$\boldsymbol{E}_0$	external field, 74
$\boldsymbol{E}_0$	homogeneous part of the Maxwell field in a dielectric, 293
$\boldsymbol{E}_c$	cavity field, 78, 81
$\boldsymbol{E}_d$	directing field, 161, 173
$\boldsymbol{E}_i$	internal field, 70, 161
$\boldsymbol{E}_l$	local field, 208
$\boldsymbol{E}_s$	field due to the surroundings of a molecule, 130
$\boldsymbol{E}_F$	Fröhlich field, 252, 299
$\boldsymbol{E}_L$	Lorentz field, 166
$\mathbf{E}_0$	3N-dimensional vector connected with the external field, 209
$\mathbf{E}_1$	3N-dimensional vector connected with the local field, 209
f	reaction field factor, 129, 134
F	free energy
$\boldsymbol{F}$	force
$\boldsymbol{\mathsf{F}}$	reaction field tensor, 130, 139
$\tilde{F}$	transformed free energy, 306
$\mathcal{F}$	transformed free energy, 108
g	chemical potential, 306
g	radial distribution function, 218
g	Kirkwood correlation factor, 249, 258
$\tilde{G}$	transformed free enthalpy, 319
h	correction factor for molar polarization, 193
$\boldsymbol{i}$	unit vector in direction of the x-axis
$\boldsymbol{\mathsf{I}}$	unit tensor
I	3N-dimensional unit tensor, 209
$\boldsymbol{j}$	unit vector in direction of the y-axis

k	Boltzmann's constant
$\boldsymbol{k}$	unit vector in direction of the z-axis
K	equilibrium constant
$\boldsymbol{l}$, l	distance
L	Langevin function
$\boldsymbol{L}$	depolarizing tensor, 80
m	dipole strength, 34
m	mass of microscopic particle, 60
$\boldsymbol{m}$	electric moment, dipole moment, 9, 10
$\mathbf{m}$	3N-dimensional vector connected with the dipole moments of the molecules, 209
M	molecular weight
$\boldsymbol{M}$, M	instantaneous dipole moment of a dielectric, 205
$\boldsymbol{M}_d$	orientational part of the instantaneous dipole moment of a dielectric in Fröhlich's model, 251
$\boldsymbol{M}_i^*$	average moment of a microscopic sphere with the central dipole held fixed, 246
n	refractive index, 2, 173
n	number of moles per cm^3, molar density
n_D	refractive index for Na$_D$-line
$\boldsymbol{n}$	unit vector in arbitrary direction
$\boldsymbol{n}$	unit vector normal to a surface
N	number of molecules in a system, 207
N	number of particles per cm^3, number density, 131
N_A	Avogadro's number
$\mathcal{N}$	number of molecules in a microscopic sphere, 207
p	pressure
p	probability, weight function
$\boldsymbol{p}$	induced molecular dipole moment, 109
P_n	Legendre polynomial, 25
$P_n^{(m)}$	associated Legendre polynomial, 58
$\boldsymbol{P}$	dipole density, electric polarization, 22, 69
$\boldsymbol{P}_{in}$	induced polarization, 251

$\boldsymbol{P}_{or}$	orientation polarization, 251
$\boldsymbol{P}_{\alpha}$	induced polarization, 160
$\boldsymbol{P}_{\mu}$	dipole polarization, 160
$\mathscr{P}$	electric polarization, 69
$[P]$	molar polarization, 170
q	quadrupole strength, 34, 45, 48
q_0	factor characterizing a quadrupole moment not leading to a potential contribution, 49
Q_n	Legendre polynomial of the second kind, 140
$\boldsymbol{Q}$	quadrupole moment, 44
$\boldsymbol{Q}'$	standardized quadrupole moment, 49
$\boldsymbol{Q}_{H}$	quadrupole moment standardized according to Hirschfelder, 52
$\boldsymbol{Q}_{norm}$	quadrupole moment standardized according to Buckingham, 51
$\mathscr{Q}$	quadrupole density, 61, 69
∂Q	amount of heat absorbed by a system
$\boldsymbol{r}, r$	radius vector
R	gas constant
R_s	reduction factor of saturation, 304
$\boldsymbol{R}$	reaction field, 129, 134
$\boldsymbol{s}, s$	distance
$\boldsymbol{ds}$, $\mathrm{d}s$	differential line element
S	surface
S	entropy
$\boldsymbol{dS}$, $\mathrm{d}S$	differential surface element
t	temperature in centigrades
T	absolute temperature
$\boldsymbol{T}$	torque
$\boldsymbol{T}$	dipole field tensor, dipole-dipole interaction tensor, 18, 116
$\mathbf{T}$	3N-dimensional tensor connected with the dipole-dipole interactions, 209
u	octupole strength, 45
U	internal energy, energy of a system

$\boldsymbol{U}$	octupole moment, 44
v	volume
dv	differential volume element
V	potential energy
V	volume
W	work
x	molar fraction, 261
y	reaction coordinate, 306
$y^{(n)}$	2^n-pole strength, 35
$\boldsymbol{Y}^{(n)}$	2^n-pole moment, 53
z	number of nearest neighbours of a molecule, 249
α	polarizability, 86
α^e	polarizability connected with electronic polarization, 173
α^a	polarizability connected with atomic polarization, 191
$\boldsymbol{\alpha}$	polarizability tensor, 87
β	compressibility, 319
β_E	electrocaloric coefficient, 103
$\boldsymbol{\beta}$	first hyperpolarizability tensor, 290, 310
γ	second hyperpolarizability tensor, 290, 310
δ	delta function, 352
δ_{ij}	Kronecker delta, 341
ε	dielectric constant, permittivity, 1, 71, 159
ε'	ratio D/E for non-linear dielectrics, 291
ε_0	field independent part of the permittivity, 289
ε_∞	dielectric constant characteristic for the induced polarization, 172
ε_i	internal dielectric constant, 88
ε_E	incremental dielectric constant, 289

$\boldsymbol{\varepsilon}$	permittivity tensor, 71
$\Delta\varepsilon$	electric saturation term, 290
$\Delta\varepsilon_a$	anomalous saturation term, 303
$\Delta\varepsilon_e$	saturation term due to electrostriction, 318
$\boldsymbol{\Theta}$	permanent molecular quadrupole moment, 109
λ	heat of vaporization, 155
μ	value of permanent molecular dipole moment, 109
μ_{gas}	permanent moment calculated from measurements on diluted gases, 179
$\boldsymbol{\mu}$	permanent molecular dipole moment, 109
$\boldsymbol{\mu}^*$	permanent molecular dipole moment enlarged by the reaction field, 135
$\boldsymbol{\mu}_d$	molecular dipole moment in Fröhlich's model, 251
$\boldsymbol{\mu}_e$	external moment, 134
$\mathbf{\mu}$	3N-dimensional vector connected with the permanent dipole moments, 209
ν	stoechiometric coefficient, 306
ξ	coefficient of non-linear dielectric behaviour, electric saturation coefficient, 71, 289
$\boldsymbol{\xi}$	tensor of non-linear dielectric behaviour, 289
ρ	volume charge density, 12
σ	surface charge density, 12
σ	diameter of rigid sphere, 227
ϕ	potential of electric field, 13
φ	volume fraction, 200
φ	molar volume of a mixture, 261
χ	dielectric susceptibility, 70
$\boldsymbol{\chi}$	dielectric susceptibility tensor, 71
$\boldsymbol{\Omega}$	permanent molecular octupole moment, 109

HISTORICAL INTRODUCTION

Electric polarization has long been a subject of investigation, different aspects being emphasized in different periods.

As a natural phenomenon, electricity was known in ancient times, but it was not until the eighteenth century that the first experimental studies were undertaken, making it possible to establish a clear distinction between positive and negative electric charges—as they were later called—and between conductors and insulators. It was discovered that large quantities of electric charge could be stored in a condensor, an apparatus consisting of two conducting plates separated by insulating materials. The condensor constructed in 1745 by Cunaeus and Musschenbroek[1] became very popular for a variety of experimental purposes under the name of Leyden jar.

Relatively little attention was paid to the properties of the insulating material until 1837, when Faraday[2] published the first numerical results of measurements on this material, which he called the dielectric. Cavendish[3] had performed comparable experiments about sixty years earlier, but his results remained unpublished until 1879.

The results of Faraday's rather crude measurements indicated that the capacity of a condensor was dependent on the nature of the material separating the conducting surfaces. For the ratio between the capacity of a condensor filled with a dielectric and the capacity of the same condensor when empty, Faraday introduced the term *specific inductive capacity*. This quantity is now generally called the permittivity or the dielectric constant, and denoted by ε.

In this period, other electric and magnetic phenomena were also investigated intensively. The results were summarized by Maxwell[4] in the middle of the 1860s in his unified theory of electromagnetic phenomena. In this theory the permittivity is conceived to be the ratio between the electric field intensity and the dielectric displacement, introduced by this author. Since light was taken as a form of electromagnetic radiation, it followed that for most dielectrics the dielectric constant should be equal to the square of the refractive index n. This simple relation, $\varepsilon = n^2$, is called the Maxwell relation.

In the last part of the nineteenth and the beginning of twentieth century the dielectric constant was determined for a number of substances, especially for the verification of the Maxwell relation. The experimental results showed that the equation held reasonably well for many solids as well as for some classes of liquids and gases, but for many other substances, which were called "associating", the dielectric constant proved to be considerably higher than the square of the refractive index measured in the visible region. After the experimental realization by Hertz[5] of electromagnetic waves of low frequency, the determination of the refractive index for these frequencies became possible. Drude's[6] experiments showed that the Maxwell relation was also confirmed for "associating" compounds when n was determined for waves of sufficiently low frequency. The same experiments revealed that for some substances adsorption of energy occurred at these frequencies, which was always accompanied by anomalous dispersion, *i.e.* the refractive index decreased with increasing frequency. Thus, it appeared possible to extend the validity of the Maxwell relation by the formal generalization of the dielectric constant to a complex frequency-dependent quantity and the use of a complex refractive index, where in both cases the imaginary part is a measure of the absorption of energy.

In addition to the development of this phenomenological approach, attempts were made as early as the middle of the nineteenth century to correlate the dielectric constant with the microscopic structure of matter. In 1847, following Faraday in considering the dielectric to be composed of conducting spheres in a non-conducting medium, Mossotti[7] succeeded in deriving a relation between the dielectric constant and the volume fraction occupied by the conducting particles in the dielectric. However, this expression remained relatively unknown until it was again derived—probably independently—in 1879 by Clausius.[8] At about the same time, a corresponding expression for the square of the refractive index was given by Lorenz.[9]

Lorentz[10] derived the validity of the expression with n^2 for a medium built up of particles containing elastically bound electrons, an assumption in agreement with J. J. Thomson's model for the distribution of the electric charge in the molecules. In his derivation Lorentz introduced the concept of the internal field, *i.e.* the average field working on an individual particle.

This concept advanced the theoretical understanding of the problem considerably and also contributed to the subsequent development of the theory. By taking the number of particles per unit volume proportional to the density of the compound, an equation accessible to experimental verification was obtained; this equation is known as the Lorenz-Lorentz equation.

For low frequencies, n^2 can be replaced by the dielectric constant, which transforms the Lorenz-Lorentz equation to a form corresponding with the expression derived by Mossotti and Clausius if in the latter expression the relative space-filling of the conducting spheres is taken proportional to the density. The resulting equation is still in use in this form and is generally called the Clausius-Mossotti equation.

Comparison with the experimental results then available, showed that the Lorenz-Lorentz equation held very well for the optical region, but that in the static case the Clausius-Mossotti equation had a very limited validity. For the compounds considered to be "associating"—partially because of the large difference observed in the liquid state between the dielectric constant and the square of the refractive index for visible light—the Clausius-Mossotti equation failed to hold even for the gaseous phase. The deviations depended strongly on pressure, temperature, and the state of aggregation.

Among the suggestions put forward to explain the discrepancy between the theory and the experimental results, the most useful proved to be the attribution of a permanent electric dipole moment to the molecules. Essentially, the concept of molecules with a permanent dipole moment goes back as far as Berzelius,[11] who assumed molecules to be composed of oppositely charged regions. In the nineteenth century this concept was discussed explicitly by several authors (*e.g.* Clausius, von Helmholtz), but the treatment remained vague and on a qualitative level until Debye[12] published a quantitative theory in 1912. Applying Langevin's calculations for the average orientation of permanent magnetic moments in an external field to the electric case and accepting Lorentz's approach for the internal field, Debye succeeded in deriving an extension of the Clausius-Mossotti equation: the Debye equation. In this equation the dielectric constant depends not only on the molecular polarizability, as had been implicit in the earlier theories, but also on the permanent moment of the molecules.

With the Debye equation the temperature dependence of the dielectric constant could be elucidated to a great extent, especially for low pressure gases, for which the agreement between theory and experimental results was very close. On the basis of the existence of permanent electric moments in the molecules, the anomalous dispersion observed by Drude could also be understood and a better insight into the nature of the intermolecular forces was obtained.

The Debye equation afforded the possibility of calculating molecular dipole moments from measurements of the dielectric constant, a calculation actually carried out for some compounds in the 1912 publication. The method did not arouse much interest until the middle of the 1920s, however. At that time, understanding of the structure of molecules had been advanced to such a degree that the value of the dipole moment could be connected with the geometrical arrangement of the atoms in the molecule. From then on, the significance of Debye's method was fully realized and an increasing number of determinations of permanent dipole moments were performed; in this connexion special mention must be made of the work of Smyth[13] and his co-workers.

Measurements performed on the gaseous phase and dilute solutions of polar compounds (*i.e.* compounds with a non-zero permanent dipole moment) in different non-polar solvents yielded comparable values of the dipole moment, which strengthened confidence in the underlying theory. Values calculated from measurements on concentrated solutions and on pure polar liquids did not, however, agree with the gas value. Since for these liquids the temperature dependence of the dielectric constant, too, did not comply with the theoretical requirements, it became clear that the Debye equation, although better than the Clausius-Mossotti equation, was also of limited validity. The deviations were often attributed to association, *i.e.* the tendency of particular types of molecules to form aggregates of two or more molecules. Several modifications of the Debye equation were proposed, but because of either restricted applicability or lack of a sound theoretical foundation, none of them gained general acceptance.

In 1936, Onsager[14] succeeded in constructing a fundamental modification of Debye's approach, which greatly enlarged the applicability. He pointed out that the internal field used by Debye included a part produced by the permanent moment of the molecule itself via the moments induced by it in the dielectric surrounding this molecule. This contribution to the internal field, the so-called reaction field, always has the direction of the permanent moment, and therefore cannot produce a torque on the molecule. Consequently, its average component in the direction of the internal field must be subtracted from the internal field to obtain the field that tends to direct the permanent dipoles, called here the directing field. In all previous theories the directing field had been taken equal to the internal field.

By considering the surroundings of a molecule as a continuum characterized by the macroscopic dielectric constant of the substance and by introducing some plausible approximations, it became possible to derive a new relation between the dielectric constant and the molecular dipole moment, the Onsager equation. Except for some specific classes of compounds, Onsager's theory accounted for most of the deviations from the Debye equation in the case of polar liquids. From this it could be concluded, in contradiction to earlier ideas, that the occurrence of association in liquids was an exception rather than a rule.

For gases and dilute solutions of polar compounds in non-polar solvents, the results of Debye's theory are only slightly affected when the reaction field is taken into account. Therefore, the values for the dipole moments based on measurements done on gases and dilute solutions retained their validity. Application of the Onsager equation made it also possible to determine the permanent dipole moments from measurements on concentrated solutions and pure polar liquids. Some further improvement was obtained when the non-spherical shape of the molecules was incorporated into Onsager's theory.

The Onsager equation generally does not hold for the liquid phase of such compounds as carboxylic acids and alcohols. In these cases real association occurs, due to the formation of H-bonds; this has also been shown by other experimental results (thermal data, IR- and NMR-spectra, etc.). A theory relating the deviations from the Onsager equation in the case of associating liquids to the orientation correlation of neighbouring molecules, as the result of short-range specific interactions, was given by Kirkwood.[15] On this basis it is possible to use the dielectric constant as a source for deriving information about specific interactions between the molecules and about the liquid structure. It is not always possible, however, to reach definite conclusions, due to a lack of additional information about the structure of the liquid.

Shortly after Debye[16] had introduced the concept of molecular permanent dipole moments into the theory of dielectrics, he used this concept to explain the anomalous dispersion of the dielectric constant observed by Drude and others. Debye pointed out that the process of orientation of the permanent moments connected with changes in the field requires a definite time interval, since it depends on the rotational movements accomplished by the molecules. From Debye's assumptions concerning the molecular reorientations, it follows that after the removal of an externally applied field the average dipole orientation decays exponentially with time; the characteristic time of this exponential decay is called the relaxation time.

For an alternating field, Debye deduced that the time lag between the average orientation of the moments and the field becomes noticeable when the frequency of the field becomes of the same order as the reciprocal relaxation time. In this way the molecular relaxation process leads to the macroscopic phenomena of dielectric relaxation, *i.e.* the anomalous dispersion of the dielectric constant and the accompanying absorption of electromagnetic energy.

For the calculation of the relaxation time, Debye considered the rotational movements of the molecules to be governed by the frictional resistance in the medium to changes of the molecular orientation. When the molecules were taken as rigid spheres and the macroscopic viscosity of the liquid was taken as an approximation for the molecular friction constant, the relaxation time proved to be roughly of the order required for the explanation of the experimental results.

Several modifications and extensions of Debye's theory for dielectric relaxation have been proposed generally leading to the replacement of the single relaxation time by a set of different relaxation times in the description of the macroscopic relaxation process. For the description of the experimental results, the introduction of continuous distributions of relaxation times was also proposed in many cases. Cole and Cole[17] gave an important graphical method, the so-called Cole-Cole plot, to distinguish these cases from that of a single relaxation time.

Experimentally, the relaxation time was found to depend exponentially on the reciprocal temperature. This led to consideration of the dielectric relaxation process as a rate phenomenon. The dipole moments do not change their direction continuously but rather in discrete "jumps" between stable orientations separated by potential energy barriers. In this conception, analyzed extensively by Kauzmann,[18] the process of molecular reorientation is governed by an activation free energy, dependent on the intermolecular forces. Variations in the local environment of the molecules give rise to a distribution of activation energies, resulting in a distribution of relaxation times.

For a long time the difficulty of performing measurements at high frequencies limited studies of dielectric relaxation behaviour to substances in which strong intermolecular forces exist, *e.g.* substances in which real association occurs. But during and after the Second World War the development of new apparatus extended the experimentally accessible frequency range to such an extent that many of the relaxation phenomena of non-associating polar compounds in the liquid state could be observed.

For the interpretation of dielectric relaxation behaviour in terms of molecular properties, two problems must be solved. First, the relationship between the macroscopic relaxation behaviour and the molecular re-orientation process must be established unequivocally. Secondly, the reorientation process must be explained in terms of the properties of the single molecule and the molecular interactions. In recent years, as a consequence of developments in irreversible statistical mechanics, considerable advances have been achieved with respect to the first problem. The final explanation in terms of molecular properties is only in its initial stage.

From this account it appears that the development of theories of dielectric polarization and relaxation has been stimulated by the development of theories in related fields, as well as by the availability of new experimental results. It must be expected that in this way in the future further advances will be achieved in the theory of electric polarization.

References

1. P. van Musschenbroek, *Introductio ad Philosophiam Naturalem*, Luchtmans, Leiden 1762.
2. M. Faraday, *Phil. Trans.* **128** (1837/38) 1, 79, 265.
3. *The Electric Researches of the Honourable Henry Cavendish*, ed. J. C. Maxwell, Cambridge 1879.
4. J. C. Maxwell, *Phil. Trans.* **155** (1865) 459, *Ibid.* **158** (1868) 643, *Treatise on Electricity and Magnetism*, Dover, New York 1954.
5. H. Hertz, *Ann. Phys.* **31** (1887) 421, *Gesammelte Werke* Vol. II, 1890.
6. P. Drude, *Z. Phys. Chem.* **23** (1897) 267.
7. P. F. Mossotti, *Bibl. Univ. Modena* **6** (1847) 193.
8. R. Clausius, *Die Mechanische Wärmetheorie* Vol. II, Braunschweich 1879.
9. L. Lorenz, *Ann. Phys. Chem.* **11** (1880) 70.
10. H. A. Lorentz, *Verh. Kon. Acad. van Wetenschappen*, Amsterdam 1879, *Ann. Phys.* **9** (1880) 641, *Theory of Electrons*, Dover, New York 1952.
11. J. J. Berzelius, *Essai sur la Théorie des Proportions Chimique et sur l'Influence Chimique de l'Electricité, Paris* 1819.

12. P. Debye, *Phys. Z.* **13** (1912) 97, *Collected Papers*, Interscience, New York 1954, p. 173, *Polar Molecules*, Dover, New York.
13. C. P. Smyth, *Dielectric Constant and Molecular Structure*, Chemical Catalog, New York 1931, *Dielectric Behavior and Structure*, McGraw-Hill, New York 1955.
14. L. Onsager, *J. Am Chem. Soc.* **58** (1936) 1486.
15. J. G. Kirkwood, *J. Chem. Phys.* **7** (1939) 911.
16. P. Debye, *Ber. D. Phys. Ges.* **15** (1913) 777, *Collected Papers*, Interscience, New York 1954, p. 158.
17. K. S. Cole and R. H. Cole, *J. Chem. Phys.* **9** (1941) 341.
18. W. Kauzmann, *Revs. Mod. Phys.* **14** (1942) 12.

CHAPTER I

ELECTRIC DIPOLES AND MULTIPOLES

§1. Electric moment and electric dipoles

For the theory of electric polarization the electric moment of a system of charges is a fundamental notion, and it will therefore be useful to start with a definition of this quantity.

The electric moment of a point charge e relative to a fixed point is defined as $e\boldsymbol{r}$, in which $\boldsymbol{r}$ is the radius vector from the fixed point to e. Consequently, the electric moment of a system of charges e_i, relative to a fixed origin is defined as:

$$\boldsymbol{m} = \sum_i e_i \boldsymbol{r}_i. \tag{1.1}$$

If the net charge of the system is zero, the electric moment is independent of the choice of the origin: when the origin is displaced over a distance $\boldsymbol{r}_0$, the change in $\boldsymbol{m}$ is, according to eqn. (1.1), given by:

$$\Delta\boldsymbol{m} = -\sum_i e_i \boldsymbol{r}_0 = -\boldsymbol{r}_0 \sum_i e_i.$$

Thus, $\Delta\boldsymbol{m}$ equals zero when the net charge is zero. Then $\boldsymbol{m}$ is independent of the choice of the origin. In this case eqn. (1.1) can be written in another way by the introduction of the electric centres of gravity of the positive and negative charges. These centres are defined by the equations:

$$\sum_{\text{positive}} e_i \boldsymbol{r}_i = \boldsymbol{r}_\mathrm{p} \sum_{\text{positive}} e_i = \boldsymbol{r}_\mathrm{p} e,$$

and:

$$\sum_{\text{negative}} e_i \boldsymbol{r}_i = \boldsymbol{r}_\mathrm{n} \sum_{\text{negative}} e_i = -\boldsymbol{r}_\mathrm{n} e,$$

in which the radius vectors from the origin to these centres are represented by $\boldsymbol{r}_\mathrm{p}$ and $\boldsymbol{r}_\mathrm{n}$ respectively and the total positive charge is called e. For a zero net charge, eqn. (1.1) can therefore be written as:

$$\boldsymbol{m} = (\boldsymbol{r}_\mathrm{p} - \boldsymbol{r}_\mathrm{n})e.$$

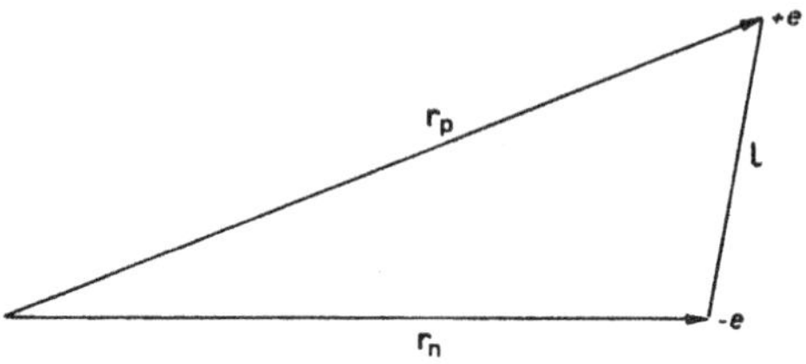

Fig. 1

The difference $\boldsymbol{r}_p - \boldsymbol{r}_n$ is equal to the vectorial distance between the centres of gravity, represented by a vector $\boldsymbol{l}$ pointing from the negative to the positive centre (Fig. 1). We then have:

$$\boldsymbol{m} = \boldsymbol{l}e. \tag{1.2}$$

Therefore, the electric moment of a system of charges with zero net charge is generally called the electric dipole moment of the system.

A simple case is a system consisting of only two point charges $+e$ and $-e$ at a distance $\boldsymbol{l}$. Such a system is called a (physical) electric dipole, its moment is equal to $e\boldsymbol{l}$, the vector $\boldsymbol{l}$ pointing from the negative to the positive charge.

In theoretical organic chemistry the dipole vector is generally taken as pointing from the positive to the negative charge. We prefer to use the physical definition given above.

A mathematical abstraction derived from the above defined physical dipole is the ideal or point dipole. Its definition is as follows: the distance $\boldsymbol{l}$ between two point charges $+e$ and $-e$ is replaced by $\boldsymbol{l}/n$ and the charge e by en. The limit approached as the number n tends to infinity is the ideal dipole. The equations derived for ideal dipoles are much simpler than those obtained for non-ideal dipoles.

Many neutral molecules are examples of charge systems with a non-ideal electric dipole moment, since in most types of molecule the centres of gravity of the positive and negative charge distributions do not coincide.

Apart from these permanent or intrinsic dipole moments, a temporary or induced dipole moment arises when a particle is brought into an external electric field. Under the influence of this field the positive and negative charges in the particle are moved apart: the particle is polarized. In general, these induced dipoles can be treated as ideal; permanent dipoles, however, may generally not be treated as ideal when the field at molecular distances is to be calculated (see section 2).

The values of molecular dipole moments are usually expressed in Debye units. The Debye unit, abbreviated as D, equals 10^{-18} electrostatic units (e.s.u.). The permanent dipole moments of non-symmetrical molecules generally lie between 0.5 and 5D. This order of magnitude can be expected, because the elementary charge e_0 is 4.4 10^{-10} e.s.u. and the distances of the charge centres in the molecules amount to about 10^{-9}–10^{-8} cm.

Since the most important quantity in the theory of electric polarization, the dipole moment, is generally expressed in Debye units, we prefer for present purposes to use electrostatic units rather than Giorgi's m.k.s. units. For readers accustomed to the m.k.s. system, we may mention that the electrostatic system is based upon a definition of the charge arising from Coulomb's law: for two equal unit charges 1 cm apart, the mutual force is 1 dyne. Thus Coulomb's law can be represented by $F = e_1 e_2/r^2$. The e.s.u. of electric charge is equal to 3.33 10^{-10} coulombs.

§2. The electric field of an ideal dipole *in vacuo*

Before considering the field of an ideal dipole *in vacuo*, the general theory of the electrostatic field *in vacuo* will be briefly summarized.*

(a) *The potentials and fields due to electric charges*

According to Coulomb's experimental inverse square law, the force between two charges e and e' with distance r is given by:

$$\boldsymbol{F} = \frac{ee'}{r^2}\frac{\boldsymbol{r}}{r}. \tag{1.3}$$

Taking one of the charges, say e', as a test charge to measure the effect of the charge e on its surroundings, we arrive at the concept of an electric field produced by e and with a field strength or intensity defined by:

$$\boldsymbol{E} = \lim_{e' \to 0} \frac{\boldsymbol{F}}{e'}. \tag{1.4}$$

The field strength due to an electric charge at a distance r is then given by:

* See for instance J. A. Stratton.[1]

$$\boldsymbol{E} = \frac{e}{r^2} \frac{\boldsymbol{r}}{r}, \tag{1.5}$$

in which r is expressed in cm, e in electrostatic units and $\boldsymbol{E}$ in dynes per charge unit, *i.e.* the e.s.u. of field intensity.

A simple vector-analytic calculation* shows that eqn. (1.5) leads to:

$$\oiint \boldsymbol{E} \cdot \boldsymbol{dS} = 4\pi e, \tag{1.6}$$

in which the integration is taken over any closed surface around the charge e, and where $\boldsymbol{dS}$ is a surface element having the vectorial direction of the outward normal.

By assuming that the electric field intensity is additively built up of the contributions of all the separate charges (a form of the principle of superposition), eqn. (1.6) can be extended to:

$$\oiint \boldsymbol{E} \cdot \boldsymbol{dS} = 4\pi \sum_i e_i. \tag{1.7}$$

This relation will still hold for the case of a continuous charge distribution, represented by a volume charge density ρ or a surface charge density σ. For the case of a volume charge density we write:

$$\oiint \boldsymbol{E} \cdot \boldsymbol{dS} = 4\pi \iiint_V \rho \, \mathrm{d}v, \tag{1.8}$$

or, using Gauss' divergence theorem:

$$\operatorname{div} \boldsymbol{E} = 4\pi\rho. \tag{1.9}$$

When surface charges are present, eqn. (1.7) transforms to:

$$\oiint \boldsymbol{E} \cdot \boldsymbol{dS} = 4\pi \iint_S \sigma \, \mathrm{d}S, \tag{1.10}$$

and taking as the closed surface of integration two surfaces having an infinitely small distance from the surface charge, we find:

$$(E_n)_2 - (E_n)_1 = 4\pi\sigma, \tag{1.11}$$

where $(E_n)_1$ and $(E_n)_2$ denote the normal components of $\boldsymbol{E}$ at two points

* See Appendix I for the meaning of the symbols and for a number of vector-analytic rules used.

P_1 and P_2 on different sides of the surface charge but with an infinitely small distance between them. In eqn. (1.10) the normal components of $\boldsymbol{E}$ are taken to point in the direction from P_1 to P_2.

Eqn. (1.9) is the first of Maxwell's well-known equations for the electrostatic field *in vacuo*; it is generally called the source equation. From its integral form as given in eqn. (1.8), eqns. (1.6), (1.7), and (1.11) can be derived by inserting the appropriate charge distributions.

The second of Maxwell's equations, necessary to derive $\boldsymbol{E}$ uniquely for a given charge distribution, is:

$$\operatorname{curl} \boldsymbol{E} = \boldsymbol{0}, \tag{1.12}$$

or, using Stokes's theorem:

$$\oint \boldsymbol{E} \cdot \boldsymbol{ds} = 0, \tag{1.13}$$

in which the integration is taken along a closed curve of which $\boldsymbol{ds}$ is a line element. From eqn. (1.13) it follows that $\boldsymbol{E}$ can be written as the gradient of a scalar field ϕ,* which is called the potential of the field:

$$\boldsymbol{E} = -\operatorname{grad} \phi. \tag{1.14}$$

The relation between the potential ϕ and the energy of the field will be discussed in Chapter III. We have introduced the potential here because it is often far easier to calculate the potential first and derive $\boldsymbol{E}$ from it by eqn. (1.14) than to calculate $\boldsymbol{E}$ directly from eqn. (1.9).

The combination of (1.9) and (1.14) leads to Poisson's equation:

$$\operatorname{div} \operatorname{grad} \phi = \Delta\phi = -4\pi\rho. \tag{1.15}$$

In charge-free parts of the field this reduces to Laplace's equation:**

$$\Delta\phi = 0. \tag{1.16}$$

Substitution of the appropriate charge distribution for the case of a single point charge e in the solution of Poisson's equation leads to:

$$\phi = \frac{e}{r}. \tag{1.17}$$

Taking the gradient, we immediately recover eqn. (1.5).

---

* See Appendix I, eqn. (A1.33).

** See for a discussion of Poisson's and Laplace's equation, Appendix II.

It is evident from the foregoing that the theory of the electrostatic field *in vacuo* can be based on two different sets of equations, *i.e.* (1.3) together with the assumption of additivity of the field quantities or (1.9) and (1.12). We have mentioned the second approach with a view to the fact that the phenomenological theory of the electric polarization is best described with the help of Maxwell's theory.

(b) *The potential due to an ideal dipole*

We will first consider a physical dipole consisting of the charges $+e$ and $-e$ at a distance $\boldsymbol{l}$, the vector $\boldsymbol{l}$ pointing from $-e$ to $+e$. The distances from these charges to a point P are $\boldsymbol{r}_+$ and $\boldsymbol{r}_-$, respectively. The negative charge is situated at the origin of a coordinate system. According to eqn. (1.17) the potential at P due to this dipole is:

$$\phi(P) = \frac{e}{r_+} - \frac{e}{r_-}. \tag{1.18}$$

To obtain an ideal dipole we now let the distance between the two charges decrease and at the same time let the amount of charge increase in such a way that the product $\boldsymbol{m} = e\boldsymbol{l}$ remains constant. Continuing this process indefinitely, we get an ideal dipole, which is situated at the origin and has a moment $\boldsymbol{m}$ in the direction of $\boldsymbol{l}$.

To calculate the potential of this ideal dipole we first rewrite (1.18) as:

$$\phi(P) = \phi_+(P) + \phi_-(P), \tag{1.19}$$

in which ϕ_+ and ϕ_- are the potentials of the two charges, considered separately. Extending the triangle formed by $-e$, $+e$, and P to a parallelogram by denoting the fourth corner as P' (see Fig. 2), we see that $\phi_-(P)$ is equal to $-\phi_+(P')$, so that:

$$\phi(P) = \phi_+(P) - \phi_+(P') = -\{\phi_+(P') - \phi_+(P)\}. \tag{1.20}$$

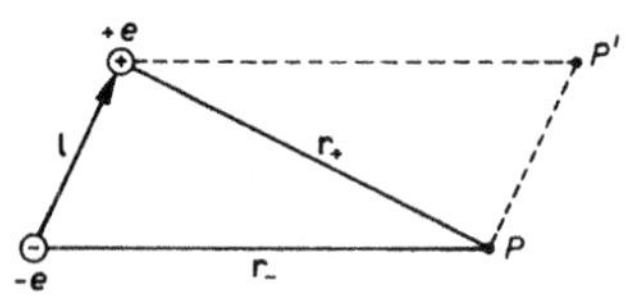

Fig. 2

The process of forming the ideal dipole at the origin is now equivalent to letting the point P' approach the point P indefinitely along the line $P'P$. According to the definition of the gradient (see eqn. (A1.48)), we may write the potential of the ideal dipole, by taking the limit in eqn. (1.20), as:

$$\phi(P) = -\boldsymbol{l} \cdot \operatorname{grad} \phi_+(P). \tag{1.21}$$

Since the charge is not affected by the gradient operation, it can be brought before the gradient. Reminding ourselves that the product of $\boldsymbol{l}$ with the charge has remained at the constant value $\boldsymbol{m}$, and that in the limit $r_+ = r_- = r$, we find for the potential of the ideal dipole:

$$\phi(P) = -\boldsymbol{m} \cdot \operatorname{grad} \frac{1}{r}. \tag{1.22}$$

Using (A1.22), eqn. (1.22) can be written as:

$$\phi(P) = \frac{\boldsymbol{m} \cdot \boldsymbol{r}}{r^3}. \tag{1.23}$$

If we take the dipole as the origin of a coordinate system with the z-axis along the dipole moment vector (Fig. 3), this equation may be written as:

$$\phi = \frac{mz}{r^3}, \tag{1.24}$$

or using the angle θ between the radius vector $\boldsymbol{r}$ and the positive z-axis:

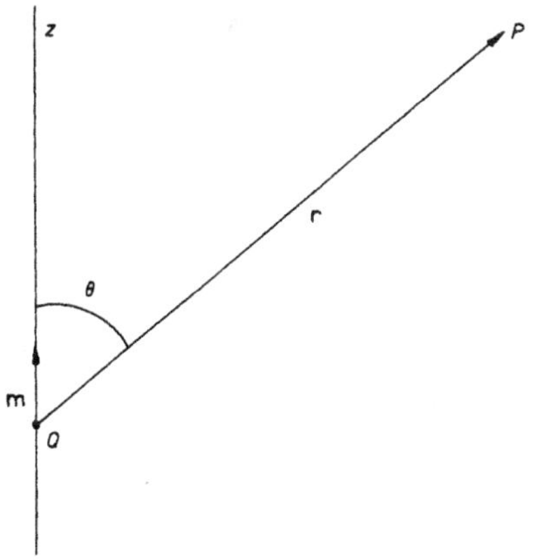

Fig. 3

$$\phi = \frac{m}{r^2} \cos \theta. \tag{1.25}$$

Remarks:

1. The four eqns. (1.22) to (1.25) for the potential due to an ideal dipole are sometimes also used for the non-ideal dipoles occurring in molecules. For this reason, we shall estimate the inaccuracy when these equations are used for the calculation of ϕ at a small distance from a non-ideal dipole.

As an extreme case we consider a point P on the positive z-axis at a distance l from the positive charge e, while the negative and positive charge of the dipole are situated at $z = -\frac{1}{2}l$ and $z = \frac{1}{2}l$ respectively. In this case the potential at P is equal to:

$$\phi(P) = \frac{e}{l} - \frac{e}{2l} = 0.50\frac{e}{l},$$

whereas eqn. (1.25) would give:

$$\phi(P) = \frac{e \cdot l}{(\frac{3}{2}l)^2} = 0.44\frac{e}{l}.$$

At this small distance we make an error of about 12% when we use the equation for the potential of the ideal dipole in the case of a non-ideal dipole.

For a point R on the positive z-axis at a distance $2l$ from the positive charge of the dipole, the potential is:

$$\phi(R) = \frac{e}{2l} - \frac{e}{3l} = 0.167\frac{e}{l}.$$

Eqn. (1.25) would give:

$$\phi(R) = \frac{e \cdot l}{(\frac{5}{2}l)^2} = 0.160\,\frac{e}{l}.$$

At this distance the inaccuracy is reduced to about 4%.

2. The potential due to a dipole consisting of charges $+e$ and $-e$ decreases more with increasing distance from the dipole centre than the potential due to one of these charges. This is illustrated by the example in Table 1, in which the potentials at distances of 2 Å and 10 Å from the centres of a positive mono-valent ion are compared with the potentials on the axis of a point

TABLE 1

THE POTENTIALS ON THE AXIS OF A DIPOLE AT DISTANCES OF 2Å AND 10Å, COMPARED WITH THE POTENTIALS AT THE SAME DISTANCES FROM A MONOVALENT ION

r (Å)	e_0/r (volt)	m/r^2 (volt)	
		$m = 5D$	$m = 0.5D$
2	7.2	3.8	0.4
10	1.4	0.15	0.02

dipole at the same distances of 2 Å and 10 Å from the dipole. The calculation is carried out for two values of the dipole moment, 0.5 and $5D$, these being respectively a rather low and rather high value for the permanent dipole moment of a molecule.

3. According to eqn. (1.24), the equipotential surfaces are given by:

$$\frac{mz}{r^3} = \text{const.}, \tag{1.26}$$

or:

$$z^2 = \text{const.}\ (x^2 + y^2 + z^2)^3. \tag{1.27}$$

(c) *The field intensity due to an ideal dipole*

From the potential $\phi(P)$, caused at P by an ideal dipole $\boldsymbol{m}$, placed at Q (Fig. 4), as given by eqn. (1.23), we find using the general relation (1.14):

$$\boldsymbol{E} = -\operatorname{grad} \phi,$$

for the field strength at the point P:

$$\boldsymbol{E} = -\operatorname{grad} \frac{\boldsymbol{m} \cdot \boldsymbol{r}}{r^3} = -\boldsymbol{m} \cdot \boldsymbol{r} \operatorname{grad} \frac{1}{r^3} - \frac{1}{r^3} \operatorname{grad} \boldsymbol{m} \cdot \boldsymbol{r}. \tag{1.28}$$

Since we have:

$$\operatorname{grad} \boldsymbol{m} \cdot \boldsymbol{r} = \boldsymbol{m},$$

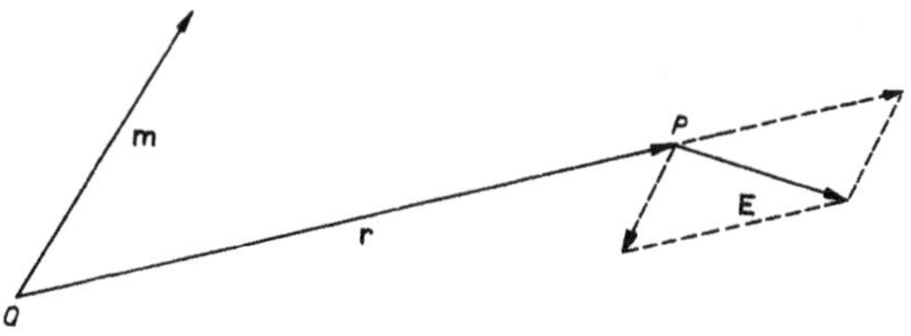

Fig. 4

and:

$$\operatorname{grad}\frac{1}{r^3} = -\frac{3}{r^4}\operatorname{grad} r = -\frac{3\boldsymbol{r}}{r^5},$$

(see Appendix I, eqns. (A1.24) and (A1.21)), eqn. (1.28) is reduced to:

$$\boldsymbol{E} = \frac{3\boldsymbol{m}\cdot\boldsymbol{r}}{r^5}\boldsymbol{r} - \frac{\boldsymbol{m}}{r^3}. \tag{1.29}$$

This important equation shows how $\boldsymbol{E}$ can be resolved into a component along the radius vector and a component parallel to the dipole vector.

From eqn. (1.29) it follows that each component of $\boldsymbol{E}$ is a linear function of the components of $\boldsymbol{m}$, the coefficients being functions of the components x, y, z of the radius vector $\boldsymbol{r}$:

$$\left.\begin{aligned} E_x &= \frac{3x^2 - r^2}{r^5}m_x + \frac{3xy}{r^5}m_y + \frac{3xz}{r^5}m_z \\ E_y &= \frac{3xy}{r^5}m_x + \frac{3y^2 - r^2}{r^5}m_y + \frac{3yz}{r^5}m_z \\ E_z &= \frac{3xz}{r^5}m_x + \frac{3yz}{r^5}m_y + \frac{3z^2 - r^2}{r^5}m_z \end{aligned}\right\}, \tag{1.30}$$

or in tensor notation:

$$\boldsymbol{E} = -\boldsymbol{T}\cdot\boldsymbol{m}. \tag{1.31}$$

The choice of the sign in this equation has been made to conform with general usage.

From eqns. (1.29) and (1.31) we derive the following expression for the tensor $\boldsymbol{T}$ (see Appendix I, Section 6):

$$\boldsymbol{T} = r^{-3}(\boldsymbol{I} - 3r^{-2}\boldsymbol{rr}). \tag{1.32}$$

This tensor may be called the dipole field tensor; its more usual name, however, is the dipole–dipole interaction tensor because the interaction energy of two dipoles is given by the product of one dipole moment and the field of the other dipole (see Chapter III, p. 116).

An alternative expression for the dipole field tensor can be derived from (1.14) and (1.22). Combining these equations we have:

$$\boldsymbol{E} = \operatorname{grad}\left[\left(\operatorname{grad}\frac{1}{r}\right)\cdot \boldsymbol{m}\right]. \tag{1.33}$$

Comparison with (1.31) and introduction of the nabla-operator notation (see Appendix I) now leads to:

$$\boldsymbol{T} = -\nabla\nabla\frac{1}{r}. \tag{1.34}$$

Remarks:

1. Eqn. (1.29) shows that if we resolve the vector $\boldsymbol{E}$ into a component along the radius vector $\boldsymbol{r}$ and a component parallel to the dipole vector, the latter is always in the direction opposite to the vector $\boldsymbol{m}$. Considering a plane through the centre of the dipole, perpendicular to the dipole vector, the product $\boldsymbol{m}\cdot\boldsymbol{r}$ is positive at the side of this plane on which the positive charge of the dipole is situated. Hence, at that "positive" side of the plane the component of $\boldsymbol{E}$ along the radius vector $\boldsymbol{r}$ is directed away from the dipole, whereas at the other side of the plane it is directed towards the dipole.

This corresponds with the fact that, according to (1.24), ϕ is positive at the "positive" side of this plane and negative at the "negative" side, simply because a point at the "positive" side of the plane is nearer to the positive charge of the dipole than to the negative charge.

2. If we take the z-axis along the dipole vector and the origin at the centre of the dipole, the set of eqns. (1.30) is reduced to:

$$\left.\begin{aligned} E_x &= \frac{3xz}{r^5}m \\ E_y &= \frac{3yz}{r^5}m \\ E_z &= \frac{3z^2 - r^2}{r^5} \end{aligned}\right\}. \tag{1.35}$$

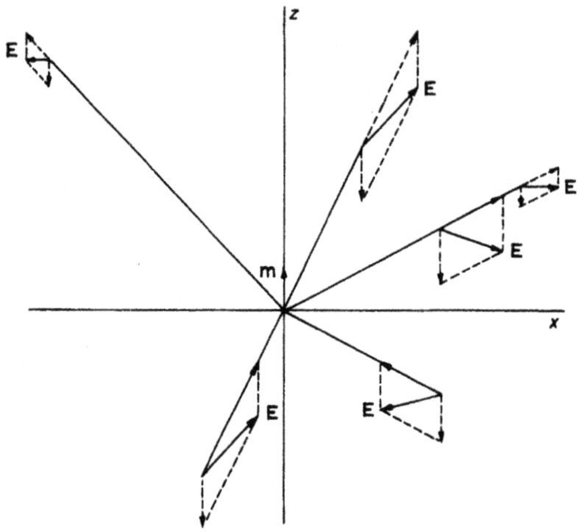

Fig. 5
The intensity of the dipole field.

For this case the relative values of $\boldsymbol{E}$ at some points in the xz-plane are shown in Fig. 5. In this figure the vectors $\boldsymbol{E}$ are resolved into a component along the radius vector and a component parallel to the dipole vector, to demonstrate the first remark made above.

In the xy-plane $\boldsymbol{m} \cdot \boldsymbol{r} = 0$, since $\boldsymbol{m}$ is perpendicular to $\boldsymbol{r}$. Hence:

$$(\boldsymbol{E})_{z=0} = -\frac{\boldsymbol{m}}{r^3}.$$

On the z-axis we have:

$$(\boldsymbol{E})_{\substack{x=0\\y=0}} = 2\frac{\boldsymbol{m}}{r^3}.$$

In Fig. 6 the relative values of $\boldsymbol{E}$ are given for some points on the z-axis and in the xy-plane.

In Fig. 7 the lines of force in the xz-plane are given for the same situation of the dipole as in Figs. 5 and 6.

For the same cases as those for which we calculated the potential due to a dipole and that due to a mono-valent ion, the calculated values of the field strength are given in Table 2.

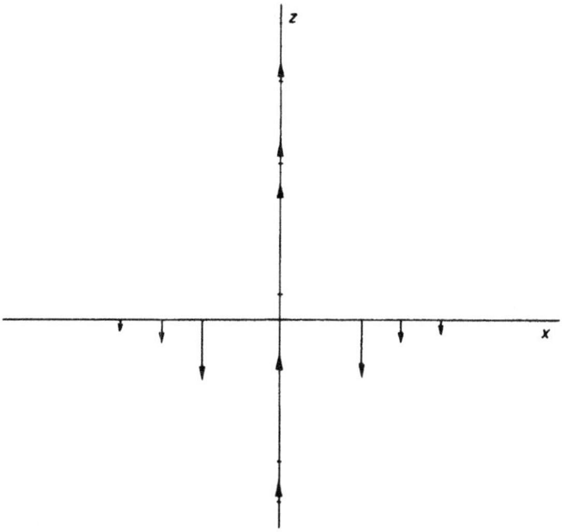

Fig. 6
Relative values of $\boldsymbol{E}$ at some points of the dipole field. The dipole is placed along the z-axis at the origin.

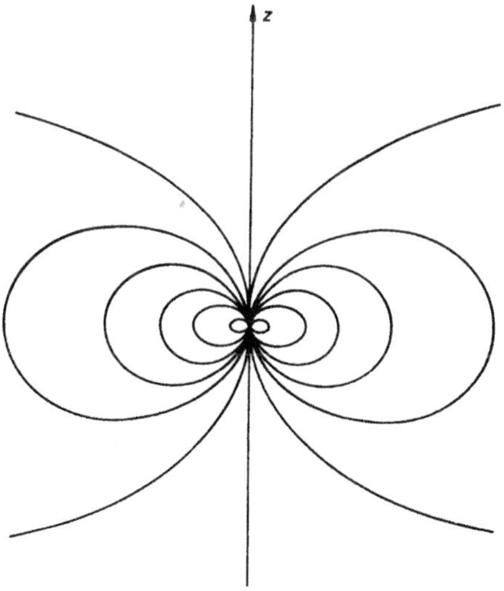

Fig. 7
Lines of force in the dipole field. The dipole is placed along the z-axis at the origin.

TABLE 2

THE FIELD STRENGTHS ON THE AXIS OF A DIPOLE AT DISTANCES OF 2Å AND 10Å COMPARED WITH THE FIELD STRENGTHS AT THE SAME DISTANCES FROM A MONOVALENT ION

r (Å)	e_0/r^2 (volts/cm)	$2m/r^3$ (volts/cm)	
		$m = 5D$	$m = 0.5D$
2	360 10^6	375 10^6	38 10^6
10	14 10^6	3 10^6	0.3 10^6

(d) *The potential due to a dipole density*

In section 2a, it has been mentioned that the function ρ, which describes the charge distribution by giving its volume density, can be derived from a distribution of point charges because for a small volume element ΔV the total charge contained within it can be represented as $\sum_{i \text{ in } \Delta V} e_i$ as well as $\iiint_{\Delta V} \rho \, \mathrm{d}v$. In the same way as the potential of a continuous charge distribution is derived from the potential of the electric point charge, we can derive from the potential of the ideal dipole the potential of a continuous dipole distribution $\boldsymbol{P}$. The equation for the potential of a discrete and a continuous charge distribution are given respectively by:

$$\phi = \sum_i \frac{e_i}{r_i}, \tag{1.36}$$

and:

$$\phi = \iiint \frac{\rho}{r} \mathrm{d}v. \tag{1.37}$$

To calculate the potential in the origin due to an arbitrary continuous dipole distribution, we shall first consider the potential in the origin due to one ideal dipole. For an ideal dipole in the origin the potential in a point with radius vector $\boldsymbol{r}$ is given by eqn. (1.22):

$$\phi = -\boldsymbol{m} \cdot \operatorname{grad} \frac{1}{r}.$$

Here the gradient is taken at the point where the potential is calculated. If this equation is transformed to a system of axes with its origin in the point where the potential is taken, we must give to the gradient the opposite sign, according to eqn. (A1.20). Hence we find for the potential in the origin due to an ideal dipole in $\boldsymbol{r}$:

$$\phi(0) = \boldsymbol{m} \cdot \operatorname{grad} \frac{1}{r}. \tag{1.38}$$

Passing to a continuous dipole distribution we then find:

$$\phi(0) = \iiint \boldsymbol{P} \cdot \operatorname{grad} \frac{1}{r} \, \mathrm{d}v. \tag{1.39}$$

For the case that the dipole distribution is limited to a certain volume V, this equation can be rewritten to a form which will be used in the next chapter. First we write as a consequence of eqn. (A1.25):

$$\operatorname{div} \frac{\boldsymbol{P}}{r} = \frac{1}{r} \operatorname{div} \boldsymbol{P} + \boldsymbol{P} \cdot \operatorname{grad} \frac{1}{r}.$$

Hence:

$$\begin{aligned} \phi(0) &= \iiint_V \operatorname{div} \frac{\boldsymbol{P}}{r} \, \mathrm{d}v - \iiint_V \frac{1}{r} \operatorname{div} \boldsymbol{P} \, \mathrm{d}v \\ &= \oiint \frac{\boldsymbol{P}}{r} \cdot \mathrm{d}\boldsymbol{S} - \iiint_V \frac{1}{r} \operatorname{div} \boldsymbol{P} \, \mathrm{d}v. \end{aligned} \tag{1.40}$$

So the potential due to a dipole distribution is equal to the superposition of the potential of a surface charge P_n on the boundary of the dipole distribution and of a charge density $-\operatorname{div} \boldsymbol{P}$ within the dipole distribution. The surface charge P_n is called the apparent charge. For a homogeneous dipole distribution, $\operatorname{div} \boldsymbol{P} = 0$, and the potential is determined solely by the apparent surface charge.

§3. Non-ideal dipoles. The description of potentials with the aid of Legendre functions

In section 2 it was shown that for a non-ideal dipole the eqns. (1.22) to (1.25), derived for an ideal dipole, are rather poor approximations for the

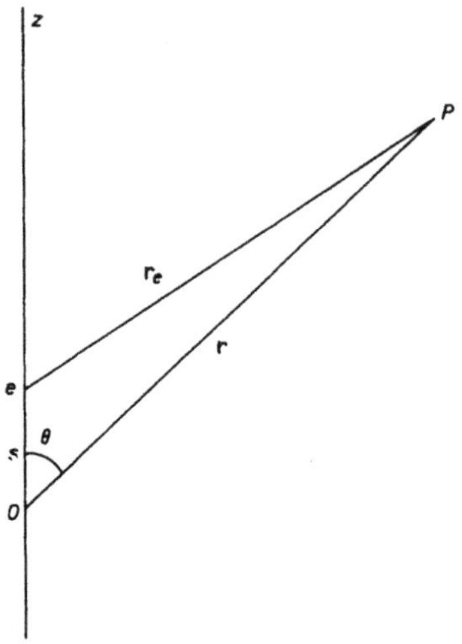

Fig. 8

potential at small distances from a non-ideal dipole. Better approximations result from developing the potential in a series of so-called Legendre functions, as will be shown in section 3a. This development of the potential will first be demonstrated for the case of a single point charge e.

We take the z-axis of the coordinate system through this point charge and the origin O at a distance s from e (Fig. 8). The potential at P due to the charge e expressed in the spherical coordinates r and θ is given by:

$$\phi(P) = \frac{e}{r_e} = \frac{e}{\sqrt{s^2 - 2sr\cos\theta + r^2}}. \tag{1.41}$$

In this equation there is no dependence on the third coordinate φ due to the axial symmetry. First, we consider the case of $r > s$. Introducing the abbreviation $\cos\theta = \tau$, eqn. (1.41) can be written as:

$$\phi(P) = \frac{e}{r}\left(1 - 2\tau\frac{s}{r} + \frac{s^2}{r^2}\right)^{-\frac{1}{2}}. \tag{1.42}$$

We use the binomial expansion:

$$(1 + x)^n = 1 + \binom{n}{1}x + \binom{n}{2}x^2 + \binom{n}{3}x^3 + \cdots,$$

where $\binom{n}{k}$ is written for $\dfrac{n!}{k!(n-k)!}$. Introduction of this expression with $n = -\frac{1}{2}$ in (1.42) gives:

$$\phi(P) = \frac{e}{r}\left[1 + \frac{-\frac{1}{2}}{1}\left(-2\tau\frac{s}{r} + \frac{s^2}{r^2}\right) + \frac{-\frac{1}{2}\cdot -\frac{3}{2}}{1\cdot 2}\left(-2\tau\frac{s}{r} + \frac{s^2}{r^2}\right)^2 + \cdots\right]$$

$$= \frac{e}{r}\left[1 + \tau\frac{s}{r} + \frac{3\tau^2 - 1}{2}\left(\frac{s}{r}\right)^2 + \frac{5\tau^3 - 3\tau}{2}\left(\frac{s}{r}\right)^3 + \cdots\right], \tag{1.43}$$

where we have collected terms with the same power of s/r. Thus, we can express $\phi(P)$ in a series of ascending powers of s/r:

$$\phi(P) = \frac{e}{r}\sum_{n=0}^{\infty}\left(\frac{s}{r}\right)^n P_n(\cos\theta). \tag{1.44}$$

The coefficients $P_n(\cos\theta)$ are called Legendre functions, Legendre polynomials or zonal surface harmonics. Comparing eqns. (1.43) and (1.44), we find:

$$P_0(\cos\theta) = 1,$$

$$P_1(\cos\theta) = \cos\theta,$$

$$P_2(\cos\theta) = \frac{3\cos^2\theta - 1}{2},$$

$$P_3(\cos\theta) = \frac{5\cos^3\theta - 3\cos\theta}{2}.$$

It can be shown that the expression (1.43) converges absolutely for $r \geqslant s$, unless $r = s$ and $\theta = 0$; this, however, is the case of the potential calculated at the point where the charge itself is situated.

For $r < s$ it is preferable to choose a series of ascending powers of r/s. In that case we write (1.41) as:

$$\phi(P) = \frac{e}{s}\left(1 - 2\tau\cdot\frac{r}{s} + \frac{r^2}{s^2}\right)^{-\frac{1}{2}}. \tag{1.45}$$

Using the binomial expansion again, we obtain:

$$\phi(P) = \frac{e}{s}\sum_{n=0}^{\infty}\left(\frac{r}{s}\right)^n P_n(\cos\theta). \tag{1.46}$$

This series converges absolutely for $r \leqslant s$ unless $r = s$ and $\theta = 0$.

A series development of a potential with the aid of Legendre functions, as represented by (1.44) or (1.46), is useful in many cases. A number of applications will be given in this section. Some properties of the Legendre functions are discussed in Appendix III, where a Table of $P_n(\cos\theta)$, up to P_7, is also given.

APPLICATIONS

(a) *The potential of a non-ideal dipole*

As an initial example of the application of eqns. (1.44) and (1.46) we will consider a non-ideal dipole, consisting of the charges $+e$ and $-e$ on the z-axis at $+s$ and $-s$ respectively. In other words: we add to the charge e in Fig. 8 a second charge $-e$ on the z-axis at the other side of the origin. For $r > s$ we develop the potential of this dipole in a series of ascending powers of s/r.

According to eqn. (1.44) the potential of the positive charge is given by:

$$\phi = \frac{e}{r}\sum_{n=0}^{\infty}\left(\frac{s}{r}\right)^n P_n(\cos\theta).$$

The potential of the negative charge is then:

$$\phi = -\frac{e}{r}\sum_{n=0}^{\infty}\left(\frac{s}{r}\right)^n P_n\{\cos(\theta+\pi)\} = -\frac{e}{r}\sum_{n=0}^{\infty}\left(\frac{s}{r}\right)^n P_n(-\cos\theta). \tag{1.47}$$

Therefore we find for the potential of the non-ideal dipole:

$$\phi = \frac{e}{r}\sum_{0}^{\infty}\left(\frac{s}{r}\right)^n\{P_n(\cos\theta) - P_n(-\cos\theta)\} =$$

$$= \frac{e}{r}\left[2\frac{s}{r}P_1(\cos\theta) + 2\frac{s^3}{r^3}P_3(\cos\theta) + 2\frac{s^5}{r^5}P_5(\cos\theta) + \cdots\right], \tag{1.48}$$

if we use the fact that $P_n(\cos\theta)$ is an even function of $\cos\theta$ if n is even, and an odd function of $\cos\theta$ if n is odd.

Introducing the dipole moment $m = 2es$ and the values of $P_n(\cos\theta)$ given above, we obtain:

$$\phi = \frac{m}{r^2}\cos\theta + \frac{ms^2}{r^4}\,\frac{5\cos^3\theta - 3\cos\theta}{2} + \cdots. \tag{1.49}$$

The first term is the value the potential would have if the dipole was ideal. The use of even just two terms leads to a much higher accuracy. This will be shown for the case treated in section 2b, remark 1. There we calculated the potential in a point P on the z-axis with $z = \frac{3}{2}l$ due to a non-ideal dipole with charges $+e$ and $-e$ situated at $z = \frac{1}{2}l$ and $z = -\frac{1}{2}l$ respectively. If we calculate $\phi(P)$, using only the first two terms of (1.49), we find:

$$\frac{m}{r^2}\cos\theta \qquad = \frac{el}{(\frac{3}{2}l)^2} \qquad = 0.4444\frac{e}{l}$$

$$\frac{ms^2}{r^4}\frac{5\cos^3\theta}{2} = \frac{el(\frac{1}{2}l)^2}{(\frac{3}{2}l)^4} = 0.0494\frac{e}{l}$$

$$\phi(P) = 0.4938\frac{e}{l}.$$

The correct value of $\phi(P)$ is $e/l - e/2l$, which is equal to $0.5000e/l$. Even at this small distance from the dipole the inaccuracy is only 1.2% by using two terms, whereas when only the first term of (1.49) is used it is 12%.

(b) *The average field inside a sphere surrounding an arbitrary system of charges*

As a second example of the application of eqn. (1.44), we shall calculate, for the case of an arbitrary system of point charges, the average field inside a sphere enclosing all the charges.

We call the radius of the sphere a and take its centre as the origin of the coordinate system. We first consider one point charge e, located, as in Fig. 8, on the z-axis at a distance s from the origin ($s < a$, see Fig. 9).

The average of the vector $\boldsymbol{E}$ is given by:

$$\bar{\boldsymbol{E}} = \frac{1}{V}\iiint_V \boldsymbol{E}\,\mathrm{d}v, \tag{1.50}$$

where $V = \frac{4}{3}\pi a^3$. Using eqns. (1.14) and (A1.34), we have:

$$\bar{\boldsymbol{E}} = -\frac{1}{V}\iiint_V \operatorname{grad}\phi\,\mathrm{d}v = -\frac{3}{4\pi a^3}\oint\!\!\oint \phi\,\boldsymbol{dS}. \tag{1.51}$$

Since there is symmetry around the z-axis, the average of the vector $\boldsymbol{E}$ is directed along the z-axis and its magnitude is given by:

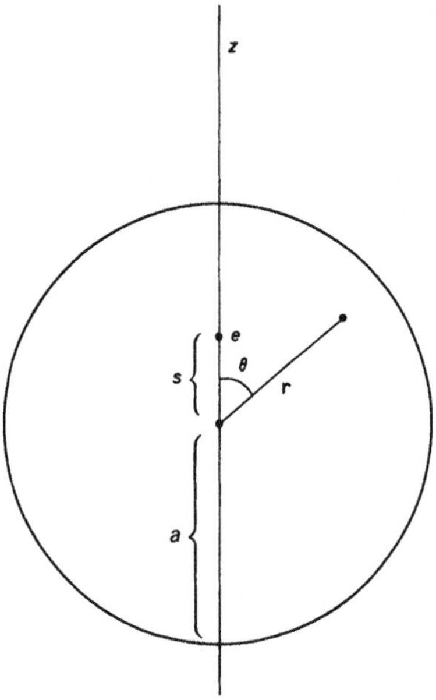

Fig. 9

$$\bar{E}_z = -\frac{3}{4\pi a^3} \oiint \phi (\mathrm{d}S)_z = -\frac{3}{4\pi a^3} \oiint \phi \cos\theta \, \mathrm{d}S. \tag{1.52}$$

The surface element of the sphere is given by $\mathrm{d}S = a^2 \sin\theta \, \mathrm{d}\theta \, \mathrm{d}\varphi$. Hence, eqn. (1.52) can be written as:

$$\bar{E} = -\frac{3}{4\pi a^3} \int_{\theta=0}^{\pi} \int_{\varphi=0}^{2\pi} \phi \cos\theta a^2 \sin\theta \, \mathrm{d}\theta \, \mathrm{d}\varphi,$$

which results in:

$$\bar{E} = -\frac{3}{2a} \int_{\cos\theta=-1}^{\cos\theta=1} \phi \cos\theta \, \mathrm{d}(\cos\theta). \tag{1.53}$$

We now write $\cos\theta = \tau$, $\tau = P_1(\tau)$, and substitute for ϕ the series development (1.44):

$$\bar{E} = -\frac{3}{2a} \int_{-1}^{+1} P_1(\tau) \frac{e}{a} \left\{ \sum_{n=0}^{\infty} \left(\frac{s}{a} \right)^n P_n(\tau) \right\} \mathrm{d}\tau. \tag{1.54}$$

Using the orthogonality of the Legendre functions, as given by eqn. (A3.27):

$$\int_{-1}^{+1} P_n(\tau)P_m(\tau)\,\mathrm{d}\tau = 0 \text{ for } n \neq m,$$

and by eqn. (A3.30):

$$\int_{-1}^{+1} P_n^2(\tau)\,\mathrm{d}\tau = \frac{2}{2n+1},$$

we find:

$$\bar{E} = -\frac{3}{2a}\frac{e}{a}\frac{s}{a}\frac{2}{3} = \frac{se}{a^3}. \tag{1.55}$$

Using the electric moment of the charge with respect to the origin $\boldsymbol{m} = se\boldsymbol{k}$, where $\boldsymbol{k}$ is the unit vector in the direction of the z-axis, we may write:

$$\bar{\boldsymbol{E}} = -\frac{\boldsymbol{m}}{a^3}. \tag{1.56}$$

This expression can easily be extended to the case of an arbitrary system of charges inside the sphere. Since the electric field is additively built up of the contributions of the individual charges, eqn. (1.56) is also valid in this case, $\boldsymbol{m}$ being the total electric moment with respect to the origin, as defined by eqn. (1.1). We recall that in the special case of a zero net charge, $\boldsymbol{m}$ is independent of the choice of the origin, being then called the dipole moment of the system. In that case the system of charges constituting the dipole can be displaced in the sphere without changing the average field $\bar{\boldsymbol{E}}$.

(c) *The potential of an ideal dipole, expressed as a function of the distance to an arbitrary origin*

As a third example of the application of eqns. (1.44) and (1.46) we consider an ideal dipole on the z-axis, at a distance s from the origin, the dipole vector $\boldsymbol{m}$ being directed along the positive z-axis. When the infinitely small distance between the negative and positive charge is $\mathrm{d}s$, the charge e is determined by the condition $m = e\,\mathrm{d}s$. The potential due to this dipole is given by the algebraic sum of the potentials due to the two charges. Thus we have:

$$\phi_m = \frac{\mathrm{d}s}{\mathrm{d}s}\{\phi_e(s + \mathrm{d}s) - \phi_e(s)\},$$

or:

$$\phi_m = \mathrm{d}s\frac{\partial \phi_e}{\partial s}, \tag{1.57}$$

where ϕ_e is the potential due to the single charge, given by (1.44) and (1.46). Since $m = e\,\mathrm{d}s$, we can write (1.57) as:

$$\phi_m = \frac{m}{e}\,\frac{\partial \phi_e}{\partial s}. \tag{1.58}$$

Combining (1.58) with (1.44) and (1.46), respectively, we have:

$$\phi_m = m\sum_{n=0}^{\infty} n\frac{s^{n-1}}{r^{n-1}}\,P_n(\cos\theta) \qquad \text{for } r > s, \tag{1.59}$$

and:

$$\phi_m = -\,m\sum_{n=0}^{\infty}(n+1)\frac{r^n}{s^{n+2}}\,P_n(\cos\theta) \qquad \text{for } r < s. \tag{1.60}$$

(d) *A conducting sphere in a homogeneous external field*

An important application of the Legendre functions is the solution of those electrostatical problems, in which certain regions contain no electric charge, so that Laplace's equation holds in these regions. This application is based on the fact that the general solution of Laplace's equation in the case of cylindrical symmetry can be expressed by an infinite series of the Legendre functions (see Appendix II).

For an uncharged conducting sphere in a homogeneous external field the solution proceeds in the following way. The potential outside the conductor is denoted by ϕ_1, the potential within the conductor by ϕ_2. Because there are no charges outside the conductor, Laplace's equation holds in this region. Further there is axial symmetry, so that according to eqn. (A2.15) we can write:

$$\phi_1 = \sum_{n=0}^{\infty} \left(A_n r^n + \frac{B_n}{r^{n+1}} \right) P_n(\cos\theta). \tag{1.61}$$

We also know that the potential in a conductor is a constant:

$$\phi_2 = C. \tag{1.62}$$

In addition to these expressions for the potential outside and inside the sphere, we must use boundary conditions to solve the problem:

1. Far removed from the conducting sphere, the potential is only determined by the external field $\boldsymbol{E}_0$:

$$(\phi_1)_{r\to\infty} = -E_0 r \cos\theta. \tag{1.63}$$

2. The potentials must be equal on both sides of the boundary:

$$(\phi_1)_{r=a} = (\phi_2)_{r=a}. \tag{1.64}$$

3. Finally, we know that the total charge of the sphere must be zero, so that according to (1.7):

$$\oint\!\!\oint \boldsymbol{E} \cdot d\boldsymbol{S} = 0, \tag{1.65}$$

if the integration is performed at the outer surface of the sphere. The problem can now be solved by combining eqns. (1.61) through (1.65).

Combining (1.61) and (1.63), and using the linear independence of the Legendre functions (see Appendix II, p. 352 and Appendix III, p. 362), we find:

$$A_n = 0 \qquad \text{if} \quad n \neq 1,$$

and:

$$A_1 = -E_0.$$

Thus:

$$\phi_1 = \sum_{n=0}^{\infty} \frac{B_n}{r^{n+1}} P_n(\cos\theta) - E_0 r \cos\theta. \tag{1.66}$$

Combining (1.66) with (1.62) and (1.64), we find:

$$\sum_{n=0}^{\infty} \frac{B_n}{a^{n+1}} P_n(\cos\theta) - E_0 a \cos\theta = C. \tag{1.67}$$

Since the Legendre polynomials are linearly independent, we obtain from (1.67):

$$B_n = 0 \quad \text{if} \quad n > 1,$$

$$\frac{B_1}{a^2} - E_0 a = 0 \quad \text{or} \quad B_1 = E_0 a^3,$$

$$\frac{B_0}{a} = C \quad \text{or} \quad B_0 = Ca.$$

Substitution of these values in eqn. (1.66) gives:

$$\phi_1 = \frac{Ca}{r} + \left(\frac{a^3 E_0}{r^2} - E_0 r\right) \cos\theta. \tag{1.68}$$

We now rewrite (1.65) as follows:

$$\oiint \left(\frac{\partial \phi_1}{\partial r}\right)_{r=a} \mathrm{d}S = 0, \tag{1.69}$$

after which substitution of (1.68) in (1.69) results in:

$$-\frac{Ca}{a^2} 4\pi a^2 = 0 \quad \text{since} \quad \oiint \cos\theta \, \mathrm{d}S = 0.$$

Hence $C = 0$ and thus, the potential outside the uncharged conducting sphere is given by:

$$\phi_1 = \left(\frac{a^3 E_0}{r^2} - E_0 r\right) \cos\theta, \tag{1.70}$$

whereas inside the sphere the potential is given by:

$$\phi_2 = 0. \tag{1.71}$$

From (1.70) we conclude that the additional field outside the sphere, due to the true surface charges on the sphere, is equal to the field that would be caused by a point dipole placed at the centre of the sphere and surrounded by a vacuum, its moment $\boldsymbol{m}$ being given by:

$$\boldsymbol{m} = a^3 \boldsymbol{E}_0. \tag{1.72}$$

According to eqn. (1.11), the surface charge density on the sphere is given by:

$$\sigma = \frac{1}{4\pi}\{(E_n)_1 - (E_n)_2\},$$

where the normal pointing away from the origin is considered to be positive. Using $(E_n)_1 = -\frac{\partial \phi_1}{\partial r}$ and $(E_n)_2 = 0$, we find:

$$\sigma = \frac{3}{4\pi} E_0 \cos\theta. \tag{1.73}$$

This charge density is illustrated in Fig. 10.

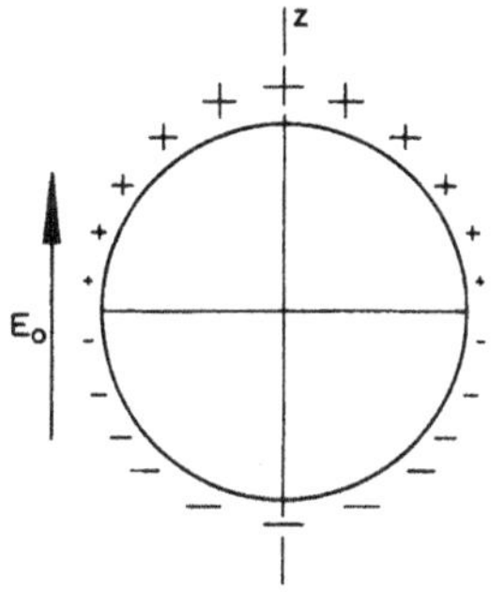

Fig. 10
Surface charge density on a conducting sphere in a uniform field.

Since the potential ϕ_2 inside the sphere is zero and the part of it due to the external field is $-E_0 z$, the potential inside the sphere due to the induced surface charges is $+E_0 z$. Hence, the field inside the sphere due to the surface charges on the sphere is uniform and equal to $-\boldsymbol{E}_0$.

§4. Axial multipoles

For the description of the potential due to a charge distribution, it is useful to extend the concept of a dipole as introduced in section 1, to the more general concept of a multipole. To this end we shall first calculate the potential due to two antiparallel ideal dipoles m, directed along the z-axis, at points $+s$ and $-s$ on the z-axis. This potential can be derived from (1.59) and (1.60):

$$\phi = m \sum_{n=0}^{\infty} n \frac{s^{n-1}}{r^{n+1}} \{P_n(\cos\theta) + P_n(-\cos\theta)\} \qquad \text{for } r > s, \tag{1.74}$$

and:

$$\phi = -m \sum_{n=0}^{\infty} (n+1) \frac{r^n}{s^{n+2}} \{P_n(\cos\theta) + P_n(-\cos\theta)\} \qquad \text{for } r < s. \quad (1.75)$$

The first terms of these series are:

$$\phi = m\left[\frac{2s}{r^3}(3\cos^2\theta - 1) + \frac{s^3}{r^5}(35\cos^4\theta - 30\cos^2\theta + 3) + \cdots\right]$$

$$\text{for } r > s, \qquad (1.76)$$

and:

$$\phi = -m\left[\frac{2}{s^2} + \frac{3r^2}{s^4}(3\cos^2\theta - 1) + \frac{5r^4}{4s^6}(35\cos^4\theta - 30\cos^2\theta + 3) + \cdots\right]$$

$$\text{for } r < s. \qquad (1.77)$$

Comparing (1.44), (1.49), and (1.76), we see that for $r > s$ the potential of a single charge, a non-ideal dipole and a combination of two dipoles are dominated by terms proportional to r^{-1}, r^{-2}, and r^{-3}, respectively. This demonstrates the increase of the compensation of the potentials due to positive and negative charges, leading to a strong decrease of the potential at large distances.

The above-mentioned combination of two anti-parallel dipoles is a fairly good model for molecules such as p-dichlorobenzene: ClCl. However, it remains an approximation, since the dipoles of the C–Cl bonds are non-ideal, although the distance between the two dipoles is much greater than the distance between the electric centres of gravity of each individual dipole. A better model for such a molecule is the combination of two non-ideal dipoles, for which the potential can also easily be derived from (1.44) and (1.46).

The system of two equal dipoles, arranged along the same axis but opposite in direction, as introduced in the above example, is called an axial quadrupole. Its quadrupole strength is defined as the product of the dipole moment* and the distance of the dipoles. Thus, in the case described above the quadrupole strength is $q = 2ms$. Similarly, an axial octupole is obtained as a

* In section 5 the distinction between the multipole moment and the corresponding multipole strength will be discussed in detail. In the case of a dipole we should speak of the dipole moment $\boldsymbol{m}$ and the corresponding dipole strength m. Since the context will always make it clear in the case of the dipole which of the two is meant, we will not always adhere to the distinction introduced here and continue to speak of a dipole moment m.

combination of two quadrupoles etc. As in the definition of the ideal dipole, we obtain an ideal axial quadrupole when we replace the dipole moment m by mn and the distance s by s/n and let n approach infinity. Axial quadrupoles and other axial multipoles are a special kind of general multipoles. The theory of the general multipoles is given in section 5.

For the ideal dipole, axial quadrupole, axial octupole etc., placed at the origin and directed along the z-axis, we have, according to the definitions:

$$\phi_m = \mathrm{d}s\frac{\partial \phi_e}{\partial s} = m\frac{\partial}{\partial z}\left(\frac{1}{r}\right),$$

$$\phi_q = \mathrm{d}s\frac{\partial \phi_m}{\partial s} = q\frac{\partial^2}{\partial z^2}\left(\frac{1}{r}\right), \text{ etc.}$$

The differentiations are performed at the origin. If we make the differentiations at the point where the potential is calculated, we have:

$$\phi_m = -m\frac{\partial}{\partial z}\left(\frac{1}{r}\right),$$

$$\phi_q = +q\frac{\partial^2}{\partial z^2}\left(\frac{1}{r}\right), \text{ etc.}$$

A multipole consisting of 2^n charges is called a 2^n-pole. According to the above calculation, the potential $\phi^{(n)}$ of an ideal axial 2^n-pole, located at the origin and directed along the z-axis, is represented by:

$$\phi^{(n)} = (-1)^n y^{(n)}\frac{\partial^n}{\partial z^n}\left(\frac{1}{r}\right), \tag{1.78}$$

where $y^{(n)}$ is the strength of the 2^n-pole.

The axial multipoles can be used for the description of the potential of an arbitrary axial charge distribution. Again we first consider the case of a single charge, represented in Fig. 8.

We call r_e the distance of the charge e to P. Then the potential at P is given by $\phi(P) = e/r_e$. For $r > s$ we develop $1/r_e$ in a Taylor series:

$$\frac{1}{r_e} = \frac{1}{r} + s\frac{\partial}{\partial z}\left(\frac{1}{r}\right) + \frac{s^2}{2!}\frac{\partial^2}{\partial z^2}\left(\frac{1}{r}\right) + \cdots,$$

where the differentiations are made at the origin. The restriction $r > s$ is introduced, since otherwise the series does not converge. Thus $\phi(P) = e/r_e$ is given by:

$$\phi(P) = \sum_{n=0}^{\infty} es^n \frac{(-1)^n}{n!} \frac{\partial^n}{\partial z^n}\left(\frac{1}{r}\right), \tag{1.79}$$

where the differentiations are made at P. Comparison with eqn. (1.78) shows that each term of this infinite series corresponds to the potential of a 2^n-pole. Thus, we find that for $r > s$ the potential of the charge e is identical to the potential of a system consisting of a point charge, an ideal dipole, an ideal axial quadrupole, and so on, all located at the origin and directed along the z-axis. The corresponding strength of the axial 2^n-pole is calculated from (1.78) and (1.79) to be:

$$y^{(n)} = \frac{es^n}{n!}. \tag{1.80}$$

In section 3, $\varphi(P) = e/r_e$ was developed in another way, which led to eqn. (1.44):

$$\phi(P) = \frac{e}{r} \sum_{n=0}^{\infty} \left(\frac{s}{r}\right)^n P_n(\cos\theta).$$

Because both (1.79) and (1.44) are valid for every s, we find:

$$P_n(\cos\theta) = \frac{(-1)^n}{n!} r^{n+1} \frac{\partial^n}{\partial z^n}\left(\frac{1}{r}\right). \tag{1.81}$$

This equation may be used for the calculation of the Legendre functions. Computation from Rodrigues' equation (eqn. (A3.37)) is much easier, however.

We now combine (1.78) and (1.81):

$$\phi^{(n)} = n! \frac{y^{(n)}}{r^{n+1}} P_n(\cos\theta). \tag{1.82}$$

In this expression for the potential of an axial 2^n-pole, the dependence on r and the dependence on θ are separated. From the factor $r^{-(n+1)}$ we conclude, as suggested at the beginning of this section, that the potentials of a charge, an ideal dipole, an ideal axial quadrupole, etc., are proportional to r^{-1}, r^{-2}, r^{-3}, etc. The appearance of the Legendre polynomial shows the close relationship between the potential of ideal axial multipoles and the corresponding Legendre polynomials.

The factor $n!$ in (1.82) vanishes if, instead of the physical definition, given at the beginning of the present section, the multipole strength of order n is defined as es^n. This definition is chosen by Stratton.[1]

For an arbitrary system of charges e_i distributed along the z-axis, we obtain, if the condition $r > s_i$ is satisfied for every i:

$$\phi(P) = \sum_i \sum_{n=0}^{\infty} e_i s_i^n \frac{(-1)^n}{n!} \frac{\partial^n}{\partial z^n}\left(\frac{1}{r}\right).$$

This expression can be rewritten as:

$$\phi(P) = \sum_{n=0}^{\infty} \frac{(-1)^n}{n!}\left[\sum_i e_i s_i^n\right] \frac{\partial^n}{\partial z^n}\left(\frac{1}{r}\right). \tag{1.83}$$

Thus, the potential of an arbitrary axial charge distribution can be written as the potential of a series of axial multipoles located at the origin, with strengths given by:

$$y^{(n)} = \frac{1}{n!}\sum_i e_i s_i^n. \tag{1.84}$$

For a continuous charge distribution along an axis, this equation must be written as:

$$y^{(n)} = \frac{1}{n!}\int \rho(s)s^n \, \mathrm{d}s. \tag{1.85}$$

Thus, we again find a potential that can be expressed in terms of potentials of ideal axial multipoles, or, equivalently, with the help of Legendre polynomials.

For the applicability of eqn. (1.83) it is not necessary that the charges be confined to the z-axis. As long as the charge distribution shows axial symmetry, we are allowed to use the general solution of Laplace's equation in terms of Legendre polynomials (see Appendix II). In this solution terms corresponding to the potentials of axial multipoles occur; these can be identified as follows. The solution is given by eqn. (A2.15):

$$\phi = \sum_{n=0}^{\infty}\left(a_n r^n + \frac{b_n}{r^{n+1}}\right)P_n(\cos\theta),$$

where a_n and b_n are constants. (This result was already used in section 3d.) In this equation the terms with b_n correspond with ideal axial 2^n-poles, as can be seen from a comparison with eqn. (1.82).

The terms with a_n have a different meaning: for $n = 0$ we have a term $\phi_0 = a_0$, corresponding to a field intensity zero; for $n = 1$ we have a term $\phi_1 = a_1 r\cos\theta = a_1 z$, corresponding to a uniform field in the direction of the z-axis; for $n = 2$ the field is a linear function of the coordinates, etc.

The difference between the two types of contributions to the potential corresponds to a difference in the sources responsible for these contributions. When in the case of a charge distribution with axial symmetry the potential at a distance r_0 from the origin is calculated, the contribution due to charges with greater distance to the origin than r_0 can be expressed as $a_n r^n P_n(\cos\theta)$. The contribution due to charges with a smaller* distance to the origin than r_0, can be expressed as $b_n r^{-n-1} P_n(\cos\theta)$. Thus the charges inside the sphere with radius r_0 need not be situated on the z-axis to give rise to axial multipole potentials, but any charge distribution showing axial symmetry will have a potential that outside the region of the charge distribution can be expressed as a sum of axial multipole potentials.

An example of a three-dimensional charge distribution whose potential is the same as the potential of an ideal dipole, was given in section 3d. It was shown there that the charge distribution on a spherical surface, represented by eqn. (1.73):

$$\sigma = \frac{3}{4\pi} E_0 \cos\theta,$$

causes outside the sphere a potential given by the first term of eqn. (1.70):

$$\phi_1 = \frac{a^3 E_0}{r^2} P_1(\cos\theta). \tag{1.86}$$

Inside the sphere the potential contribution of the surface charge on the sphere is (*cf.* p. 33):

$$\phi_2 = E_0 r P_1(\cos\theta). \tag{1.87}$$

These potentials are the terms with $n = 1$ of the general solution (A2.15) of Laplace's equation for the case of axial symmetry.

It will be shown now that for each value of n, a charge distribution on a spherical surface can be constructed in such a way that the potentials ϕ_1 and ϕ_2 are given by the corresponding terms of this general solution. This charge distribution is:

$$\sigma = \frac{C}{4\pi} \frac{2n+1}{a^{n+2}} P_n(\cos\theta), \tag{1.88}$$

where C is a constant and a is the radius of the sphere.

---

* For charges at a distance r_0, either of the series can be chosen arbitrarily.

To calculate ϕ_1 and ϕ_2 for this charge distribution, we use the boundary conditions:

1. $4\pi\sigma = \left(\frac{\partial\phi_2}{\partial r}\right)_{r=a} - \left(\frac{\partial\phi_1}{\partial r}\right)_{r=a}$, according to (1.11),
2. $(\phi_2)_{r=a} = (\phi_1)_{r=a}$,
3. $(\phi_2)_{r=0}$ must be finite,
4. $(\phi_1)_{r\to\infty} = 0$.

The charge distribution (1.88) shows axial symmetry about the z-axis. Thus the solution of Laplace's equation given in eqn. (A2.15) can be used. According to the third and fourth boundary conditions, we conclude from (A2.15) that ϕ_1 and ϕ_2 have the form:

$$\phi_1 = \sum_{m=0}^{\infty} \frac{B_m}{r^{m+1}} P_m(\cos\theta), \tag{1.89}$$

$$\phi_2 = \sum_{m=0}^{\infty} A_m r^m P_m(\cos\theta). \tag{1.90}$$

According to the first and second boundary conditions we have for all values of θ:

$$\sum_{m=0}^{\infty}\left[A_m m a^{m-1} + (m+1)\frac{B_m}{a^{m+2}}\right]P_m(\cos\theta) = C\frac{2n+1}{a^{n+2}}P_n(\cos\theta), \tag{1.91}$$

$$\sum_{m=0}^{\infty}\left[A_m\, a^m - \frac{B_m}{a^{m+1}}\right]P_m(\cos\theta) = 0. \tag{1.92}$$

Since the Legendre functions are linearly independent, it follows from (1.91) and (1.92) that

$$A_m = 0 \quad \text{for} \quad m \neq n,$$

$$B_m = 0 \quad \text{for} \quad m \neq n,$$

$$A_n = \frac{C}{a^{2n+1}},$$

$$B_n = C.$$

Thus the potential functions ϕ_1 and ϕ_2, related to the surface charge distribution (1.88), are:

$$\phi_1 = \frac{C}{r^{n+1}} P_n(\cos\theta), \tag{1.93}$$

$$\phi_2 = \frac{Cr^n}{a^{2n+1}} P_n(\cos\theta). \tag{1.94}$$

Comparing (1.93) with (1.82), we see that ϕ_1 can be identified with the potential of an ideal axial 2^n-pole placed at the centre of the sphere, directed along the z-axis, with axial multipole strength $y^{(n)} = C/n!$. This means that $C/n!$ is the axial multipole strength $y^{(n)}$ of the applied charge distribution. Thus, we can write eqns. (1.88), (1.93), and (1.94) as:

$$\sigma = \frac{n!y^{(n)}}{4\pi} \frac{2n+1}{a^{n+2}} P_n(\cos\theta), \tag{1.95}$$

$$\phi_1 = \frac{n!y^{(n)}}{r^{n+1}} P_n(\cos\theta), \tag{1.96}$$

$$\phi_2 = \frac{n!y^{(n)}r^n}{a^{2n+1}} P_n(\cos\theta). \tag{1.97}$$

1. For $n = 0$, the axial multipole strength equals the total charge e of the charge distribution. Then the above equations can be simplified to:

$$\sigma = \frac{e}{4\pi a^2}, \tag{1.98}$$

$$\phi_1 = \frac{e}{r}, \tag{1.99}$$

$$\phi_2 = \frac{e}{a}. \tag{1.100}$$

These results are in accordance with the well-known fact that a charge e, evenly distributed over a spherical surface ($\sigma = e/4\pi a^2$), gives inside the sphere a constant potential $\phi_2 = e/a$ (zero field), whereas outside the sphere we have the Coulomb-potential $\phi_1 = e/r$.

2. For $n = 1$, we can replace $y^{(n)}$ by the dipole strength m of the system. We then have:

$$\sigma = \frac{3m}{4\pi a^3} \cos\theta, \tag{1.101}$$

$$\phi_1 = \frac{m}{r^2} \cos\theta, \tag{1.102}$$

$$\phi_2 = \frac{mr}{a^3}\cos\theta. \tag{1.103}$$

Inside the sphere the field is uniform. Outside the sphere we have the field of an ideal dipole.

For an uncharged conducting sphere in a uniform field we have, according to (1.73), a surface charge distribution of the type (1.101), and the dipole moment is given by eqn. (1.72):

$$m = a^3 E_0.$$

Using this value of m, we can check that the potential eqns. (1.86) and (1.87) are in accordance with (1.102) and (1.103).

3. For $n = 2$ we calculate from (1.95), (1.96), and (1.97):

$$\sigma = \frac{5q}{4\pi a^4}(3\cos^2\theta - 1), \tag{1.104}$$

$$\phi_1 = \frac{q}{r^3}(3\cos^2\theta - 1), \tag{1.105}$$

$$\phi_2 = \frac{qr^2}{a^5}(3\cos^2\theta - 1), \tag{1.106}$$

where q is a constant, equal to the axial quadrupole strength of the surface charge distribution.

The charge distribution given by (1.104) is illustrated in Fig. 11. Since $P_2(\cos\theta) = 0$ for $\cos\theta_1 = 1/\sqrt{3}$ and $\cos\theta_2 = -1/\sqrt{3}$, the sphere is divided into three zones by the parallels $\theta = \theta_1$ and $\theta = \theta_2$. On these parallels the surface charge density is zero. In the zone between the parallels it is negative and in the other two zones it is positive.

The field outside the sphere is the field of an ideal axial quadrupole placed at the centre of the sphere and directed along the z-axis. Inside the sphere the potential is, according to (1.106), proportional to $(z^2 - \frac{1}{2}x^2 - \frac{1}{2}y^2)$. Therefore the field strength is a linear function of the coordinates.

For each value of n the surface charge distribution, given by (1.88), consists of $(n + 1)$ zones, separated by parallels where $P_n(\cos\theta) = 0$. The fact that the parallels $P_n(\cos\theta) = 0$ divide a spherical surface into zones explains why the Legendre functions are often called zonal surface harmonics.

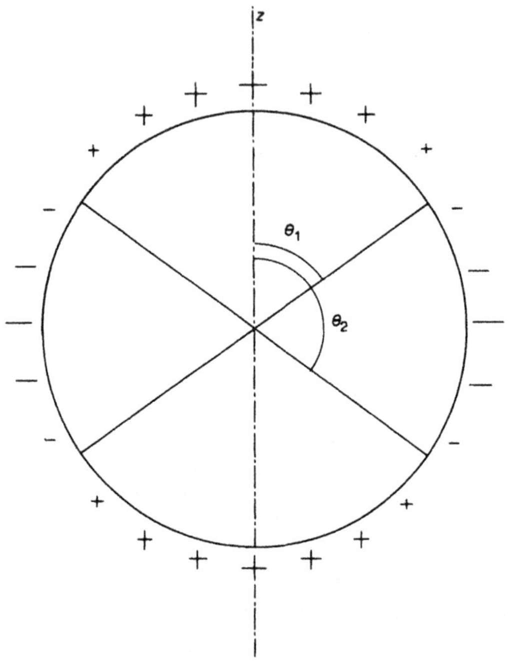

Fig. 11
Charge distribution on spherical surface, corresponding with an ideal axial quadrupole at the centre of the sphere.

§5. General multipoles

If we do not limit ourselves to charge distributions with axial symmetry, we will need general multipoles to describe the potential due to the charge distribution.

We will consider a system of point charges e_i at distances $r_i(x_i y_i z_i)$ from an origin O, chosen somewhere in between the charges. We shall calculate the potential due to this system at a point P with coordinates x, y, z (Fig. 12). We assume $r > r_i$ for all values of i. Thus, we calculate the potential outside a sphere containing all the charges. We call s_i the distance of a charge e_i to P. The potential at P due to this system of charges is then given by:

$$\phi(P) = \sum_i \frac{e_i}{s_i}. \tag{1.107}$$

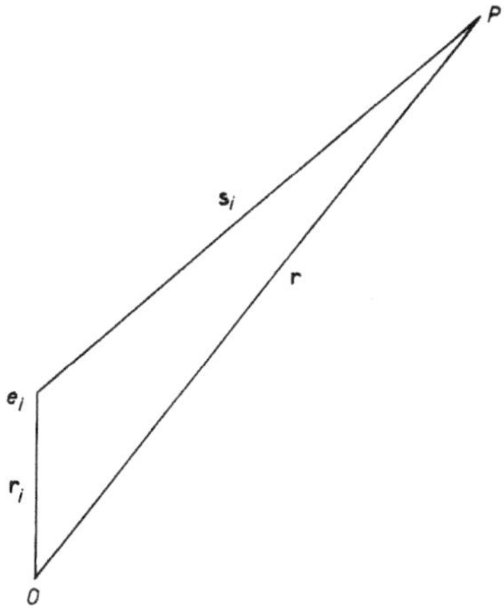

Fig. 12

We develop $1/s_i$ in a Taylor series:

$$\frac{1}{s_i} = \frac{1}{r} + x_i\frac{\partial}{\partial x}\left(\frac{1}{r}\right) + y_i\frac{\partial}{\partial y}\left(\frac{1}{r}\right) + z_i\frac{\partial}{\partial z}\left(\frac{1}{r}\right) + \\ + \tfrac{1}{2}\left[x_i^2\frac{\partial^2}{\partial x^2}\left(\frac{1}{r}\right) + 2x_iy_i\frac{\partial^2}{\partial x\partial y}\left(\frac{1}{r}\right) + \cdots\right] + \cdots,$$

where the differentiations are made at the origin. Thus we have:

$$\phi(P) = \sum_i\frac{e_i}{r} - \sum_i\left[e_ix_i\frac{\partial}{\partial x}\left(\frac{1}{r}\right) + e_iy_i\frac{\partial}{\partial y}\left(\frac{1}{r}\right) + e_iz_i\frac{\partial}{\partial z}\left(\frac{1}{r}\right)\right] + \\ + \tfrac{1}{2}\sum_i\left[e_ix_i^2\frac{\partial^2}{\partial x^2}\left(\frac{1}{r}\right) + 2e_ix_iy_i\frac{\partial^2}{\partial x\partial y}\left(\frac{1}{r}\right) + \cdots\right] - \cdots, \tag{1.108}$$

where the differentiations are made at P.

The first term of this series is the potential that would be caused at P by a charge $e = \sum_i e_i$, placed at the origin.

Introducing the electric moment $\boldsymbol{m}$ of the system of charges relative to the origin, which is, according to eqn. (1.1), given by:

$$\boldsymbol{m} = \sum_i e_i\boldsymbol{r}_i,$$

we can write the second term of the series in (1.108) as $-\boldsymbol{m}\cdot\text{grad}\,1/r$, where the gradient is taken at P. According to (1.22), this is the potential at P

caused by a point dipole $\boldsymbol{m}$ at the origin. Thus the second term of the series is the potential that would be caused at P by an ideal dipole at the origin, its moment being equal to the electric moment of the system of charges relative to the origin.

Similarly, we can describe the third and subsequent terms of (1.108) by introducing, besides the concepts single charge (unipole) and dipole, the concepts quadrupole, octupole and multipoles of higher order. In section 4 we defined the axial multipoles, connected with an axial charge distribution. For the description of the fields due to an arbitrary charge distribution, we must use the general multipoles.

Analogously to the definition of the vectorial dipole moment $\boldsymbol{m}$, the quadrupole moment $\boldsymbol{Q}$ is defined as a tensor:

$$\boldsymbol{Q} = \frac{1}{2!}\sum_i e_i \boldsymbol{r}_i \boldsymbol{r}_i. \tag{1.109}$$

In the same way, the octupole moment $\boldsymbol{U}$ is defined as a tensor of the third degree:

$$\boldsymbol{U} = \frac{1}{3!}\sum_i e_i \boldsymbol{r}_i \boldsymbol{r}_i \boldsymbol{r}_i, \tag{1.110}$$

and the hexadecapole moment* and higher multipole moments are defined analogously as tensors of the fourth degree, of the fifth degree, etc. Using these multipole moments, eqn. (1.108) can be rewritten as:

$$\phi(P) = e\frac{1}{r} - \boldsymbol{m}\cdot\nabla\frac{1}{r} + \boldsymbol{Q}:\nabla\nabla\frac{1}{r} - \boldsymbol{U}\vdots\nabla\nabla\nabla\frac{1}{r} + \cdots. \tag{1.111}$$

So far the quadrupole, octupole, and higher moments have been simply abbreviations for different summations. We shall now, as in section 1 for the dipole moment, introduce non-ideal and ideal multipoles. It will be shown that the potential due to an ideal multipole has the same form as the corresponding term in the multipole expansion for the potential given by eqn. (1.111). For the non-ideal multipole the potential will be dominated by this term.

---

* The names *dipole*, *quadrupole*, *octupole*, and *hexadecapole* have been chosen in agreement with common usage. In principle, it is preferable to derive all names in the same way, *i.e.* from Latin ordinals. Thus we have *unipole*, instead of *monopole*, as used in magnetic theory. Since *hexadecapole* is more common than the correct *sedecipole*, we have used the former.

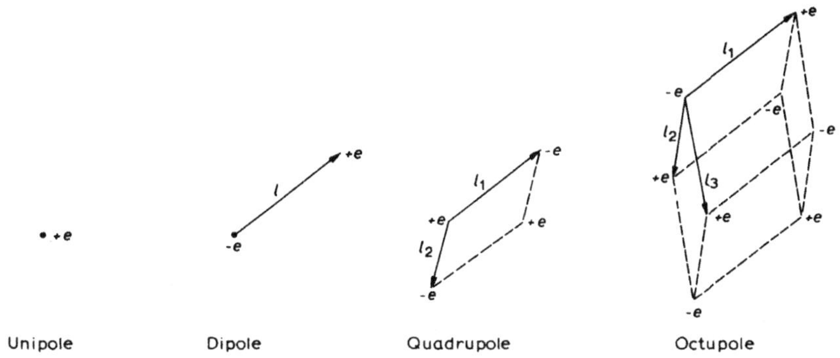

Fig. 13
Multipoles.

The electric quadrupole is obtained from a dipole in the same way as the dipole is obtained from a single charge: to construct the quadrupole from a dipole we imagine a second dipole, parallel to the original one, that would be obtained by displacing the latter over a certain distance in a direction arbitrary to that of the dipole vector. Then we interchange the charges of the original dipole. The combination of two anti-parallel dipoles thus obtained is a quadrupole (Fig. 13). The higher multipoles are similarly defined, *e.g.* an octupole is obtained by displacing a quadrupole over a certain distance in an arbitrary direction and changing the sign of all the charges of the original quadrupole (Fig. 13).

Where convenient we shall abbreviate the terms dipole, quadrupole, etc. to 2-pole, 2^2-pole, 2^3-pole, ... 2^n-pole, etc.

It must be emphasized that the increase in spatial dimensions stops at the octupole. All 2^n-poles with $n \geqslant 3$ can be constructed in ordinary three-dimensional space.

By analogy with the dipole strength m, we define the quadrupole strength:

$$q = el_1 l_2, \tag{1.112}$$

and the octupole strength:

$$u = el_1 l_2 l_3, \tag{1.113}$$

where l_1, l_2, and l_3 are the sides of the parallelogram and the parallelopiped respectively (Fig. 13).

The multipole strength, which for every kind of multipole is a scalar quantity, must be distinguished from the multipole moment, as defined above, which is a tensorial quantity. Only in the first two cases is there a simple relation between multipole moment and multipole strength: for a charge, the unipole moment and the unipole strength are identical; for a dipole the dipole strength is the absolute value of the dipole moment. For a quadrupole this relation is more complicated; it will be derived below (eqn. (1.122)).

Ideal multipoles are defined in the same way as ideal dipoles. We let the distances approach zero and at the same time increase the charges to such an extent that the strength remains constant. An ideal 2^n-pole is fixed by $(2n + 1)$ quantities: the strength and n directions, given by $2n$ angles.

All the above definitions are in accordance with the definitions given in section 4 for the special case of axial multipoles.

We shall now calculate the potentials due to the ideal multipoles defined above. In section 2 it was shown that at a point P at a distance r from an ideal dipole $\boldsymbol{m}$, the potential due to the dipole is (eqn. (1.22)):

$$\phi(P) = -\boldsymbol{m} \cdot \operatorname{grad} \frac{1}{r},$$

where the gradient is taken at the point P. Using the unit vector $\boldsymbol{n}$ in the direction of the dipole vector (see Appendix I, section 2d):

$$\boldsymbol{n} = \alpha \boldsymbol{i} + \beta \boldsymbol{j} + \gamma \boldsymbol{k},$$

we can write this in the form:

$$\phi_m = -m(\boldsymbol{n} \cdot \nabla)\frac{1}{r}$$

or:

$$\phi_m = -m\left(\alpha\frac{\partial}{\partial x} + \beta\frac{\partial}{\partial y} + \gamma\frac{\partial}{\partial z}\right)\frac{1}{r}. \tag{1.114}$$

This expression can easily be extended to the case of an ideal multipole. The potential of an ideal dipole was derived from the potential of a unipole (single charge) by using the fact that the dipole is obtained by displacing a unipole over a distance l_1 in a direction $\boldsymbol{n}_1$, and combining it with a unipole of the opposite sign, placed at the point where the displaced unipole was originally located. Combining this with the definition of the gradient we have:

$$\phi_m = -l_1 \boldsymbol{n}_1 \cdot \operatorname{grad} \phi_e, \tag{1.115}$$

which leads to eqns. (1.22) and (1.114).

Since the quadrupole is similarly obtained by displacing a dipole in a direction $\boldsymbol{n}_2$ over a distance l_2, we have, as in eqn. (1.115):

$$\phi_q = -l_2 \boldsymbol{n}_2 \cdot \operatorname{grad} \phi_m. \tag{1.116}$$

Using (1.115) and the nabla-notation we have:

$$\phi_q = q(\boldsymbol{n}_2 \cdot \nabla)(\boldsymbol{n}_1 \cdot \nabla)\frac{1}{r}, \tag{1.117}$$

where q is the strength of the quadrupole.

Similarly:

$$\phi_u = -u(\boldsymbol{n}_3 \cdot \nabla)(\boldsymbol{n}_2 \cdot \nabla)(\boldsymbol{n}_1 \cdot \nabla)\frac{1}{r}, \tag{1.118}$$

where u is the strength of the octupole.

Working out the products $\boldsymbol{n} \cdot \nabla$ in (1.117), we obtain:

$$\phi_q = q\left(\alpha_2 \frac{\partial}{\partial x} + \beta_2 \frac{\partial}{\partial y} + \gamma_2 \frac{\partial}{\partial z}\right)\left(\alpha_1 \frac{\partial}{\partial x} + \beta_1 \frac{\partial}{\partial y} + \gamma_1 \frac{\partial}{\partial z}\right)\frac{1}{r},$$

or:

$$\phi_q = q\left[\alpha_2\alpha_1 \frac{\partial^2}{\partial x^2} + (\alpha_2\beta_1 + \alpha_1\beta_2)\frac{\partial^2}{\partial x \partial y} + \beta_2\beta_1 \frac{\partial^2}{\partial y^2} + \cdots\right]\frac{1}{r}. \tag{1.119}$$

The coefficients of this equation, $q\alpha_2\alpha_1$, $q(\alpha_2\beta_1 + \alpha_1\beta_2)$, $q\beta_2\beta_1$, etc. are related to the components of the quadrupole moment $\boldsymbol{Q}$ of the ideal quadrupole. According to (1.109) and Fig. 13, this moment is given by:

$$\begin{aligned} \boldsymbol{Q} &= \tfrac{1}{2}e[\boldsymbol{r}_0\boldsymbol{r}_0 - (\boldsymbol{r}_0 + \boldsymbol{n}_1 l_1)(\boldsymbol{r}_0 + \boldsymbol{n}_1 l_1) - (\boldsymbol{r}_0 + \boldsymbol{n}_2 l_2)(\boldsymbol{r}_0 + \boldsymbol{n}_2 l_2) + \\ &\quad + (\boldsymbol{r}_0 + \boldsymbol{n}_1 l_1 + \boldsymbol{n}_2 l_2)(\boldsymbol{r}_0 + \boldsymbol{n}_1 l_1 + \boldsymbol{n}_2 l_2)] \\ &= \tfrac{1}{2}e[\boldsymbol{n}_1\boldsymbol{n}_2 l_1 l_2 + \boldsymbol{n}_2\boldsymbol{n}_1 l_1 l_2] = \tfrac{1}{2}q[\boldsymbol{n}_1\boldsymbol{n}_2 + \boldsymbol{n}_2\boldsymbol{n}_1]. \end{aligned} \tag{1.120}$$

Hence, the components of $\boldsymbol{Q}$ are given by:

$$\left.\begin{matrix} q_{11} = q\alpha_1\alpha_2 & q_{12} = q_{21} = \tfrac{1}{2}q(\alpha_1\beta_2 + \alpha_2\beta_1) \\ q_{22} = q\beta_1\beta_2 & q_{13} = q_{31} = \tfrac{1}{2}q(\alpha_1\gamma_2 + \alpha_2\gamma_1) \\ q_{33} = q\gamma_1\gamma_2 & q_{23} = q_{32} = \tfrac{1}{2}q(\beta_1\gamma_2 + \beta_2\gamma_1) \end{matrix}\right\}. \tag{1.121}$$

With the help of the conditions on the direction cosines:

$$\alpha_1^2 + \beta_1^2 + \gamma_1^2 = 1,$$

$$\alpha_2^2 + \beta_2^2 + \gamma_2^2 = 1,$$

we can solve for q. We then find the following relation between the quadrupole strength q and the components of the quadrupole moment $\boldsymbol{Q}$:

$$q^2 = \sum_{i=1}^{3} \sum_{j=1}^{3} (2q_{ij}^2 - q_{ii}q_{jj}). \tag{1.122}$$

Combining (1.120) with (1.117), we find for the potential of the ideal quadrupole:

$$\phi_q = \boldsymbol{Q}:\nabla\nabla\frac{1}{r}. \tag{1.123}$$

Comparing (1.123) with (1.111), we see that the potential of an ideal quadrupole is indeed wholly given by the quadrupole term in the multipole expansion.

Although it is easy to find the components of the quadrupole moment of the ideal quadrupole, it is generally impossible to construct an ideal quadrupole with a given quadrupole moment $\boldsymbol{Q}$. For a dipole and a dipole moment, both procedures are possible; the reason for the difficulties with quadrupoles is the following.

For both the quadrupole moment of an arbitrary charge distribution and the quadrupole moment of an ideal quadrupole it holds, that there are six quantities in these tensors. For an ideal quadrupole they are given by (1.121).

From the way the direction cosines occur in eqn. (1.121) it follows that the six components of an ideal quadrupole, as given in (1.121), are not independent of each other. This is in accordance with the fact that the ideal quadrupole should be fixed by five quantities. After some calculations, the following relation between them can be established:

$$q_{11}q_{22}q_{33} + 2q_{12}q_{13}q_{23} - q_{11}q_{23}^2 - q_{22}q_{13}^2 - q_{33}q_{12}^2 = 0. \tag{1.124}$$

Using the determinant notation we can write (1.124) as

$$|Q| = 0 \tag{1.125}$$

From the theory of matrix algebra we know that such an equation expresses the fact that the rows of the matrix are linearly dependent or, equivalently, that they represent a two-dimensional (plane) quantity.

Eqn. (1.124) is a condition on the six components of a quadrupole moment that will not be fulfilled for an arbitrary quadrupole moment.

Therefore for an arbitrary quadrupole moment $\boldsymbol{Q}$, it is in general impossible to construct an ideal quadrupole with the same quadrupole moment.

However, it is always possible to find an ideal quadrupole that gives rise to the same potential as the contribution of the quadrupole moment $\boldsymbol{Q}$ to the potential (*cf.* eqn. (1.111)). This quadrupole should have a quadrupole moment $\boldsymbol{Q}'$, such that:

$$\boldsymbol{Q}:\nabla\nabla\frac{1}{r} = \boldsymbol{Q}':\nabla\nabla\frac{1}{r}. \tag{1.126}$$

If the products in eqn. (1.126) are written out, we obtain terms of the type $q_{11}\dfrac{\partial^2}{\partial x^2}\left(\dfrac{1}{r}\right)$ as well as mixed terms like $q_{12}\dfrac{\partial^2}{\partial x\partial y}\left(\dfrac{1}{r}\right)$. From the fact that $\Delta\dfrac{1}{r} = 0$, or:

$$\left(\frac{\partial^2}{\partial x^2} + \frac{\partial^2}{\partial y^2} + \frac{\partial^2}{\partial z^2}\right)\frac{1}{r} = 0,$$

we now infer that a quadrupole moment $\boldsymbol{Q}''$, given by:

$$\boldsymbol{Q}'' = q_0\,\boldsymbol{I} \tag{1.127}$$

will not lead to a contribution to the potential. For in this case all mixed terms in the expression for the potential will disappear, and only the diagonal terms are left, *i.e.*:

$$q_0\left(\frac{\partial^2}{\partial x^2} + \frac{\partial^2}{\partial y^2} + \frac{\partial^2}{\partial z^2}\right)\frac{1}{r} = 0. \tag{1.128}$$

Thus, we find that we may take:

$$\boldsymbol{Q}' = \boldsymbol{Q} - \boldsymbol{Q}'' = \boldsymbol{Q} - q_0\boldsymbol{I}. \tag{1.129}$$

According to (1.128), such a quadrupole moment satisfies eqn. (1.126). Since q_0 is as yet undetermined, we may take for q_0 that value which causes the components of $\boldsymbol{Q}'$ to satisfy eqn. (1.124). Substituting the components of $\boldsymbol{Q}'$ in eqn. (1.124) we obtain an equation of the third degree in q_0, so that we will have three, possibly different, solutions.

Using the determinant eqn. (1.125) we now have:

$$|\boldsymbol{Q}'| = |\boldsymbol{Q} - q_0\boldsymbol{I}| = 0.$$

Consequently we must solve the secular equation for Q; the solutions for q_0 will be the three eigenvalues of the matrix Q.

Only one of the solutions for q_0 will reduce the original quadrupole moment $\boldsymbol{Q}$ to the quadrupole moment $\boldsymbol{Q}'$ of an ideal quadrupole. The other solutions will also effect a reduction to a charge distribution in a plane; these charge distributions, however, cannot be constructed by a simple displacement and inversion of a dipole.

The structure of these charge distributions is best found after first transforming $\boldsymbol{Q}$ to principal axes (see Appendix I, section 6). In the new coordinate system $\boldsymbol{Q}$ will be given by the diagonal matrix Q^*, with diagonal elements q_{11}^*, q_{22}^*, q_{33}^*, and all non-diagonal elements equal to zero. In this coordinate system the reduced quadrupole moment $\boldsymbol{Q}'$ will be given by another diagonal matrix:

$$Q'^* = Q^* - q_0 I. \tag{1.131}$$

Eqn. (1.124) should still hold; upon substitution we now find:

$$(q_{11}^* - q_0)(q_{22}^* - q_0)(q_{33}^* - q_0) = 0. \tag{1.132}$$

Thus, the three solutions for q_0 are equal to the diagonal elements of Q^*, that is, the eigenvalues of Q.

We now compare in the new coordinate system (ξ, η, ζ) the quadrupole contribution to the potential given by:

$$\phi_q = \left(q_{11}^* \frac{\partial^2}{\partial \xi^2} + q_{22}^* \frac{\partial^2}{\partial \eta^2} + q_{33}^* \frac{\partial^2}{\partial \zeta^2}\right)\frac{1}{r}, \tag{1.133}$$

with the potential of an axial quadrupole for instance directed along the ξ-axis. According to eqn. (1.78) this potential is given by:

$$\phi^{(2)} = y^{(2)} \frac{\partial^2}{\partial \xi^2}\left(\frac{1}{r}\right). \tag{1.134}$$

It appears that the quadrupole contribution to the potential may be considered as the superposition of the potentials of three axial quadrupoles along the principle axes of the quadrupole moment tensor. From (1.131) it

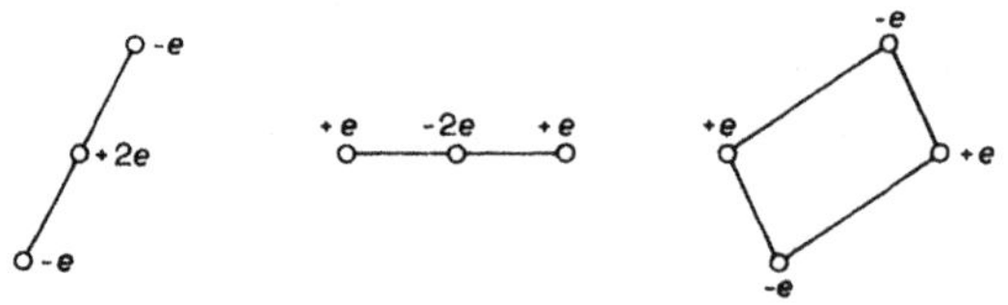

Fig. 14

follows that if we subtract from all three axial quadrupoles the value of one of them, the potential of the system will not change and the new quadrupole will obey relation (1.124).

We have characterized a quadrupole as consisting of four point charges at the corners of a parallelogram. It can therefore be considered as the superposition of two axial quadrupoles with opposite signs along the diagonals of the parallelogram in which the central charges in the intersecting point of the diagonals cancel each other (Fig. 14).

Not every combination of two non-ideal axial quadrupoles will result in a general quadrupole. This will only be the case, if the axial quadrupoles have a central charge, and if these central charges are of opposite signs and equal magnitudes. If the axial quadrupoles have opposite signs, however, it is always possible to replace them, by changing the charges and the distances between the charges, with axial quadrupoles having the same strengths and directed along the same axes as the original quadrupoles, but with central charges compensating each other. The potential of the general quadrupole obtained in this way differs from the potential of the original combination of the two axial quadrupoles only by its octupole and higher terms. In the case of ideal quadrupoles these terms do not exist. Therefore two ideal axial quadrupoles together always form an ideal general quadrupole, provided that the signs of the axial quadrupoles are opposite.

Returning to the question of the three values of q_0, we see that if we subtract the lowest value found for q_0 we find two axial quadrupoles of positive strength, and subtracting the highest value, we find two axial quadrupoles of negative strength. Only if we use the third (middle) value, will we find two axial quadrupoles with opposite signs, together forming a general quadrupole. So there is always one and only one way by which the quadrupole moment of an arbitrary distribution can be reduced to the moment of an ideal quadrupole.

If we want to compare the quadrupole moments of different charge distributions (*e.g.* molecules), they must have been previously standardized. This can be done by reducing them to the moments of ideal quadrupoles, as pointed out above. Most authors, for instance Buckingham,[2] Kielich,[3] Stogryn and Stogryn,[4] and Krishnaji and Prakash,[5] choose another way of normalization, however. They define the quadrupole moment as a tensor:

$$\boldsymbol{Q}_{\text{norm}} = \tfrac{1}{2}\sum_i e_i(3\boldsymbol{r}_i\boldsymbol{r}_i - r_i^2\boldsymbol{I}), \qquad (1.135)$$

in which:

$$r_i^2 = x_i^2 + y_i^2 + z_i^2.$$

From this definition it follows that the trace is zero (*cf.* eqn. (A1.27)):

$$\mathrm{Tr}\,(\boldsymbol{Q}_{\text{norm}}) = 0. \tag{1.136}$$

This tensor can be derived from the quadrupole moment as defined in (1.109) by choosing for q_0 in eqn. (1.129) the value $q_0 = \frac{1}{3}\mathrm{Tr}\,(\boldsymbol{Q})$ and afterwards multiplying all elements by 3. The advantage of this normalization to a traceless tensor or deviator is that it is much easier to perform than solving the secular equation to find the appropriate value for q_0.

In the special case of an axial quadrupole, however, the definition of (1.135) will result in three non-zero elements after diagonalization. Normalization according to condition (1.124) will under the same circumstances effect a reduction to a matrix with only one non-zero element, equal to the quadrupole strength as given in eqn. (1.84). If eqn. (1.135) is used for systems with an axial quadrupole moment and the z-axis is taken along the axis of this quadrupole moment, one finds:

$$q_{33} = -2q_{11} = -2q_{22}. \tag{1.137}$$

The value of q_{33} is often used as the scalar quadrupole strength; it coincides with the quadrupole strength according to Stratton's definition. If the definition of (1.135) is used, eqn. (1.111) for the potential no longer holds; the quadrupole contribution to the potential is then given by $\frac{1}{3}\boldsymbol{Q}_{\text{norm}}:\nabla\nabla\frac{1}{r}$.

Hirschfelder *et al.*[6] define the quadrupole moment as a tensor with elements twice as large as in eqn. (1.135):

$$\boldsymbol{Q}_{\mathrm{H}} = \sum_i e_i(3\boldsymbol{r}_i\boldsymbol{r}_i - r_i^2\boldsymbol{I}). \tag{1.138}$$

In this case the quadrupole contribution to the potential is given by $\frac{1}{6}\boldsymbol{Q}_{\mathrm{H}}:\nabla\nabla\frac{1}{r}$.

Apart from the different definitions of the quadrupole moment, another reason why different values are given by different authors to characterize the molecular quadrupole moment, is that some authors give the average value of the quadrupole strength of rotating molecules. Clearly, great caution must be used in comparing values given by different authors.

The quadrupole moment of an ideal quadrupole is given by eqn. (1.120):

$$\boldsymbol{Q} = \tfrac{1}{2}q(\boldsymbol{n}_1\boldsymbol{n}_2 + \boldsymbol{n}_2\boldsymbol{n}_1).$$

Using the symmetrant notation (see Appendix I, section 6), this can be written as:

$$\boldsymbol{Q} = q \operatorname{Sym}(\boldsymbol{n}_1\boldsymbol{n}_2). \tag{1.139}$$

We shall prove that the moment of an ideal 2^n-pole is given analogously by:

$$\begin{aligned}\boldsymbol{Y}^{(n)} &= \frac{1}{n!}y^{(n)}(\boldsymbol{n}_1\boldsymbol{n}_2\cdots\boldsymbol{n}_n + \boldsymbol{n}_1\boldsymbol{n}_2\cdots\boldsymbol{n}_n\boldsymbol{n}_{n-1} + \cdots + \boldsymbol{n}_n\boldsymbol{n}_{n-1}\cdots\boldsymbol{n}_1)\\ &= y^{(n)}\operatorname{Sym}(\boldsymbol{n}_1\boldsymbol{n}_2\cdots\boldsymbol{n}_n),\end{aligned} \tag{1.140}$$

where $y^{(n)}$ is the multipole strength.

To this end we use the method of mathematical induction: supposing that eqn. (1.140) holds for a 2^n-pole, we prove that it then also holds for a 2^{n+1}-pole. A 2^{n+1}-pole is formed from a 2^n-pole in the following way. The charges of a second 2^n-pole equal to the original one are placed at a distance l_{n+1} in the direction of a unit vector $\boldsymbol{n}_{n+1}$. Then the signs of the charges of the original 2^n-pole are interchanged. Hence, the moment of the 2^{n+1}-pole, defined in analogy to eqns. (1.109) and (1.110), is:

$$\boldsymbol{Y}^{(n+1)} = \frac{1}{(n+1)!}\sum_{i=1}^{2^{n+1}} e_i(\boldsymbol{r}_i)^{n+1}, \tag{1.141}$$

where $(\boldsymbol{r}_i)^{n+1}$ denotes a $(n+1)$-times repeated tensor product. Eqn. (1.141) can be written as:

$$\boldsymbol{Y}^{(n+1)} = \frac{1}{(n+1)!}\sum_{i=1}^{2^n}\{-e_i(\boldsymbol{r}_i)^{n+1} + e_i(\boldsymbol{r}_i + l_{n+1}\boldsymbol{n}_{n+1})^{n+1}\}. \tag{1.142}$$

Using the binomial expansion (A1.76) we get:

$$\begin{aligned}\boldsymbol{Y}^{(n+1)} = \frac{1}{(n+1)!}\sum_{i=1}^{2^n}\Bigg\{&-e_i(\boldsymbol{r}_i)^{n+1} +\\ &+ e_i\sum_{m=0}^{n+1}\binom{n+1}{m}\operatorname{Sym}[(\boldsymbol{r}_i)^m(l_{n+1}\boldsymbol{n}_{n+1})^{n+1-m}]\Bigg\}.\end{aligned} \tag{1.143}$$

The terms in $(\boldsymbol{r}_i)^{n+1}$ cancel each other, so after interchanging the order of summation we obtain:

$$\boldsymbol{Y}^{(n+1)} = \frac{1}{(n+1)!} \sum_{m=0}^{n} \binom{n+1}{m} (l_{n+1})^{n+1-m} \mathrm{Sym}\left[(\boldsymbol{n}_{n+1})^{n+1-m} \sum_{i=1}^{2^n} e_i(\boldsymbol{r}_i)^m \right] \cdot \quad (1.144)$$

We now use the fact that the charges $i = 1, 2, \ldots, 2^n$ form a 2^n-pole. Hence all lower multipole moments are zero, so that for $m < n$:

$$\sum_{i=1}^{2^n} e_i(\boldsymbol{r}_i)^m = m!\boldsymbol{Y}^{(m)} = \boldsymbol{0}. \quad (1.145)$$

We further use the supposition (1.140):

$$\boldsymbol{Y}^{(n)} = \frac{1}{n!} \sum_{i=1}^{2^n} e_i(\boldsymbol{r}_i)^n = y^{(n)} \,\mathrm{Sym}\,(\boldsymbol{n}_1 \boldsymbol{n}_2 \ldots \boldsymbol{n}_n), \quad (1.146)$$

and find from (1.144):

$$\begin{aligned} \boldsymbol{Y}^{(n+1)} &= \frac{1}{(n+1)!} \binom{n+1}{n} l_{n+1} \,\mathrm{Sym}\,[n! y^{(n)} \boldsymbol{n}_{n+1} \,\mathrm{Sym}\,(\boldsymbol{n}_1 \boldsymbol{n}_2 \ldots \boldsymbol{n}_n)] \\ &= l_{n+1} y^{(n)} \,\mathrm{Sym}\,[\boldsymbol{n}_{n+1} \,\mathrm{Sym}\,(\boldsymbol{n}_1 \boldsymbol{n}_2 \ldots \boldsymbol{n}_n)] \\ &= y^{(n+1)} \,\mathrm{Sym}\,(\boldsymbol{n}_1 \boldsymbol{n}_2 \ldots \boldsymbol{n}_{n+1}). \end{aligned} \quad (1.147)$$

Thus, we have proved that if eqn. (1.140) holds for n, it is also valid for $n + 1$. We further know that eqn. (1.140) holds for $n = 2$ (eqn. (1.139)); hence eqn. (1.140) holds for every value of n.

From eqn. (1.140) it appears that the 2^n-pole moment of a 2^n-pole is characterized by n directions and the multipole strength, *i.e.* by $2n + 1$ independent quantities. The multipole moment tensor, however, contains $(n + 1)(\frac{1}{2}n + 1)$ different components. Hence, in the case of an ideal 2^n-pole, there are $(n + 1)(\frac{1}{2}n + 1) - (2n + 1) = \frac{1}{2}n(n - 1)$ relations between the components of the 2^n-pole moment of the 2^n-pole; the one relation for the case $n = 2$ has been given in (1.124).

It must be always possible to reduce an arbitrary 2^n-pole moment to the moment of a 2^n-pole in the same way as we have demonstrated above for the quadrupole moment, because for the higher multipoles too, potential contributions due to three tensor components can cancel each other out. For the octupole, three relations comparable with (1.128) exist, of the type of:

$$u_0 \left(\frac{\partial^3}{\partial x^3} + \frac{\partial}{\partial x} \frac{\partial^2}{\partial y^2} + \frac{\partial}{\partial x} \frac{\partial^2}{\partial z^2} \right) \frac{1}{r} = u_0 \frac{\partial}{\partial x} \Delta \frac{1}{r} = 0. \quad (1.148)$$

In the general case of the 2^n-pole, the number of relations which can be derived in this way from $\Delta\frac{1}{r} = 0$ amounts to $\frac{1}{2}n(n-1)$. This number agrees with the number of relations between the components of the moment of a 2^n-pole. Therefore it must be always possible to reduce the 2^n-pole moment of an arbitrary charge distribution to the 2^n-pole moment of an ideal 2^n-pole.

A definition of the octupole moment, analogous to (1.135), can be given by defining the components as:

$$u_{\alpha\beta\gamma} = \tfrac{1}{2}\sum_i e_i\{5r_{i\alpha}r_{i\beta}r_{i\gamma} - r_i^2(r_{i\alpha}\delta_{\beta\gamma} + r_{i\beta}\delta_{\gamma\alpha} + r_{i\gamma}\delta_{\alpha\beta})\}, \qquad (1.149)$$

where $\delta_{\beta\gamma}$, $\delta_{\gamma\alpha}$, and $\delta_{\alpha\beta}$ are Kronecker deltas. If this definition is used, the octupole moment is standardized, *i.e.* all charge distributions with the same octupole contribution to the potential have also the same octupole moment. For the 2^n-pole, the generalization of (1.149) is given by:

$$y^{(n)}_{\alpha\beta\cdots\gamma} = \frac{(-1)^n}{n!}\sum_i e_i r_i^{2n+1}\frac{\partial^n}{\partial r_{i\alpha}\partial r_{i\beta}\cdots\partial r_{i\gamma}}\left(\frac{1}{r_i}\right). \qquad (1.150)$$

Using this definition, we find for the potential development:

$$\begin{aligned}\phi &= \sum_{n=0}^{\infty}\frac{(-1)^n 2^n n!}{(2n)!}\boldsymbol{Y}^{(n)} \overset{(n)}{\cdot} \boldsymbol{\nabla}^n\frac{1}{r} \\ &= e\frac{1}{r} - \boldsymbol{m}\cdot\boldsymbol{\nabla}\frac{1}{r} + \tfrac{1}{3}\boldsymbol{Q}:\boldsymbol{\nabla\nabla}\frac{1}{r} - \tfrac{1}{15}\boldsymbol{U}\vdots\boldsymbol{\nabla\nabla\nabla}\frac{1}{r} + \cdots, \end{aligned} \qquad (1.151)$$

where $\overset{(n)}{\cdot}$ denotes an n times repeated inner multiplication, and the exponent n of the nabla operator denotes an n times repeated tensor product. $\boldsymbol{Q}$ and $\boldsymbol{U}$ are defined according to eqns. (1.135) and (1.149).

In section 1 we found the dipole moment of a system whose net charge is zero, to be independent of the origin. This rule may be generalized for the higher multipole moments to the following form: the 2^n-pole moment $\boldsymbol{Y}^{(n)}$ of a charge distribution is independent of the choice of the origin if all lower moments are zero. To prove this we shall calculate the 2^n-pole moment $\boldsymbol{Y}'^{(n)}$ of this charge distribution in a new coordinate system with radius vector $\boldsymbol{r}'$, in which $\boldsymbol{r}'_0$ is the origin of the old coordinate system with radius vector $\boldsymbol{r}$:

$$\begin{aligned}\boldsymbol{Y}'^{(n)} &= \frac{1}{n!}\sum_i e_i(\boldsymbol{r}'_i)^n = \frac{1}{n!}\sum_i e_i(\boldsymbol{r}_i + \boldsymbol{r}'_0)^n \\ &= \frac{1}{n!}\sum_i e_i \sum_{m=1}^{n} \binom{n}{m} (\boldsymbol{r}_i)^m (\boldsymbol{r}'_0)^{n-m} \\ &= \sum_{m=1}^{n} \frac{1}{(n-m)!} (\boldsymbol{r}'_0)^{n-m} \frac{1}{m!} \sum_i e_i (\boldsymbol{r}_i)^m \\ &= \sum_{m=1}^{n} \frac{1}{(n-m)!} (\boldsymbol{r}'_0)^{n-m} \boldsymbol{Y}^{(m)}. \end{aligned} \tag{1.152}$$

Since $\boldsymbol{Y}^{(m)} = \mathbf{0}$ if $m < n$, this gives:

$$\boldsymbol{Y}'^{(n)} = \boldsymbol{Y}^{(n)}. \tag{1.153}$$

When not all moments $\frac{1}{m!}\sum_i e_i(\boldsymbol{r}_i)^m$ are zero for $m < n$, the multipole moment $\boldsymbol{Y}^{(n)}$ will depend on the choice of the origin. When the lower moments make no contribution to the potential, however, the 2^n-pole moment standardized according to some definition will still be independent of the origin.

This can be shown as follows. The charge distribution is considered to be built up of two parts:

1. a charge distribution ρ_1 with multipole moments given by $\boldsymbol{Y}_1^{(m)} = \boldsymbol{Y}_{\text{norm}}^{(m)}$,
2. a charge distribution ρ_2 with multipole moments given by $\boldsymbol{Y}_2^{(m)} = \boldsymbol{Y}^{(m)} - \boldsymbol{Y}_1^{(m)}$.

From these assumptions it follows that $\boldsymbol{Y}_{2_{\text{norm}}}^{(m)} = \mathbf{0}$ for every m. Therefore, the charge distribution ρ_2 does not contribute to the potential.

For the charge distribution under consideration, it has been assumed that $\boldsymbol{Y}_{\text{norm}}^{(m)} = \mathbf{0}$ for $m < n$. Hence, for $m < n$ we have $\boldsymbol{Y}_1^{(m)} = \mathbf{0}$, and ρ_1 fulfills the requirements for eqn. (1.153) to be valid. So if we choose another origin, we find $(\boldsymbol{Y}_1^{(m)})' = \boldsymbol{Y}_1^{(m)}$. Since per definition, $\boldsymbol{Y}_1^{(m)}$ is a standardized moment, $(\boldsymbol{Y}_1^{(m)})'$ will also be standardized: $(\boldsymbol{Y}_{1_{\text{norm}}}^{(m)})' = (\boldsymbol{Y}_1^{(m)})' = \boldsymbol{Y}_{\text{norm}}^{(m)}$. Furthermore, ρ_2 gives no contribution to the potential, so that $(\boldsymbol{Y}_{2_{\text{norm}}}^{(m)})' = \mathbf{0}$. Hence, $(\boldsymbol{Y}_{\text{norm}}^{(n)})' = (\boldsymbol{Y}_{1_{\text{norm}}}^{(m)})' + (\boldsymbol{Y}_{2_{\text{norm}}}^{(m)})'$, is equal to $\boldsymbol{Y}_{\text{norm}}^{(n)}$, *i.e.* the standardized moment does not depend on the choice of the origin.

Consequently in the potential development as given in eqn. (1.111) the first non-zero reduced moment is independent of the choice of the origin.

For most non-polar molecules the quadrupole term dominates the potential development as given in (1.111). Most quadrupole moments of non-polar molecules can be reduced to the moments of axial quadrupoles whose strengths are on the order of 10^{-26}–10^{-25} e.s.u. Considering the axial quadrupole to be built up from two dipoles, these quadrupole strengths correspond to dipole strengths of 0.5–5*D* and mutual distances of the order of 2 Å. A large number of quadrupole moments are given by Stogryn and

TABLE 3

QUADRUPOLE STRENGTHS OF SOME NON-POLAR MOLECULES FOR WHICH THE CORRESPONDING QUADRUPOLE IS AXIAL

Compound	q (10^{-26} e.s.u.)
H_2	+0.331
D_2	+0.325
Li_2	+6.9
N_2	−0.76
O_2	−0.20
F_2	+0.44
CO_2	−2.2
$(CH_3)_2$	−0.33

Stogryn[4] and by Krishnaji and Prakash.[5] A comparison of quadrupole strengths determined with different experimental and theoretical methods indicates that these values are rather crude. In Table 3 we give for a number of molecules with axial quadrupole moments the value of the quadrupole strength, as recommended by Stogryn and Stogryn and adapted to our definitions.

For molecules such as CH_4 and CCl_4, the quadrupole moment vanishes after reduction. For these molecules, the potential is dominated by the octupole term. For CH_4, the octupole strength amounts to $1.8\ 10^{-34}$ e.s.u.* For SF_6, the octupole is also zero, and the potential is dominated by the hexadecapole term.**

In many cases it is possible to draw conclusions concerning the multipole moments from the symmetry of the molecules. For instance, it is easily seen from our definitions that if the system has a centre of symmetry, the dipole

* This value is derived from the value $4.5 \cdot 10^{-34}$ e.s.u. given by Stogryn and Stogryn,[4] by multiplication with a conversion factor 0.4. This factor is obtained by considering an octupole consisting of charges at the vertices of a cube, *i.e.* with charges $+e$ at the points (1, 0, 0), (0, 1, 0), (0, 0, 1), and (1, 1, 1), and charges $-e$ at the points (0, 0, 0), (1, 1, 0), (1, 0, 1), and (0, 1, 1); this octupole has the same symmetry as the CH_4 molecule. Clearly the strength of this octupole according to eqn. (1.113) amounts to $+e$. According to Stogryn's definition (1.149) only the xyz-elements in the moment of this octupole are non-zero, and they are all equal to $+\frac{5}{2}e$. This common value is used by Stogryn to characterize the octupole, so that a conversion factor 0.4 must be applied to obtain the octupole strength from the quantity given by Stogryn.

** The vanishing of the dipole, quadrupole, or octupole moment generally holds only for the ground state of the molecule.

moment, the octupole moment, and any 2^{2n+1}-pole moment with respect to the centre of symmetry will be zero. Furthermore, in section 4 we found that if the system has axial symmetry, all multipole moments can be characterized by a scalar quantity. This was generalized by Stogryn and Stogryn[4] to the following theorem: if a molecule has an axis of n-fold symmetry, all 2^m-pole moments with $m < n$ can be characterized by a scalar quantity.

In section 3 the potential of a charge distribution with axial symmetry was expressed by using Legendre functions. These functions give the solution of Laplace's equation in the case of axial symmetry. In section 4 they appeared to be closely related to the axial multipoles. General multipoles are related in an analogous way to the so-called associated Legendre functions, which give the general solution of Laplace's equation:

$$\phi = \sum_{n=0}^{\infty} \sum_{m=-n}^{n} \left(a_n^{(m)} r^l + \frac{b_n^{(m)}}{r^{n+1}} \right) P_n^{(m)}(\cos \theta) e^{im\varphi}. \tag{1.152}$$

In this equation the terms $b_n^{(m)}/r^{n+1}$ give the 2^n-pole contributions, determined by $2n + 1$ independent quantities. An extensive treatment of these associated Legendre functions is given by MacRoberts.[7]

References

1. J. A. Stratton, *Electromagnetic Theory*, McGraw-Hill, New York 1941.
2. A. D. Buckingham, *Quart. Rev.* **13** (1959) 183.
3. S. Kielich, *Physica* **31** (1965) 444.
4. D. E. Stogryn and A. P. Stogryn, *Mol. Phys.* **11** (1966) 371.
5. Krishnaji and V. Prakash, *Rev. Mod. Phys.* **38** (1966) 690.
6. J. O. Hirschfelder, C. F. Curtiss, and R. B. Bird, *Molecular Theory of Gases and Liquids*, Wiley, New York 1954.
7. T. M. MacRoberts, *Spherical Harmonics*, Revised with the assistance of J. N. Sneddon, Pergamon, Oxford 1967.

CHAPTER II

SOME CONCEPTS AND PROBLEMS OF ELECTROSTATICS

§6. The vector fields *E* and *D*

For measurements inside matter, the definition of ***E*** *in vacuo*, as given in eqn. (1.4), cannot be used, unless some further specifications are given about the way in which the test charge should be inserted into the piece of matter. These specifications depend upon the particular model of matter which is used.

In general two different approaches to the solution of the problem how to measure ***E*** inside matter can be distinguished. Historically, the first method is the one used by Kelvin[1] and Maxwell,[2] who pictured matter as a continuum in which, by a sort of thought experiment, virtual cavities were made. Inside these cavities it would be possible to use the original definition of ***E***, eqn. (1.4). On the other hand Lorentz,[3] and after him Rosenfeld,[4] as well as Mazur[5] and de Groot,[6] tried to take the molecular structure of matter into account by picturing matter as a collection of point charges *in vacuo* forming clusters of various types. Here, if one applies the original definition (1.4), one arrives at a so-called microscopic field. If this microscopic field is averaged, one obtains the macroscopic or Maxwell field ***E***.

Both approaches show the difficulties of passing from a microscopic description in terms of electrons, nuclei, atoms, molecules, and ions, to a macroscopic or phenomenological description. Lorentz and Rosenfeld average the microscopic field over small intervals of space and time to get rid of the rapid fluctuations. Mazur and de Groot use the methods of statistical mechanics to pass from a detailed to an over-all description. In Maxwell's approach the difficulties are partially hidden; they appear in full force, however, when one asks to what extent it is actually possible to represent matter as a continuum.

The characteristic of matter making it possible to picture it as a continuum in many problems, is the fact that typical lengths with respect to molecules and their effects (of the order 1–1000 Å) may be neglected compared with typical macroscopic lengths (of the order of 0.1–100 cm). In a macroscopic problem such molecular distances may be treated as physically infinitesimally small, and a volume element of about 1000 Å^3 may be used as if it were a point. Although these distances and volumes are not mathematically infinitesimal in the exact sense of the word, all the operations of the differential calculus can be performed without risk of appreciable errors.

The difference in scale between macroscopic and microscopic phenomena is used in the approach of Mazur and de Groot in the following way. We label all point charges with two indices i and k, denoting the i-th charge element of the k-th charge cluster as e_{ki}, *e.g.* the i-th electron in the k-th molecule. In the same way we denote its mass as m_{ki} and its radius vector as $\boldsymbol{r}_{ki}$. Introducing the centre of gravity $\boldsymbol{r}_k$ of the k-th molecule with the help of:

$$\boldsymbol{r}_k \sum_i m_{ki} = \sum_i m_{ki}\boldsymbol{r}_{ki}, \tag{2.1}$$

we can rewrite the coordinates of the point charges, expressing them in terms of vectors $\boldsymbol{l}_{ki}$ given by:

$$\boldsymbol{l}_{ki} = \boldsymbol{r}_{ki} - \boldsymbol{r}_k. \tag{2.2}$$

We can now also define the moments of the charge clusters with respect to their centres of gravity:

$$e_k = \sum_i e_{ki}, \tag{2.3}$$

$$\boldsymbol{m}_k = \sum_i e_{ki}\boldsymbol{l}_{ki}, \tag{2.4}$$

$$\boldsymbol{Q}_k = \tfrac{1}{2}\sum_i e_{ki}\boldsymbol{l}_{ki}\boldsymbol{l}_{ki}, \tag{2.5}$$

respectively the charge, dipole moment, and quadrupole moment of the molecule or ion. The potential in a point with radius vector $\boldsymbol{r}$ due to this distribution of point charges is given by:

$$\phi(\boldsymbol{r}) = \sum_k \sum_i \frac{e_{ki}}{|\boldsymbol{r}_{ki} - \boldsymbol{r}|}, \tag{2.6}$$

according to eqn. (1.17). The denominators in eqn. (2.6) can be expressed as:

$$|\boldsymbol{r}_{ki} - \boldsymbol{r}| = |(\boldsymbol{r}_k - \boldsymbol{r}) + \boldsymbol{l}_{ki}|, \tag{2.7}$$

where (2.2) is used. If we now develop all terms of the potential in a Taylor series with respect to $1/(|\boldsymbol{r}_k - \boldsymbol{r}|)$, the inverse of the distance between the centre of gravity and the point of observation, we get an expression of the form given in section 5 (eqn. (1.108)). In this expression the differentiations may be combined to form nabla-operators. This leads to the expression:

$$\phi(\boldsymbol{r}) = \sum_k \sum_i \left\{ e_{ki}\frac{1}{|\boldsymbol{r}_k - \boldsymbol{r}|} + \boldsymbol{l}_{ki}\cdot\nabla_{r_k}\frac{1}{|\boldsymbol{r}_k - \boldsymbol{r}|} + \tfrac{1}{2}\boldsymbol{l}_{ki}\boldsymbol{l}_{ki}:\nabla_{r_k}\nabla_{r_k}\frac{1}{|\boldsymbol{r}_k - \boldsymbol{r}|} + \cdots \right\}, \tag{2.8}$$

where the symbol ∇_{r_k} denotes the gradient taken at $\boldsymbol{r}_k$ (*cf.* Appendix I, section 2a). Performing the summations over the index i in the same way as going from eqn. (1.108) to eqn. (1.111), we obtain:

$$\phi(\boldsymbol{r}) = \sum_k \left\{ \frac{e_k}{|\boldsymbol{r}_k - \boldsymbol{r}|} + \boldsymbol{m}_k \cdot \nabla_{\boldsymbol{r}_k} \frac{1}{|\boldsymbol{r}_k - \boldsymbol{r}|} + \boldsymbol{Q}_k : \nabla_{\boldsymbol{r}_k} \nabla_{\boldsymbol{r}_k} \frac{1}{|\boldsymbol{r}_k - \boldsymbol{r}|} + \cdots \right\}, \tag{2.9}$$

where e_k, $\boldsymbol{m}_k$, and $\boldsymbol{Q}_k$ are the charge, dipole moment and quadrupole moment of the k-th molecule, respectively. If $\boldsymbol{r}_k - \boldsymbol{r}$, the difference of two macroscopic vectors, is also of macroscopic length, the microscopic vector $\boldsymbol{l}_{ki}$ may be neglected with respect to it and the series development may be broken off after a few terms.

Thus we find that the potential can be written in terms of the potentials of the charges, dipole moments, and quadrupole moments of the separate molecules or ions. In general, however, the values, or even the statistical distribution of the radius vectors $\boldsymbol{r}_k$ in space and time, are not known. Some information can only be obtained for gases at low densities and solids at low temperatures. Therefore we are forced to leave the detailed description and shift to a macroscopic description in terms of densities of charges, of dipoles, of quadrupoles, etc.

This is best shown after first transforming eqn. (2.9) by operating on both sides with $-\Delta$, the negative of the Laplace operator. From Appendix II we use the result (A2.20):

$$-\Delta \frac{1}{|\boldsymbol{r}_k - \boldsymbol{r}|} = 4\pi\delta(\boldsymbol{r}_k - \boldsymbol{r}),$$

so that we find:

$$-\Delta\phi(\boldsymbol{r}) = \operatorname{div} \boldsymbol{E}(\boldsymbol{r}) = 4\pi \left\{ \sum_k e_k \delta(\boldsymbol{r}_k - \boldsymbol{r}) + \sum_k \boldsymbol{m}_k \cdot \nabla_{\boldsymbol{r}_k} \delta(\boldsymbol{r}_k - \boldsymbol{r}) + \sum_k \boldsymbol{Q}_k : \nabla_{\boldsymbol{r}_k} \nabla_{\boldsymbol{r}_k} \delta(\boldsymbol{r}_k - \boldsymbol{r}) \right\}. \tag{2.10}$$

From eqn. (A1.20) we see that the nabla-operator with respect to $\boldsymbol{r}_k$ may be transformed to the nabla-operator with respect to $\boldsymbol{r}$. This leads to:

$$\operatorname{div} \boldsymbol{E}(\boldsymbol{r}) = 4\pi \left\{ \sum_k e_k \delta(\boldsymbol{r}_k - \boldsymbol{r}) - \nabla_{\boldsymbol{r}} \cdot \sum_k \boldsymbol{m}_k \delta(\boldsymbol{r}_k - \boldsymbol{r}) + \nabla_{\boldsymbol{r}} \nabla_{\boldsymbol{r}} : \sum_k \boldsymbol{Q}_k \delta(\boldsymbol{r}_k - \boldsymbol{r}) \right\}. \tag{2.11}$$

From the properties of the δ-function it follows that a summation $\sum_k e_k \delta(\boldsymbol{r}_k - \boldsymbol{r})$ represents the density of a discrete distribution of charges e_k. In the same way dipole and quadrupole densities are represented by summations over δ-functions.

Denoting the charge density $\sum e_k \delta(\boldsymbol{r}_k - \boldsymbol{r})$ by a function $\rho(\boldsymbol{r})$, the dipole density by $\boldsymbol{P}(\boldsymbol{r})$, and the quadrupole density by $\mathcal{Q}(\boldsymbol{r})$, we can write:

$$\operatorname{div} \boldsymbol{E} = 4\pi\rho - 4\pi \operatorname{div} (\boldsymbol{P} - \operatorname{div} \mathcal{Q}). \tag{2.12}$$

Although this equation is derived for discrete distributions of point charges, point dipoles, and point quadrupoles, it should also be valid when the discrete distributions are replaced by continuous densities as long as one does not come too close to one of the point singularities. By a complete statistical mechanical treatment, this procedure can be justified rigorously.

From eqn. (2.12) we see that Maxwell's source equation, eqn. (1.9), has to be corrected for the presence of matter with the help of terms corresponding to dipole densities and higher multipole densities.

For the solution of the problem of how to determine the electric field inside matter, it is also possible first to introduce a new vector field $\boldsymbol{D}$ in such a way that for this field the source equation, eqn. (1.9), will be valid, and then to derive a relationship between $\boldsymbol{D}$ and $\boldsymbol{E}$. This phenomenological method is the one originally followed in the development of electromagnetic theory. In the rest of this section we shall treat the conditions on $\boldsymbol{D}$ and the determination of $\boldsymbol{E}$ and $\boldsymbol{D}$ inside virtual cavities in matter. In section 7 the relation between $\boldsymbol{D}$ and $\boldsymbol{E}$, defined in this way, will be derived.

As already mentioned, in Maxwell's theory matter is regarded as a continuum. Consequently, to use the definition of the field vector $\boldsymbol{E}$, a cavity has to be made around the point where the field is to be determined. However, the force acting upon a test charge in this cavity will generally depend on the shape of the cavity, since this force is at least partly determined by effects due to the walls of the cavity.

Therefore, the shape of the cavity has to be specified, preferably in such a way that the field vector thus defined, satisfies both of Maxwell's equations of electrostatics *in vacuo*, eqns. (1.9) and (1.12). As can be seen from the following, however, it is impossible to choose a cavity of such a shape that both fundamental equations are satisfied. Therefore, we will define two vector fields instead of one, namely the electric field strength $\boldsymbol{E}$, satisfying curl $\boldsymbol{E} = \boldsymbol{0}$, and the dielectric displacement $\boldsymbol{D}$, satisfying div $\boldsymbol{D} = 4\pi\rho$.

The electric field strength $\boldsymbol{E}$ at a point P inside matter is defined as follows. We imagine a relatively long cylindrical cavity with a diameter that is physically infinitesimally small. The point P is the centre of this needle-shaped cavity, and the axis points in the direction of the x-axis of a Cartesian

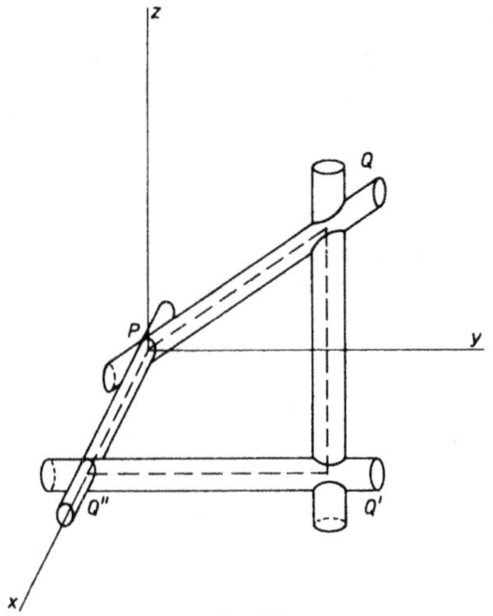

Fig. 15

coordinate system (Fig. 15). In the vacuum of this cavity a field vector $\boldsymbol{E}^*$ at P can be determined according to the definition of the electric field intensity given in eqn. (1.4). Then E_x, the component of the electric field strength inside matter in the direction of the x-axis, is defined by:

$$E_x = E_x^*. \tag{2.13}$$

The components in the y- and z-directions are similarly defined, using needle-shaped cavities with their axes in the y- and z-directions. If we assume for the moment that these three components are indeed the components of a vector, we may write:

$$\boldsymbol{E} = E_x\boldsymbol{i} + E_y\boldsymbol{j} + E_z\boldsymbol{k}, \tag{2.14}$$

by which equation the electric field strength $\boldsymbol{E}$ is given.

$\boldsymbol{E}$ is a vector when it is independent of the choice of the coordinate system. To prove this, it is sufficient to prove that the component of $\boldsymbol{E}$, as given by (2.14), in an arbitrary direction $\boldsymbol{u}$ is equal to the value determined with the help of (2.13).

We take a point Q at a distance $\boldsymbol{du}$ from P (see Fig. 15) and project Q on the xy-plane, obtaining the point Q'. We then project Q' on the x-axis, obtaining the point Q''. The distances PQ'', $Q''Q'$, and $Q'Q$ are then equal to $\mathrm{d}x$, $\mathrm{d}y$, and $\mathrm{d}z$, the components of $\boldsymbol{du}$ in the Cartesian coordinate system. So we may always write:

$$E_u \mathrm{d}u = \boldsymbol{E} \cdot \boldsymbol{du} = E_x\,\mathrm{d}x + E_y\,\mathrm{d}y + E_z\,\mathrm{d}z. \tag{2.15}$$

We now imagine four intersecting cylindrical cavities of infinitesimally small diameter around PQ'', $Q''Q'$, $Q'Q$, and PQ. In the vacuum of these cavities we may apply eqn. (1.13) to the closed circuit $PQ''Q'QP$, obtaining:

$$\oint \boldsymbol{E}^* \cdot \boldsymbol{ds} = 0. \tag{2.16}$$

According to (2.13), however, we have $E_x = E_x^*$ in the cavity in the direction of the x-axis. Using the same argument for E_y and E_z, we find for (2.16):

$$E_x\,\mathrm{d}x + E_y\,\mathrm{d}y + E_z\,\mathrm{d}z - E_u^*\,\mathrm{d}u = 0. \tag{2.17}$$

From (2.17) and (2.15) it follows that in the cavity along PQ we have:

$$E_u = E_u^*. \tag{2.18}$$

Since we have obtained the original definition again, $\boldsymbol{E}$ may indeed be treated as a vector quantity.

Next, we shall prove that $\boldsymbol{E}$ satisfies the condition curl $\boldsymbol{E} = \boldsymbol{0}$. For any closed path in the dielectric, we imagine a continuous cavity along this path with infinitesimally small diameter. Such a cavity may be approximated as closely as necessary with the help of intersecting needle-shaped cavities. In the continuous cavity, as well as in the combined intersecting needle-shaped cavities, we have:

$$\oint E_s^* \, \mathrm{d}s = 0. \tag{2.19}$$

For any needle-shaped cavity, E_s^* equals E_s (see eqns. (2.13) and (2.18)). Therefore (2.19) can be written as:

$$\oint E_s \, \mathrm{d}s = \oint \boldsymbol{E} \cdot \boldsymbol{ds} = 0, \tag{2.20}$$

or, using Stokes's theorem:

$$\operatorname{curl} \boldsymbol{E} = \boldsymbol{0}. \tag{2.21}$$

Thus, the vector $\boldsymbol{E}$, defined with the help of virtual needle-shaped cavities, satisfies one of Maxwell's equations for the electric field *in vacuo*.

From the fact that eqn. (1.12) is also valid for dielectrics, as shown by eqn. (2.21), we may derive two important corollaries. In the first place, we can apply (2.20) to a closed circuit $PQQ'P'P$ across the boundary between two dielectrics, P and Q being two points of the one dielectric at a distance l and at infinitesimally small distance from the boundary, whereas P' and Q' are the corresponding points at the other side of the boundary. Writing E_t for the component of $\boldsymbol{E}$ tangent to the boundary, eqn. (2.20) now reduces to:

$$(E_t)_1 l - (E_t)_2 l = 0, \tag{2.22}$$

when l is sufficiently small for E_t to be almost equal at P and at Q. Hence:

$$(E_t)_1 = (E_t)_2, \tag{2.23}$$

that is, the tangential components of $\boldsymbol{E}$ are continuous across a boundary.

A second corollary resulting from the validity of (2.20) and (2.21) for a dielectric, is that the potential ϕ may be used also for points within matter. The definition of the potential given in eqn. (1.14),

$$\boldsymbol{E} = -\operatorname{grad} \phi,$$

determines ϕ up to an additive constant. This constant may be fixed by requiring ϕ to be zero at an infinite distance from true and apparent charges, a requirement which may be derived from the relation between ϕ and energy (see section 11). In some idealized problems, however, it is impossible to meet this requirement, namely when there is a charge distribution at infinity (compare section 9). From this it follows that in such problems only differences between potentials will have a physical meaning.

By substitution of the definition of the potential (1.14) into (2.20) and using the closed circuit $PQQ'P'P$ again, it is seen that (2.23) is equivalent to the requirement that ϕ be continuous across a boundary.*

For the definition of the dielectric displacement $\boldsymbol{D}$, the vector field which should satisfy the source equation div $\boldsymbol{D} = 4\pi\rho$, a cavity having a different shape than the one required for the definition of $\boldsymbol{E}$ has to be used. To find the component D_x of $\boldsymbol{D}$ at a point P, we imagine P to be the centre of a cylindrical cavity with a relatively large diameter and infinitesimally small height. The axis of the cavity is taken in the direction of the x-axis. In this cavity a field $\boldsymbol{E}^\dagger$ can be measured. Then D_x is defined by the equation:

$$D_x = E_x^\dagger. \tag{2.24}$$

The components of $\boldsymbol{D}$ in the directions of the y- and z-axes are similarly defined, using disc-shaped cavities with their axes in the y- and z-directions. These components are used to form a vector $\boldsymbol{D}$ according to:

$$\boldsymbol{D} = D_x\boldsymbol{i} + D_y\boldsymbol{j} + D_z\boldsymbol{k}. \tag{2.25}$$

The proof of the vectorial character of $\boldsymbol{D}$ as defined in (2.25) is similar to the proof of the vectorial character of $\boldsymbol{E}$. A disc-shaped cavity with its axis in the arbitrary direction $\boldsymbol{u}$ is constructed; in the median plane of this disc the triangular element QRT having area $\mathrm{d}A$ (see Fig. 16) is constructed. The distance between P and the median plane is $\mathrm{d}u$. The components of the vector $\boldsymbol{dA} = \boldsymbol{u}\,\mathrm{d}A$ in the Cartesian coordinate system can now be given as:

$$\boldsymbol{dA} = \boldsymbol{i}\,dA_x + \boldsymbol{j}\,\mathrm{d}A_y + \boldsymbol{k}\,\mathrm{d}A_z, \tag{2.26}$$

where $\mathrm{d}A_x$, $\mathrm{d}A_y$, and $\mathrm{d}A_z$ are the areas of the planes PRT, PTQ, and PQR, respectively.

Around the surface elements $\mathrm{d}A_x$, $\mathrm{d}A_y$, and $\mathrm{d}A_z$, disc-shaped cavities are constructed with their axes in the direction of the coordinate axes. In the vacuum of the four cavities, eqn. (1.8) is valid:

$$\oiint \boldsymbol{E}^\dagger \cdot \boldsymbol{dS} = 4\pi \iiint_{\mathrm{d}V} \rho\,\mathrm{d}v, \tag{2.27}$$

where $\mathrm{d}V$ is the volume of the tetrahedron $PQRT$. Substituting the definition of the components of $\boldsymbol{D}$ as given in eqn. (2.24), we find:

$$D_x\,\mathrm{d}A_x + D_y\,\mathrm{d}A_y + D_z\,\mathrm{d}A_z - E_u^\dagger\,\mathrm{d}A = 4\pi\rho\,\mathrm{d}V. \tag{2.28}$$

If $\boldsymbol{D}$ is indeed a vector, we may write:

* This is only true when there is no electric double layer at the boundary. This case will not be treated.

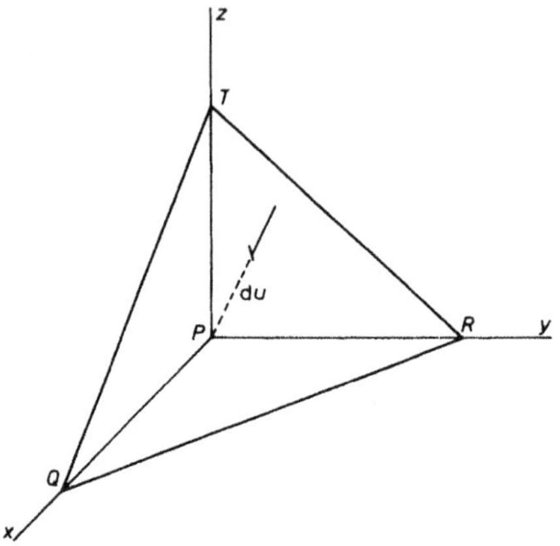

Fig. 16

$$\boldsymbol{D} \cdot \boldsymbol{dA} = D_u \, dA = D_x \, dA_x + D_y \, dA_y + D_z \, dA_z. \tag{2.29}$$

Since $\boldsymbol{du}$ and $\boldsymbol{dA}$ are parallel, dV is proportional to $du \, dA$, and thus:

$$D_u = E_u^\dagger + 4\pi\rho \, dV/dA. \tag{2.30}$$

The second term is infinitesimally small, and can be neglected. In this way we obtain the original definition (2.24), which proves that $\boldsymbol{D}$ is a vector.

The vector field $\boldsymbol{D}$, defined in this way, satisfies the condition $\operatorname{div} \boldsymbol{D} = 4\pi\rho$. This can be shown easily by taking any closed surface in the dielectric. This surface can be approximated as closely as one wishes by a set of intersecting disc-shaped cavities. In these cavities eqn. (1.8) holds; substitution of (2.24) and similar appropriate equalities now leads immediately to:

$$\operatorname{div} \boldsymbol{D} = 4\pi\rho. \tag{2.31}$$

From the validity of this condition for dielectrics we can derive a requirement on the continuity of $\boldsymbol{D}$ across a boundary between two media. The integral equivalent of eqn. (2.31) is:

$$\oiint \boldsymbol{D} \cdot \boldsymbol{dS} = 4\pi \iiint_V \rho \, dv. \tag{2.32}$$

If this relation is applied to a pill-box shaped volume whose top and bottom surfaces are parallel to the boundary on each side of it, then as long as the height of the pill-box is infinitesimally small, we find:

$$A\{(D_n)_2 - (D_n)_1\} = 4\pi \iiint_V \rho \, dv. \tag{2.33}$$

Here A is the area of the boundary enclosed by the pill-box, and D_n is the component of $\boldsymbol{D}$ normal to the boundary, where the direction from medium 1 to medium 2 is considered to be positive. If the charge density remains finite at the boundary, the volume integral will tend to zero when the volume V goes to zero. This is the case since the height of the pill-box is infinitesimally small. Thus we have:

$$(D_n)_2 = (D_n)_1. \tag{2.34}$$

When the charge density becomes infinite at the boundary, as is for instance the case when a surface charge σ is present, this argument has to be modified. We then find:

$$(D_n)_2 - (D_n)_1 = 4\pi\sigma. \tag{2.35}$$

So we see that the normal components of $\boldsymbol{D}$ are continuous across a boundary between two dielectrics, as long as no surface charge is present. When there is a surface charge σ, the normal component jumps in value by an amount $4\pi\sigma$.

§7. The electric polarization

In the Maxwell approach, in which matter is treated as a continuum, we must in many cases ascribe a dipole density to matter. In the following we want to compare the vector fields $\boldsymbol{D}$ and $\boldsymbol{E}$ for the case that only a dipole density is present.

Differences between the values of the field vectors arise from differences in their sources. Both the external charges and the dipole density of the piece of matter act as sources of these vectors. The external charges contribute to $\boldsymbol{D}$ and to $\boldsymbol{E}$ in the same manner. Because of the different cavities in which the field vectors are measured, the contributions of the dipole density to $\boldsymbol{D}$ and $\boldsymbol{E}$ are not the same.

Using eqn. (1.40), we find for the potential contribution of the dipole density in the case of a needle-shaped cavity:

$$\phi_{\text{needle}} = \oint\!\!\oint_{\text{ext}} \frac{\boldsymbol{P}}{r} \cdot d\boldsymbol{S} + \oint\!\!\oint_{\text{needle}} \frac{\boldsymbol{P}}{r} \cdot d\boldsymbol{S} - \iiint_{V} \frac{1}{r} \operatorname{div} \boldsymbol{P} \, dv + \iiint_{\text{needle}} \frac{1}{r} \operatorname{div} \boldsymbol{P} \, dv, \quad (2.36)$$

and in the case of a disc-shaped cavity:

$$\phi_{\text{disc}} = \oint\!\!\oint_{\text{ext}} \frac{\boldsymbol{P}}{r} \cdot d\boldsymbol{S} + \oint\!\!\oint_{\text{disc}} \frac{\boldsymbol{P}}{r} \cdot d\boldsymbol{S} - \iiint_{V} \frac{1}{r} \operatorname{div} \boldsymbol{P} \, dv + \iiint_{\text{disc}} \frac{1}{r} \operatorname{div} \boldsymbol{P} \, dv. \quad (2.37)$$

In these equations, the first term denotes the potential contribution of the apparent charges on the external surface of the piece of matter considered. This term is the same in both equations. The third term denotes the contribution of div $\boldsymbol{P}$ integrated over V, the volume of the whole piece of matter if no cavities had been made. This term, too, is the same in both equations. The fourth term is a correction of the third term for the volume of the cavity. In both cases this term vanishes if the volume of the cavity becomes infinitesimally small.

Therefore, the difference between the potential contributions of the dipole density is caused by the difference between the second terms. For the calculation of these terms at least one dimension of the cavities is taken to be infinitesimally small, so that we may assume $\boldsymbol{P}$ to be constant in the corresponding direction.

We first consider the needle-shaped cavity. Because the ratio between the length and the radius of the cylinder goes to infinity, we have only to reckon with the apparent charge on the cylindrical surface. We take the axis of the cylinder as the x-axis and subdivide the cylindrical surface into an infinite number of infinitesimally small rings parallel with the yz-plane. The integration is performed in two steps: first we consider the potential contribution of one ring, and after that the potential contribution of the whole surface.

All points of each ring have the same distance to the point on the axis in which we want to calculate the potential. Therefore, the contribution of each ring is proportional to the total apparent charge on the ring and does not depend on the distribution of the charge over the ring. For reasons of symmetry, however, the total apparent charge of each ring is zero. So the contribution to the potential due to each ring is zero. Therefore:

$$\oint\!\!\oint_{\text{needle}} \frac{\boldsymbol{P}}{r} \cdot d\boldsymbol{S} = 0, \quad (2.38)$$

that is, for a needle-shaped cavity the apparent charges on the walls of the cavity do not contribute to the potential.

In the case of a disc-shaped cavity, the ratio between the length and the radius of the cylinder goes to zero, and it is the apparent charge on the curved surface that may be neglected. Taking again the axis of the cylinder as the x-axis, we find for the contribution to the potential due to the apparent charge on the flat surfaces, given by $x = a$ and $x = -a$ respectively:

$$\begin{aligned}\oint\!\!\!\oint_{\text{disc}} \frac{\boldsymbol{P}}{r}\cdot d\boldsymbol{S} &= \lim_{a/R\to 0}\int_0^R \mathrm{d}r\, 2\pi r\left\{\frac{P_x}{\sqrt{r^2+(a+x)^2}}+\frac{-P_x}{\sqrt{r^2+(a-x)^2}}\right\}\\ &= 2\pi P_x \lim_{a/R\to 0}\left[\sqrt{r^2+(a+x)^2}-\sqrt{r^2+(a-x)^2}\right]_0^R\\ &= 2\pi P_x\{-(a+x)+(a-x)\}\\ &= -4\pi P_x x. \end{aligned} \tag{2.39}$$

From this it follows, according to eqn. (1.14), that the contribution of the apparent charges on the walls of the cavity to the field along the x-axis will be $4\pi P_x$. Since for the needle-shaped cavity this contribution is zero, we find:

$$D_x - E_x = 4\pi P_x. \tag{2.40}$$

A similar relation holds for the components in the direction of the y- and z-axes. So we have:

$$\boldsymbol{D} - \boldsymbol{E} = 4\pi \boldsymbol{P}. \tag{2.41}$$

In general the dipole distribution is not the only cause of differences between $\boldsymbol{D}$ and $\boldsymbol{E}$. When distributions of higher multipoles are present, eqn. (2.41) has to be modified (*e.g.* eqn. (2.12)). If we also take into account the quadrupole density $\mathcal{Q}$, we find:

$$\boldsymbol{D} - \boldsymbol{E} = 4\pi(\boldsymbol{P} - \nabla\cdot\mathcal{Q}). \tag{2.42}$$

In general one may write:

$$\boldsymbol{D} - \boldsymbol{E} = 4\pi\mathcal{P}, \tag{2.43}$$

with:

$$\mathcal{P} = \boldsymbol{P} - \nabla\cdot\mathcal{Q} + \cdots, \tag{2.44}$$

$\mathcal{P}$ being called the polarization.

In most cases the contribution due to the higher multipole densities is very small, so that these contributions may be neglected. The polarization is then identical with the dipole density, and is also denoted by $\boldsymbol{P}$.

§8. The relation between E and P

Generally, the polarization $\boldsymbol{P}$ depends on the electric field strength $\boldsymbol{E}$. This can be explained in the following way. All matter is built up of electrically charged particles: negative electrons and positive nuclei, sometimes combining to neutral atoms and molecules but in other cases to particles with a net charge, such as ions in solutions and in crystals.

When we bring a material into an external electric field, for instance between the plates of a charged condenser, every particle of the material is subjected to an "internal field" $\boldsymbol{E}_i$, proportional to the electric field $\boldsymbol{E}$ as defined in section 6. This internal field $\boldsymbol{E}_i$ will be calculated in Chapter V.

Under the influence of the internal field $\boldsymbol{E}_i$, in some materials particles are moved over long distances, for instance electrons in metals or ions in solutions. In these materials, which are called conductors, an equilibrium position is not reached before the field strength has become zero at all points inside the material.

In the case of insulators (dielectrics), however, only very small displacements of the charges occur. When an electric field is applied the forces acting upon the charges bring about a small displacement of the electrons relative to the nuclei, because the field tends to shift the positive and negative charges in opposite directions. This displacement is limited, since the electrons are bound to the nuclei. The reactive forces that arise in this way are proportional to the displacements for not too high field intensities. Furthermore the applied field tends to direct the permanent dipoles. In both cases the electric field gives rise to a dipole density; the electric field polarizes the dielectric.

The dependence of $\boldsymbol{P}$ on $\boldsymbol{E}$ can take several forms. We shall give them starting from simple scalar proportionality:

$$1.\ \boldsymbol{P} = \chi \boldsymbol{E}. \tag{2.45}$$

In many cases the polarization is proportional to the field strength and eqn. (2.45) can be used. The proportionality factor χ is called the dielectric susceptibility. From this proportionality we derive:

$$D = E + 4\pi P = (1 + 4\pi\chi)E = \varepsilon E, \tag{2.46}$$

in which ε is called the permittivity. Often it is also called the dielectric constant, because it is independent of the field strength. It is, however, dependent on the temperature, the density (or the pressure) and the chemical composition of the system. This will be elaborated in Chapters V and VI.

2. $P = \chi E + \xi E^2 E.$ (2.47)

For very high field intensities the proportionality no longer holds; a correction term proportional to E^3 must be added. Here one speaks of electric saturation; this subject will be dealt with in Chapter VII.

3. $P = \chi \cdot E.$ (2.48)

For non-isotropic dielectrics, like most solids, the scalar susceptibility must be replaced by a tensor. Hence, the permittivity ε must also be replaced by a tensor:

$$\left.\begin{aligned} D_x &= \varepsilon_{11}E_x + \varepsilon_{12}E_y + \varepsilon_{13}E_z \\ D_y &= \varepsilon_{21}E_x + \varepsilon_{22}E_y + \varepsilon_{23}E_z \\ D_z &= \varepsilon_{31}E_x + \varepsilon_{32}E_y + \varepsilon_{33}E_z \end{aligned}\right\} . \tag{2.49}$$

4. In some cases there is a permanent electric polarization, independent of the applied electric field. Such materials are called electrets, a term which is analogous to "magnet". The phenomenon is called seignette-electricity. Because the polarization is a vector quantity, electrets are always anisotropic. They will be dealt with in Volume II.

5. $P(t) = \int_{-\infty}^{t} f(t - t')E(t')\,dt'.$ (2.50)

In the case of rapidly changing fields, the polarization is no longer proportional to the field strength, but depends on the values of E at all moments before the time t at which P is considered. Relaxation and resonance phenomena, which are covered by this formula, will be treated in Volume II.

The most important dependence of P on E is the one given by expression (2.45). In this case the relation $D = \varepsilon E$ holds, and the eqns. (2.31) and (1.14):

$$\operatorname{div} \boldsymbol{D} = 4\pi\rho,$$
$$\boldsymbol{E} = -\operatorname{grad} \phi,$$

may be combined to:

$$\Delta\phi = -\frac{4\pi\rho}{\varepsilon}. \tag{2.51}$$

This is the extension of Poisson's equation for the case of a homogeneous dielectric, provided $\boldsymbol{D} = \varepsilon\boldsymbol{E}$. In charge-free parts of the dielectric it is reduced to Laplace's equation, $\Delta\phi = 0$ (eqn. (1.16)). When the condition $\boldsymbol{D} = \varepsilon\boldsymbol{E}$ is not satisfied, eqn. (2.51) is not valid.

Coulomb's law can also be extended to a dielectric. We consider a point charge e surrounded by a dielectric with dielectric constant ε. No other charges are present. When $\boldsymbol{D} = \varepsilon\boldsymbol{E}$, eqn. (2.32) can be written as:

$$\oint\!\!\oint_S \boldsymbol{E} \cdot \boldsymbol{dS} = \frac{4\pi e}{\varepsilon}. \tag{2.52}$$

We choose as the closed surface S a spherical surface with e as centre and with radius r. For reasons of symmetry, $\boldsymbol{E}$ must be directed along the radius vector, whereas the magnitude E of $\boldsymbol{E}$ will be the same at all points with the same distance r from e. Using eqn. (2.52) we obtain:

$$E4\pi r^2 = \frac{4\pi e}{\varepsilon},$$

or:

$$E = \frac{e}{\varepsilon r^2}.$$

Therefore $\boldsymbol{E}$ is given by:

$$\boldsymbol{E} = \frac{e}{\varepsilon r^3}\boldsymbol{r}. \tag{2.53}$$

Comparing (2.53) with (1.5) we see that the electric field strength in a dielectric is reduced by a factor $1/\varepsilon$. Therefore the force working between two charges is reduced by the same factor and Coulomb's law, eqn. (1.3), is changed accordingly.

The potential due to a point charge in a dielectric can be calculated by integrating the expression for the field strength (*cf.* eqn. (A1.33)):

$$\phi = \frac{e}{\varepsilon r}. \tag{2.54}$$

Comparing this with eqn. (1.17) we see that again a factor $1/\varepsilon$ appears.

In section 6 the conditions that must be fulfilled at the boundary between two dielectrics were derived. According to (2.23), it holds that:

$$(E_t)_1 = (E_t)_2,$$

and according to (2.35):

$$(D_n)_2 - (D_n)_1 = 4\pi\sigma.$$

If there are no true surface charges, this results in (2.34):

$$(D_n)_1 = (D_n)_2.$$

When $\boldsymbol{D} = \varepsilon \boldsymbol{E}$, the conditions (2.23) and (2.34) lead to a simple refraction law for the lines of force. When at a certain point of the boundary surface the lines of force on the two sides make angles θ_1 and θ_2 with the normal to the surface, the dielectric constants being ε_1 and ε_2 respectively, eqns. (2.23) and (2.34) lead to:

$$E_2 \sin \theta_2 = E_1 \sin \theta_1,$$

and:

$$\varepsilon_2 E_2 \cos \theta_2 = \varepsilon_1 E_1 \cos \theta_1.$$

Hence:

$$\frac{\tan \theta_1}{\varepsilon_1} = \frac{\tan \theta_2}{\varepsilon_2}. \tag{2.55}$$

In passing from smaller values of ε to larger ones, the lines are bent away from the normal.

An essential difference between this refraction law and Snell's law of optical refraction is that eqn. (2.55) precludes the possibility of total reflection.

In Table 4 the values of the dielectric constants of some materials are given as numerical examples. These values were obtained by averaging those given by several authors. We conclude from this Table that the dielectric constant is highly dependent on the structure of the material.

TABLE 4

THE DIELECTRIC CONSTANTS OF SOME GASES AND LIQUIDS AT 20°C AND 1 ATM.

Compound	ε
Hydrogen	1.00024
Air	1.00054
Carbon dioxide	1.00092
Hexane	1.91
Benzene	2.28
3-ethylpentanol-3	3.20
Propionic acid	3.34
Chloroform	4.78
Bromoethane	9.39
Heptanol-1	11.8
Methanol	33.6
Cyanomethane	37.5
Water	81
Formamide	111

§9. Some electrostatical problems

In section 3d the problem of a conducting sphere in a homogeneous external field was solved by using the general solution of Laplace's equation for the potential in charge-free regions in the case of axial symmetry. The special conditions of the problem were transformed to boundary conditions on the potential ϕ.

In this section a number of problems in which dielectrics with $\boldsymbol{D} = \varepsilon\boldsymbol{E}$ play a role, are solved in a similar way.

(a) *A dielectric sphere in a dielectric of different dielectric constant*

A well-known electrostatical problem is that of a dielectric sphere of radius a and dielectric constant ε_2, immersed in a dielectric extending to infinity, with dielectric constant ε_1, to which an external electric field is applied, caused by a fixed external charge distribution with a configuration such that if $\varepsilon_2 = \varepsilon_1$ the field in the dielectric would be a uniform field $\boldsymbol{E}_0$ (Fig. 17).

Outside the sphere the potential satisfies Laplace's equation $\Delta\phi = 0$, since no charges are present except the charges at a great distance required

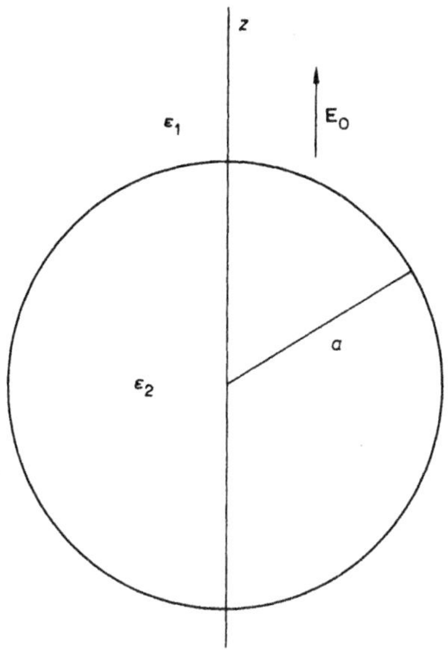

Fig. 17

to maintain the external field. On the surface of the sphere Laplace's equation is not valid, since there is an apparent surface charge. Inside the sphere, however, Laplace's equation can be used again. Therefore, for the description of ϕ, we use two different functions, ϕ_1 and ϕ_2, outside and inside the sphere, respectively.

We take the centre of the sphere as the origin of the coordinate system, we choose the z-axis in the direction of the uniform field and we introduce spherical polar coordinates r, θ, φ. In Appendix II it is shown that in the case of symmetry about the z-axis the general solution of Laplace's equation is represented by eqn. (A2.15), so that we may write:

$$\phi_1 = \sum_{n=0}^{\infty} \left(A_n r^n + \frac{B_n}{r^{n+1}} \right) P_n(\cos\theta),$$

$$\phi_2 = \sum_{n=0}^{\infty} \left(C_n r^n + \frac{D_n}{r^{n+1}} \right) P_n(\cos\theta).$$

The boundary conditions are for this problem:

1. $$(\phi_1)_{r\to\infty} = -E_0 z = -E_0 r\cos\theta, \tag{2.56}$$

since far from the origin the field approaches $\boldsymbol{E}_0$.

2. $$(\phi_1)_{r=a} = (\phi_2)_{r=a}, \tag{2.57}$$

since ϕ is continuous across a boundary (see section 6).

3. $$\varepsilon_1 \left(\frac{\partial \phi_1}{\partial r} \right)_{r=a} = \varepsilon_2 \left(\frac{\partial \phi_2}{\partial r} \right)_{r=a}, \tag{2.58}$$

since, according to (2.35), the normal component of $\boldsymbol{D}$ must be continuous at the surface of the sphere.

4. At the centre of the sphere ($r = 0$) ϕ_2 must not have a singularity.

On account of the first boundary condition and the fact that the Legendre functions are linearly independent, all coefficients A_n are zero except A_1, which has the value $A_1 = -E_0$. On account of the fourth boundary condition, all coefficients D_n are zero. Thus, we have:

$$\phi_1 = \sum_{n=0}^{\infty} \frac{B_n}{r^{n+1}} P_n(\cos\theta) - E_0 r \cos\theta, \tag{2.59}$$

$$\phi_2 = \sum_{n=0}^{\infty} C_n r^n P_n(\cos\theta). \tag{2.60}$$

Applying the second and third boundary conditions to (2.59) and (2.60), we have for any value of n except $n = 1$:

$$\frac{B_n}{a^{n+1}} = C_n a^n, \tag{2.61}$$

and:

$$-\varepsilon_1 (n+1) \frac{B_n}{a^{n+2}} = \varepsilon_2 n C_n a^{n-1}. \tag{2.62}$$

From these equations it follows that $B_n = 0$ and $C_n = 0$ for all values of n except $n = 1$. When $n = 1$, we have:

$$\frac{B_1}{a^2} - E_0 a = C_1 a,$$

$$\varepsilon_1 \left(\frac{2B_1}{a^3} + E_0 \right) = -\varepsilon_2 C_1.$$

Hence:

$$B_1 = \frac{\varepsilon_2 - \varepsilon_1}{\varepsilon_2 + 2\varepsilon_1} a^3 E_0,$$

$$C_1 = -\frac{3\varepsilon_1}{2\varepsilon_1 + \varepsilon_2} E_0.$$

Substitution of these values in (2.59) and (2.60) gives:

$$\phi_1 = \left(\frac{\varepsilon_2 - \varepsilon_1}{2\varepsilon_1 + \varepsilon_2} \frac{a^3}{r^3} - 1 \right) E_0 z, \tag{2.63}$$

and:

$$\phi_2 = -\frac{3\varepsilon_1}{2\varepsilon_1 + \varepsilon_2} E_0 z. \tag{2.64}$$

Since the potential due to the external charges is given by $\phi = -E_0 z$, it follows from (2.63) and (2.64) that the contributions ϕ'_1 and ϕ'_2 due to the apparent surface charges are given by:

$$\phi'_1 = \frac{\varepsilon_2 - \varepsilon_1}{2\varepsilon_1 + \varepsilon_2} \frac{a^3}{r^3} E_0 z, \tag{2.65}$$

and:

$$\phi'_2 = \frac{\varepsilon_2 - \varepsilon_1}{2\varepsilon_1 + \varepsilon_2} E_0 z. \tag{2.66}$$

The expression (2.65) is identical to that for the potential due to an ideal dipole at the centre of the sphere, surrounded by a vacuum, the dipole vector being directed along the z-axis and given by:

$$\boldsymbol{m} = \frac{\varepsilon_2 - \varepsilon_1}{2\varepsilon_1 + \varepsilon_2} a^3 \boldsymbol{E}_0. \tag{2.67}$$

The part of the field $\boldsymbol{E}_2$ inside the sphere due to the apparent charges is, according to (2.66), parallel to $\boldsymbol{E}_0$ and represented by:

$$\boldsymbol{E}'_2 = -\frac{\varepsilon_2 - \varepsilon_1}{2\varepsilon_1 + \varepsilon_2} \boldsymbol{E}_0. \tag{2.68}$$

The total field $\boldsymbol{E}_2$ inside the sphere is, according to (2.64), given by:

$$\boldsymbol{E}_2 = \frac{3\varepsilon_1}{2\varepsilon_1 + \varepsilon_2} \boldsymbol{E}_0. \tag{2.69}$$

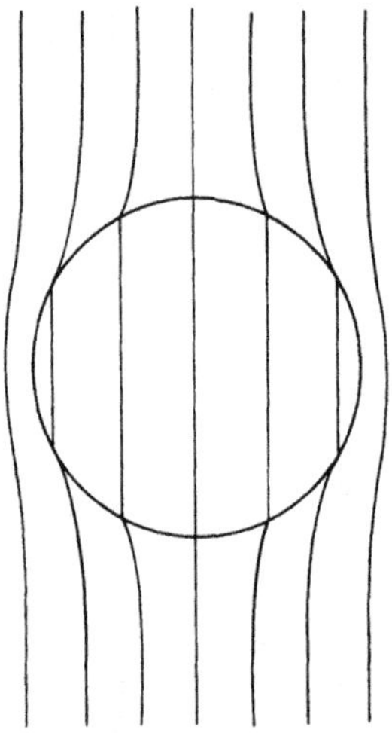

Fig. 18a
Lines of dielectric displacement.
Spherical cavity in a dielectric ($\varepsilon = 1.8$)
with uniform field.

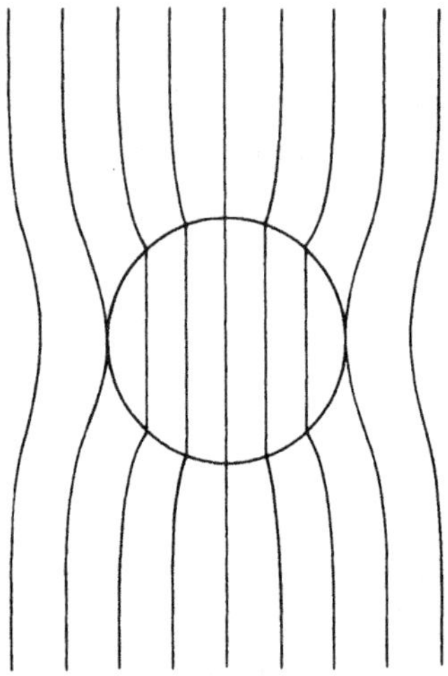

Fig. 18b
Lines of dielectric displacement.
Dielectric sphere ($\varepsilon = 3$) in a vacuum
with uniform field.

(b) *A spherical cavity in a dielectric*

In the special case of a spherical cavity in a dielectric ($\varepsilon_1 = \varepsilon$; $\varepsilon_2 = 1$), eqn. (2.69) is reduced to:

$$\boldsymbol{E}_c = \frac{3\varepsilon}{2\varepsilon + 1}\boldsymbol{E}_0. \tag{2.70}$$

This field is called the "cavity field". The lines of dielectric displacement given by $\boldsymbol{D}_c = 3\boldsymbol{D}_0/(2\varepsilon + 1)$ are shown in Fig. 18a. The density of the lines is lower in the cavity than in the surrounding dielectric, since $\boldsymbol{D}$ is larger in the dielectric than in the cavity.

(c) *A dielectric sphere in a vacuum*

For a dielectric sphere in a vacuum ($\varepsilon_1 = 1$, $\varepsilon_2 = \varepsilon$), eqn. (2.69) is reduced to:

$$\boldsymbol{E} = \frac{3}{\varepsilon + 2}\boldsymbol{E}_0, \tag{2.71}$$

where $\boldsymbol{E}$ is the field inside the sphere.

The lines of dielectric displacement are shown in Fig. 18b. The density of the lines is higher in the sphere than in the surrounding vacuum, since inside the sphere $\boldsymbol{D}_s$ equals $3\varepsilon\boldsymbol{E}_0/(\varepsilon + 2)$. Consequently, it is larger than $\boldsymbol{E}_0$.

According to eqn. (2.67), the field outside the sphere due to the apparent surface charges is the same as the field that would be caused by a dipole $\boldsymbol{m}$ at the centre of the sphere, surrounded by a vacuum, and given by:

$$\boldsymbol{m} = \frac{\varepsilon - 1}{\varepsilon + 2} a^3 \boldsymbol{E}_0. \tag{2.72}$$

This is the electric moment of the dielectric sphere. Therefore (2.72) can be checked in an interesting way. The uniform field $\boldsymbol{E}_s$ in the dielectric sphere, given by (2.71), causes a homogeneous polarization of the sphere. The induced dipole moment per cm^3 is, according to the definition of $\boldsymbol{P}$, given by:

$$\boldsymbol{P} = \frac{\varepsilon - 1}{4\pi} \boldsymbol{E}_s = \frac{\varepsilon - 1}{\varepsilon + 2} \frac{3}{4\pi} \boldsymbol{E}_0. \tag{2.73}$$

Hence the total moment of the sphere is:

$$\boldsymbol{m} = \tfrac{4}{3}\pi a^3 \boldsymbol{P} = \frac{\varepsilon - 1}{\varepsilon + 2} a^3 \boldsymbol{E}_0, \tag{2.74}$$

which is in accordance with (2.72).

(d) *An ellipsoidal body in a dielectric*

The field $\boldsymbol{E}_2$ can also be calculated inside an ellipsoidal body with dielectric constant ε_2, immersed in a dielectric with dielectric constant ε_1. We shall confine the discussion of this subject to a summary of the results.

A Cartesian coordinate system (x, y, z) is chosen with its axes in the direction of the principal axes $2a$, $2b$, and $2c$ of the ellipsoid. The potential ϕ_2 inside the ellipsoid, caused by an external field $\boldsymbol{E}_0$, can now be calculated by solving Laplace's equation in ellipsoidal harmonics. The result[7] of this calculation is:

$$\phi_2 = -\frac{x(E_0)_x}{1 + \dfrac{\varepsilon_2 - \varepsilon_1}{\varepsilon_1} A_a} - \frac{y(E_0)_y}{1 + \dfrac{\varepsilon_2 - \varepsilon_1}{\varepsilon_1} A_b} - \frac{z(E_0)_z}{1 + \dfrac{\varepsilon_2 - \varepsilon_1}{\varepsilon_1} A_c}. \tag{2.75}$$

In this expression $(E_0)_x$, $(E_0)_y$, and $(E_0)_z$ are the components of the external field $\boldsymbol{E}_0$ in the direction of the x-, y-, and z-axes. The factors A_a, A_b, and A_c depend on the form of the ellipsoid and are given by:

$$A_\alpha = \frac{abc}{2} \int_0^\infty \frac{ds}{(s + \alpha^2)R} \qquad (\alpha = a, b, c), \tag{2.76}$$

with $R^2 = (s + a^2)(s + b^2)(s + c^2)$. The integral cannot be evaluated in closed form, but extensive tabulations[8,9] of A_α as a function of a, b, and c are available. It is possible to calculate by transforming to $\frac{\mathrm{d}}{\mathrm{d}s}\left(\frac{1}{R^2}\right)$ and partial integration the sum of the factors A_α, resulting in:

$$A_a + A_b + A_c = 1. \tag{2.77}$$

When the ellipsoid degenerates into a sphere, we have $a = b = c$ and therefore all factors A_α then have the value 1/3. Substituting this value in eqn. (2.75) and taking $(E_0)_x = (E_0)_y = 0$, eqn. (2.64) is obtained again.

From the potential ϕ_2 the field $\boldsymbol{E}_2$ inside the ellipsoid can be calculated by taking the gradient (*cf.* eqn. (1.14)). Defining a tensor $\boldsymbol{\Lambda}$ as a diagonal tensor with diagonal components $1 + \frac{\varepsilon_2 - \varepsilon_1}{\varepsilon_1} A_\alpha$, this field can be expressed with the help of the inverse tensor $\boldsymbol{\Lambda}^{-1}$ as:

$$\boldsymbol{E}_2 = \boldsymbol{\Lambda}^{-1} \cdot \boldsymbol{E}_0. \tag{2.78}$$

Thus, when the external field is homogeneous, the field inside the ellipsoid will also be homogeneous; in general the fields will not have the same direction. When the external field is in the direction of, for instance, the $2c$ principal axis, eqn. (2.78) reduces to:

$$\boldsymbol{E}_2 = \frac{1}{1 + \dfrac{\varepsilon_2 - \varepsilon_1}{\varepsilon_1} A_c} \boldsymbol{E}_0 = \frac{\varepsilon_1}{\varepsilon_1 + (\varepsilon_2 - \varepsilon_1) A_c} \boldsymbol{E}_0. \tag{2.79}$$

When the ellipsoid degenerates into a sphere, we have $A_c = \frac{1}{3}$, and (2.79) reduces to (2.69).

In many cases, *e.g.* a dielectric *in vacuo*, we have $\varepsilon_2 > \varepsilon_1$. Since then the diagonal elements of $\boldsymbol{\Lambda}$ are all greater than one, the field $\boldsymbol{E}_2$ will be smaller than the external field $\boldsymbol{E}_0$. The polarization $\boldsymbol{P}_2$ of the ellipsoid tends to decrease the external field. The extent to which the polarization decreases the external field is given by the depolarizing tensor $\boldsymbol{L}$:

$$\boldsymbol{E}_2 = \boldsymbol{E}_0 - \boldsymbol{L} \cdot \boldsymbol{P}_2. \tag{2.80}$$

By combining (2.80) with (2.78) the components of $\boldsymbol{L}$ can be found. Substituting $\boldsymbol{P}_2 = (\varepsilon_2 - 1)\boldsymbol{E}_2/4\pi$ (*cf.* eqn (2.46)) into (2.80) and rearranging, we have:

$$\left(\boldsymbol{I} + \frac{\varepsilon_2 - 1}{4\pi}\boldsymbol{L}\right) \cdot \boldsymbol{E}_2 = \boldsymbol{E}_0, \tag{2.81}$$

and thus:

$$\boldsymbol{I} + \frac{\varepsilon_2 - 1}{4\pi}\boldsymbol{L} = \boldsymbol{\Lambda}. \tag{2.82}$$

From (2.82) we see that $\boldsymbol{L}$ is a diagonal tensor, with diagonal components L_α given by:

$$L_\alpha = 4\pi\frac{\varepsilon_2 - \varepsilon_1}{\varepsilon_1(\varepsilon_2 - 1)}A_\alpha. \tag{2.83}$$

The factors L_α, depending on the dielectric constants of the ellipsoid and the surrounding medium as well as on the shape of the ellipsoid, are called depolarizing factors. When the surrounding medium is a vacuum ($\varepsilon_1 = 1$), L_α is equal to $4\pi A_\alpha$. For this reason the name depolarizing factors is sometimes also given to the factors A_α.

(e) *An ellipsoidal cavity in a dielectric*

When the external field is in the direction of a principal axis α, we calculate for the cavity field inside an ellipsoidal cavity from eqn. (2.79) with $\varepsilon_1 = \varepsilon$ and $\varepsilon_2 = 1$:

$$\boldsymbol{E}_c = \frac{\varepsilon}{\varepsilon + (1 - \varepsilon)A_\alpha}\boldsymbol{E}_0. \tag{2.84}$$

From (2.84) we obtain (2.70) for the special case of a spherical cavity ($A_\alpha = \frac{1}{3}$).

(f) *A dielectric sphere in the field of a point charge within it*

In section 39 we shall be confronted with the problem of a dielectric sphere in the field of an arbitrary charge distribution within it. To solve this problem we shall first consider the problem of a dielectric sphere in the field of a single point charge within it.

We choose the coordinate system such that the centre of the sphere is situated in the origin and the point charge e on the positive z-axis. The radius of the sphere is called a, the distance between the point charge and the centre of the sphere s.

For the potential ϕ_1 in the region $r > a$ and the potential ϕ_2 in the region $s < r < a$ Laplace's equation is valid. Since there is axial symmetry the solutions can according to eqn. (A2.15) be written as:

$$\phi_1 = \sum_{n=0}^{\infty} \left(A_n r^n + \frac{B_n}{r^{n+1}} \right) P_n(\cos\theta),$$

$$\phi_2 = \sum_{n=0}^{\infty} \left(C_n r^n + \frac{D_n}{r^{n+1}} \right) P_n(\cos\theta).$$

The boundary conditions are in this case:

1. $$(\phi_1)_{r\to\infty} = 0, \tag{2.85}$$

since there are no sources of the electric field at infinity.

2. $$(\phi_1)_{r=a} = (\phi_2)_{r=a}, \tag{2.86}$$

3. $$\left(\frac{\partial\phi_1}{\partial r}\right)_{r=a} = \varepsilon\left(\frac{\partial\phi_2}{\partial r}\right)_{r=a}. \tag{2.87}$$

In considering the influence of the point charge on the potentials, it must be remembered that according to section 4 the contribution due to charges with a distance to the origin smaller than r, is expressed with the help of terms proportional to $r^{-n-1}P_n(\cos\theta)$, whereas the contribution due to charges with greater distance to the origin is expressed with the help of terms proportional to $r^n P_n(\cos\theta)$. In the region $s < r < a$ it holds that the only charge leading to terms $D_n r^{-n-1} P_n(\cos\theta)$ in ϕ_2 is the point charge e. The potential due to this point charge in a medium with dielectric constant ε is given by:

$$\phi = \frac{e}{\varepsilon r} \sum_{n=0}^{\infty} \left(\frac{s}{r}\right)^n P_n(\cos\theta), \tag{2.88}$$

as follows from (1.44) and (2.54). Therefore:

$$D_n = \frac{es^n}{\varepsilon}.$$

From the first boundary condition and the fact that the Legendre functions are linearly independent we derive $A_n = 0$ for every value of n. From the second and third boundary conditions it follows that:

$$\frac{B_n}{a^{n+1}} = C_n a^n + \frac{es^n}{\varepsilon a^{n+1}},$$

$$(-n-1)\frac{B_n}{a^{n+2}} = \varepsilon n C_n a^{n-1} + (-n-1)\frac{es^n}{a^{n+2}}.$$

Hence:

$$B_n = \frac{2n+1}{\varepsilon n + n + 1} es^n,$$

$$C_n = \frac{(\varepsilon - 1)(n+1)}{\varepsilon(\varepsilon n + n + 1)} \frac{es^n}{a^{2n+1}}.$$

Substituting A_n, B_n, C_n, and D_n in the solutions of Laplace's equation we find:

$$\phi_1 = \sum_{n=0}^{\infty} \frac{2n+1}{\varepsilon n + n + 1} \frac{es^n}{r^{n+1}} P_n(\cos\theta), \tag{2.89}$$

$$\phi_2 = \sum_{n=0}^{\infty} \left\{ \frac{(\varepsilon - 1)(n+1)}{\varepsilon(\varepsilon n + n + 1)} \frac{es^n r^n}{a^{2n+1}} + \frac{es^n}{\varepsilon r^{n+1}} \right\} P_n(\cos\theta). \tag{2.90}$$

Comparing eqn. (2.89) with eqn. (1.44) we see that the terms in the potential outside the dielectric sphere differ by a factor $(2n+1)/(\varepsilon n + n + 1)$ from the corresponding terms of the potential of a point charge not imbedded in a dielectric sphere.

For $n = 0$ we find $(2n+1)/(\varepsilon n + n + 1) = 1$: the term in $1/r$ of the potential outside the sphere is not influenced by the dielectric sphere. This result could be expected, because this term is proportional to the total electric charge within the sphere, which is equal to the value of the point charge, since the dielectric is electrically neutral.

For $n = 1$, however, the factor $(2n+1)/(\varepsilon n + n + 1)$ amounts to $3/(\varepsilon + 2)$. So the dipole term is reduced by the dielectric sphere. This is only possible when the electric moment (with respect to the origin) of the combination of dielectric sphere and point charge is smaller than the electric moment (with respect to the origin) of the point charge alone. Thus, by the point charge e the dielectric is polarized in such a way, that its electric moment amounts to:

$$\boldsymbol{m}_{\text{diel}} = \frac{3}{\varepsilon + 2}e\boldsymbol{s} - e\boldsymbol{s} = -\frac{\varepsilon - 1}{\varepsilon + 2}\boldsymbol{m}, \tag{2.91}$$

if $\boldsymbol{m} = e\boldsymbol{s}$ is the electric moment of the point charge with respect to the centre of the sphere.

For $n = 2$ we find, in the same way, that the quadrupole moment of the electric point charge is reduced by the quadrupole moment of the polarized matter:

$$\boldsymbol{Q}_{\text{diel}} = \frac{1}{2!}\left\{\frac{5}{2\varepsilon + 3}\boldsymbol{ess} - \boldsymbol{ess}\right\} = -\frac{2(\varepsilon - 1)}{2\varepsilon + 3}\boldsymbol{Q}. \tag{2.92}$$

For an arbitrary value of n this is generalized to:

$$\boldsymbol{Y}^{(n)}_{\text{diel}} = \frac{1}{n!}\left\{\frac{2n + 1}{\varepsilon n + n + 1}e(\boldsymbol{s})^n - e(\boldsymbol{s})^n\right\} = -\frac{n(\varepsilon - 1)}{\varepsilon n + n + 1}\boldsymbol{Y}^{(n)}. \tag{2.93}$$

(g) *A dielectric sphere in the field of an arbitrary charge distribution within it*

Every discrete or continuous charge distribution can be approximated to any degree of accuracy by a finite number of point charges. According to the superposition principle, the electric moment induced in the sphere by all the charges will be the sum of the moments induced by each point charge separately. Therefore:

$$\boldsymbol{m}_{\text{diel}} = \sum_i (\boldsymbol{m}_{\text{diel}})_i = \sum_i -\frac{\varepsilon - 1}{\varepsilon + 2}\boldsymbol{m}_i = -\frac{\varepsilon - 1}{\varepsilon + 2}\sum_i \boldsymbol{m}_i, \tag{2.94}$$

in which $\boldsymbol{m}_{\text{diel}}$ is the moment induced by all the charges, $(\boldsymbol{m}_{\text{diel}})_i$ the moment induced by the i-th charge, and $\boldsymbol{m}_i$ the electric moment of this charge with respect to the centre of the sphere. Hence for the electric moment induced by an arbitrary charge distribution within a dielectric sphere the same relation holds between the moment induced in the dielectric sphere and the electric moment of the charge distribution as for a single point charge.

The relations (2.92) and (2.93) can be generalized in the same way.

(h) *A dielectric sphere in the field of an external point charge*

This problem can be dealt with in a manner analogous to the case of a dielectric sphere in the field of an internal point charge. Again the coordinate system is chosen such that the centre of the sphere is situated in the origin, and the point charge on the positive z-axis at a distance s from the origin ($s > a$, the radius of the sphere). Laplace's equation is valid in the regions $a < r < s$ and $r < a$. Since there is axial symmetry the solutions can according to eqn. (A2.15) be written as:

$$\phi_1 = \sum_{n=0}^{\infty}\left(A_n r^n + \frac{B_n}{r^{n+1}}\right)P_n(\cos\theta) \quad (a < r < s),$$

$$\phi_2 = \sum_{n=0}^{\infty} \left(C_n r^n + \frac{D_n}{r^{n+1}} \right) P_n(\cos\theta) \quad (r < a).$$

The boundary conditions are:

1. $$(\phi_1)_{r=a} = (\phi_2)_{r=a}, \tag{2.95}$$

2. $$\left(\frac{\partial \phi_1}{\partial r}\right)_{r=a} = \varepsilon \left(\frac{\partial \phi_2}{\partial r}\right)_{r=a}, \tag{2.96}$$

3. ϕ_2 has no singularity for $r = 0$.

In ϕ_1 the terms $A_n r^n P_n(\cos\theta)$ are due to charges with greater distance to the origin than the point where the potential is calculated (compare section 4). In this case the only source is the point charge situated at s on the z-axis. The potential due to this point charge in the region $a < r < s$ is given by eqn. (1.46):

$$\phi = \frac{e}{s} \sum_{n=0}^{\infty} \left(\frac{r}{s}\right)^n P_n(\cos\theta),$$

so that:

$$A_n = \frac{e}{s^{n+1}}. \tag{2.97}$$

According to the third boundary condition $D_n = 0$ for every value of n. From the other boundary conditions it follows that:

$$\frac{ea^n}{s^{n+1}} + \frac{B_n}{a^{n+1}} = C_n a^n,$$

$$\frac{nea^{n-1}}{s^{n+1}} + (-n-1)\frac{B_n}{a^{n+2}} = \varepsilon n C_n a^{n-1}.$$

Hence:

$$B_n = -\frac{n(\varepsilon - 1)}{\varepsilon n + n + 1} \frac{ea^{2n+1}}{s^{n+1}},$$

$$C_n = \frac{2n+1}{\varepsilon n + n + 1} \frac{e}{s^{n+1}}.$$

Collecting the results, we find:

$$\phi_1 = \sum_{n=0}^{\infty} \left\{ \frac{er^n}{s^{n+1}} - \frac{n(\varepsilon - 1)}{\varepsilon n + n + 1} \frac{ea^{2n+1}}{r^{n+1}s^{n+1}} \right\} P_n(\cos\theta), \tag{2.98}$$

$$\phi_2 = \sum_{n=0}^{\infty} \frac{2n+1}{\varepsilon n + n + 1} \frac{er^n}{s^{n+1}} P_n(\cos\theta). \tag{2.99}$$

This last equation will be applied in section 16.

§10. The polarizability α

When a body is placed in a uniform electric field $\boldsymbol{E}_0$ *in vacuo*, caused by a fixed charge distribution, its dipole moment will in general change. The difference between the dipole moments before and after the application of the field $\boldsymbol{E}_0$ is called the induced dipole moment $\boldsymbol{m}$. If a body shows an induced dipole moment differing from zero upon application of a uniform field $\boldsymbol{E}_0$, it is said to be polarizable.

In most cases polarizable bodies are polarized linearly, that is, the induced moment $\boldsymbol{m}$ is proportional to $\boldsymbol{E}_0$. In such a case we have:

$$\boldsymbol{m} = \alpha \boldsymbol{E}_0, \tag{2.100}$$

where α is called the (scalar) polarizability of the body.

From the dimensions of the dipole moment, $[e][l]$, and the field intensity, $[e][l]^{-2}$, it follows that the polarizability has the dimension of a volume. Using the above definition of the polarizability, we conclude from eqn. (2.67) that a dielectric sphere of dielectric constant ε and with radius a has a polarizability:

$$\alpha = \frac{\varepsilon - 1}{\varepsilon + 2} a^3. \tag{2.101}$$

From (1.72) it follows that a conducting sphere with radius a has a polarizability:

$$\alpha = a^3. \tag{2.102}$$

Although this equation may be obtained by taking the limit $\varepsilon \to \infty$ in (2.101) there is a fundamental difference between the two types of polarization. In the case of a dielectric sphere every volume element is polarized, whereas in the case of a conducting sphere the induced dipole moment arises from true surface charges.

It is in general not true that a scalar polarizability is sufficient to describe the induced polarization. In general we have to use a polarizability tensor $\boldsymbol{\alpha}$, provided the effects remain linear. We then have:

$$\boldsymbol{m} = \boldsymbol{\alpha} \cdot \boldsymbol{E}_0. \tag{2.103}$$

In such a case the induced dipole moment need not have the same direction as the applied field. In general this direction will depend on the position of the body relative to the polarizing field.

In the case of an ellipsoidal body of dielectric constant ε *in vacuo*, with principal axes $2a$, $2b$, and $2c$ in the direction of the x-, y-, and z-axes, we find with the help of eqn. (2.78) and the relations $\boldsymbol{P}_2 = (\varepsilon - 1)\boldsymbol{E}_2/4\pi$ and $\boldsymbol{m} = (4\pi/3)abc\boldsymbol{P}_2$:

$$\left.\begin{array}{llll} m_x = & \alpha_a(E_0)_x & & \\ m_y = & & \alpha_b(E_0)_y & \\ m_z & & & \alpha_c(E_0)_z \end{array}\right\}, \tag{2.104}$$

where α_a, α_b, and α_c are given by:

$$\alpha_a = \frac{\varepsilon - 1}{3\{1 + (\varepsilon - 1)A_a\}} abc, \qquad \text{etc.} \tag{2.105}$$

A_a is given in (2.76) when $\alpha = a$. Thus the polarizability tensor is diagonal in this coordinate system:

$$\boldsymbol{\alpha}_{\text{principal axes}} = \begin{pmatrix} \alpha_a & 0 & 0 \\ 0 & \alpha_b & 0 \\ 0 & 0 & \alpha_c \end{pmatrix}. \tag{2.106}$$

When the ellipsoid degenerates into a sphere, we have $A_a = A_b = A_c = 1/3$, and eqn. (2.104) reduces to eqn. (2.100).

When an accurate description of the behaviour of molecules and ions in an external electric field is required, it is often necessary to attribute a scalar or tensorial polarizability to them. This is to be expected because a molecule may be considered as a cluster of positive and negative point charges. Under the influence of the electric field the negative charges will be attracted and the positive charges repelled. These charge displacements lead to a net dipole moment, which will be a linear function of the external field as long as the field strength is not too high. Therefore the behaviour of the molecule can be characterized with the help of a scalar or tensorial polarizability.

According to the macroscopic theory, the polarizability is a function of the dielectric constant of the polarizable body. Therefore, if we want to use macroscopic theory in the description of molecules, we have to picture the molecules as small spheres or ellipsoids with internal dielectric constant ε_i. The value of ε_i can be calculated with the help of eqn. (2.101) or eqn. (2.105) when the dimensions of the molecule are known. This device is often useful in model calculations. It should always be remembered, however, that the dielectric constant is a macroscopic quantity, so that the internal dielectric constant cannot be applied to give a complete characterization of the behaviour of a molecule in an electric field.

The molecular polarizability is not a scalar but a tensor when the molecule or ion is not completely spherical. For practical purposes, however, the effect of this so-called anisotropy of the polarizability is small, and the tensor $\boldsymbol{\alpha}$ may be replaced by a scalar polarizability α given by:

$$\alpha = \tfrac{1}{3}\,\mathrm{Tr}\,(\boldsymbol{\alpha}). \tag{2.107}$$

In Table 5 some values of the polarizability are given as numerical examples.

TABLE 5

THE POLARIZABILITY OF SOME SIMPLE PARTICLES

Molecule or ion	α (Å^3)
He	0.2
Na^+	0.2
Cl^-	3.0
NH_3	2.3
CO_2	2.7
Ethane	4.5
Benzene	10.3
Cyclohexane	10.9
Hexane	11.8
Dodecane	22.8

References

1. Sir W. Thomson Lord Kelvin, *Reprints of Papers on Electrostatics and Magnetism*, MacMillan and Co., London 1872, sections 479 *ff*.
2. J. C. Maxwell, *A Treatise on Electricity and Magnetism* (3rd revised edition of 1891), Dover, New York 1954, Vol. II, sections 395–400.
3. H. A. Lorentz, *Proc. Roy. Acad. Amsterdam* (1902) 254, H. A. Lorentz, *Enc. Math. Wissensch.* **2** (1) 100.
4. L. Rosenfeld, *Theory of Electrons*, North-Holland, Amsterdam 1951.
5. P. Mazur and B. R. A. Nijboer, *Physica* **19** (1953) 971, P. Mazur, *Adv. Chem. Phys.* **1** (1958) 310.
6. S. R. de Groot and J. Vlieger, *Physica* **31** (1965) 125, 254, S. R. de Groot, *The Maxwell Equations* (Studies in Statistical Mechanics, Vol. IV), North-Holland, Amsterdam 1969.
7. J. A. Stratton, *Electromagnetic Theory*, McGraw-Hill, New York 1941, Ch. 3.
8. J. A. Osborn, *Phys. Rev.* **67** (1945) 351.
9. E. C. Stoner, *Phil. Mag.* **36** (1945) 803.

CHAPTER III

POLARIZATION AND ENERGY

§11. The relation between potential and energy

As a result of the Coulomb forces, which act between point charges or, generally, between the elements of a charge distribution, a certain amount of work is necessary to change the relative position of the charge elements.

According to the definition of the electric field strength, as given by eqn. (1.4), we may write for the force $\boldsymbol{F}$ working on a test charge e:

$$\boldsymbol{F} = e\boldsymbol{E}. \tag{3.1}$$

Here, $\boldsymbol{E}$ is the electric field strength due to all charges except the test charge e itself. If the test charge undergoes a small displacement $\boldsymbol{ds}$, the amount of work $\mathrm{d}W$ necessary to bring about this displacement is given by:

$$\mathrm{d}W = -\boldsymbol{F} \cdot \boldsymbol{ds}. \tag{3.2}$$

If we now displace the test charge from its original position at a point P to its final position at a point Q, we have:

$$W = \int_P^Q \mathrm{d}W = -\int_P^Q e\boldsymbol{E} \cdot \boldsymbol{ds}. \tag{3.3}$$

From the definition of the potential (compare eqn. (1.14) and (A1.33)) it follows immediately that W is given by:

$$W = e\{\phi(Q) - \phi(P)\}. \tag{3.4}$$

If there is no loss of energy due to radiation, collisions with other particles, etc., the test charge will traverse the path from P to Q by going through a sequence of stationary states, *i.e.* in a reversible manner. The amount of work W expended in moving the test charge against the electric field is then equal to the change in potential energy V_e of the test charge in the electric field. This can be written as:

$$W = V_e(Q) - V_e(P).$$

Comparing this expression with (3.4) we infer that the potential energy is determined up to an arbitrary additive constant:

$$V_e(Q) = e\phi(Q) + \text{constant}. \tag{3.5}$$

The same situation has already been discussed for the potential itself in section 6. The arbitrary constant can be fixed by requiring the potential energy to be zero for the test charge situated at an infinite distance from all true and apparent charges. This corresponds to the physical requirement that there can be no interaction over an infinite distance. When the point Q is at infinity, expression (3.5) for $V_e(Q)$ should give zero. To make the additive constant zero the potential $\phi(Q)$ has to go to zero when the point Q goes to infinity (compare section 6, p. 64). In this way the potential energy and the potential are determined uniquely, and their relation becomes:

$$V_e(Q) = e\phi(Q), \tag{3.6}$$

i.e. the potential is the potential energy of a unit charge in the field.

From this argument we see why the potential loses its meaning as a physical quantity when there is a charge distribution at infinity, *e.g.* to create a homogeneous field. In such a case the additive constant cannot be fixed in this way, and only differences between potentials are physically meaningful quantities.

The potential energy of a dipole in an external field can now easily be calculated from the energy of two charges in a field. A non-ideal dipole, consisting of charges $+e$ at $\boldsymbol{r}_+$ and $-e$ at $\boldsymbol{r}_-$, has a potential energy given by:

$$V = e\{\phi(\boldsymbol{r}_+) - \phi(\boldsymbol{r}_-)\}. \tag{3.7}$$

Developing $\boldsymbol{r}_+ = \boldsymbol{r}_- + \boldsymbol{l}$ in a Taylor series around $\boldsymbol{r}_-$ we have:

$$V = e\boldsymbol{l}\cdot\nabla\phi(\boldsymbol{r}_-) + \tfrac{1}{2}e\boldsymbol{ll}:\nabla\nabla\phi(\boldsymbol{r}_-) + \cdots. \tag{3.8}$$

If we take the limit $e \to \infty$, $\boldsymbol{l} \to \boldsymbol{0}$, keeping $\boldsymbol{m} = e\boldsymbol{l}$ constant, the dipole becomes ideal, and we find for the potential energy:

$$V_m = \boldsymbol{m}\cdot\nabla\phi(\boldsymbol{r}) = -\boldsymbol{m}\cdot\boldsymbol{E}(\boldsymbol{r}). \tag{3.9}$$

The potential due to the dipole itself is not included in the potential ϕ. If one should take the total potential and insert it in eqn. (3.6) or (3.9), one would obtain expressions containing a self-energy term that is infinitely large. Obviously, it is meaningless to speak about the energy of a charge or a dipole in its own field. When it is not clear from the context that an external field is meant, we will use a subscript 0 to denote the external field ($\boldsymbol{E}_0$), or the corresponding potential (ϕ_0).

The calculation leading to the expression for the potential energy of a dipole in an external field, eqn. (3.9), can be extended to the case of a multipole of the n-th order with multipole moment $\boldsymbol{Y}^{(n)}$.

For a quadrupole, consisting of charges $+e$ at $\boldsymbol{r}$ and $\boldsymbol{r} + \boldsymbol{l}_1 + \boldsymbol{l}_2$, $-e$ at $\boldsymbol{r} + \boldsymbol{l}_1$ and $\boldsymbol{r} + \boldsymbol{l}_2$, we obtain, by developing all potentials in a Taylor series around $\phi(\boldsymbol{r})$ and taking the limit $e \to \infty$, $\boldsymbol{l}_1 \to \boldsymbol{0}$, $\boldsymbol{l}_2 \to \boldsymbol{0}$ while $\boldsymbol{Q} = \frac{1}{2}e(\boldsymbol{l}_1\boldsymbol{l}_2 + \boldsymbol{l}_2\boldsymbol{l}_1)$ (compare eqn. (1.120)) is kept constant:

$$V_q = \boldsymbol{Q}:\nabla\nabla\phi(\boldsymbol{r}). \tag{3.10}$$

For a general multipole with moment $\boldsymbol{Y}^{(n)}$ we find analogously:

$$V_{y^{(n)}} = \boldsymbol{Y}^{(n)} \overset{(n)}{\cdot} \nabla^n \phi(\boldsymbol{r}), \tag{3.11}$$

where as in eqn. (1.151) $\overset{(n)}{\cdot}$ denotes an n times repeated inner multiplication, and the exponent n of the nabla operator denotes an n times repeated tensor product. For $n = 0, 1, 2, 3$ we obtain from the general expression (3.11):

$$n = 0 \quad V_e = e\phi, \qquad \text{(compare (3.6))}$$

$$n = 1 \quad V_m = \boldsymbol{m}\cdot\nabla\phi = -\boldsymbol{m}\cdot\boldsymbol{E}, \qquad \text{(compare (3.9))}$$

$$n = 2 \quad V_q = \boldsymbol{Q}:\nabla\nabla\phi = -\boldsymbol{Q}:\nabla\boldsymbol{E}, \tag{3.12}$$

$$n = 3 \quad V_u = \boldsymbol{U}\,\vdots\,\nabla\nabla\nabla\phi = -\boldsymbol{U}\,\vdots\,\nabla\nabla\boldsymbol{E}. \tag{3.13}$$

From this we may conclude, for instance, that in a homogeneous external field $\boldsymbol{E}$, where $\nabla\boldsymbol{E} = \boldsymbol{0}$, the potential energy of a quadrupole and of all higher multipoles is zero. Thus, in a homogeneous field quadrupoles and all higher multipoles will have no direct physical effects.

It is also possible to derive an expression for the potential energy of an arbitrary charge distribution in an external field in terms of the multipole moments of this charge distribution. We may always take a volume enclosing the charge distribution, and denote the potential due to the charges outside this volume by ϕ. Inside the volume the potential ϕ will satisfy Laplace's equation, since its sources are all outside the volume. According to (3.6), the potential energy of the system of n charges e_i will then be given by:

$$V = \sum_{i=1}^{n} e_i\phi(\boldsymbol{r}_i), \tag{3.14}$$

where $\boldsymbol{r}_i$ is the point at which the i-th charge e_i is situated. All potentials $\phi(\boldsymbol{r}_i)$ may now be developed in a Taylor series around the origin. This leads to:

$$V = \sum_{i=1}^{n} e_i[\phi(\mathbf{0}) + \boldsymbol{r}_i \cdot \nabla\phi(\mathbf{0}) + \tfrac{1}{2}\boldsymbol{r}_i\boldsymbol{r}_i : \nabla\nabla\phi(\mathbf{0}) + \cdots], \tag{3.15}$$

where the gradients are taken at the origin.

The coefficients of $\phi(\mathbf{0})$ and its successive gradients are now equal to the multipole moments of the charge distribution, as defined in eqns. (1.109), (1.110), and (1.111).

Thus, we may write:

$$V = e\phi(\mathbf{0}) + \boldsymbol{m} \cdot \nabla\phi(\mathbf{0}) + \boldsymbol{Q} : \nabla\nabla\phi(\mathbf{0}) + \cdots. \tag{3.16}$$

From this relation we see that the multipole moments introduced to describe the potential of an arbitrary set of charges can also be used to describe the potential energy of the same set of charges in an external field. In section 5 reductions have been carried out on the multipole moment, which did not change the contribution to the potential due to the multipole moment. If these reductions, when substituted into eqn. (3.16), would change the value of the potential energy, the reductions would lead to a multipole moment physically distinguishable from the original multipole moment. Since in eqn. (3.16) the same gradients of a function satisfying Laplace's equation are involved as in the multipole expansion of the potential (where the function had the form $1/r$), the reduced multipole moments will have the same potential energy in an external field as the original ones. Therefore they will be physically indistinguishable.

§12. The work required to assemble a charge distribution

In section 11 the potential energy of a system of point charges in an external field has been calculated. From the definition of the potential energy, as given by eqn. (3.6), we see that the potential energy is the amount of work necessary to bring the charges from infinity to their actual positions without changing their relative configuration. We may also ask what amount of work is needed to assemble a set of point charges to a prescribed configuration. The work W_1 necessary to bring a charge e_1 from infinity to its actual position $\boldsymbol{r}_1$ when no other charges are present, is clearly zero. When the potential due to the charge e_1 is given by $\phi_1(\boldsymbol{r})$, the amount of work W_2 necessary to bring a charge e_2 from infinity to $\boldsymbol{r}_2$ will be given by:

$$W_2 = e_2\phi_1(\boldsymbol{r}_2), \tag{3.17}$$

according to eqn. (3.6). The potential due to the charges e_1 and e_2 will be denoted by $\phi_{12}(\boldsymbol{r})$. Since potentials may be superposed, we have:

$$\phi_{12}(\boldsymbol{r}) = \phi_1(\boldsymbol{r}) + \phi_2(\boldsymbol{r}). \tag{3.18}$$

The amount of work W_3 required to bring the charge e_3 from infinity to $\boldsymbol{r}_3$, will now be given by:

$$W_3 = e_3\phi_{12}(\boldsymbol{r}_3) = e_3\phi_1(\boldsymbol{r}_3) + e_3\phi_2(\boldsymbol{r}_3). \tag{3.19}$$

Continuing this procedure up to the n-th charge, we find for the work of assembly of the charge distribution:

$$W = \sum_{i=1}^{n} W_i = \sum_{i=1}^{n} \sum_{j=1}^{i-1} e_i\phi_j(\boldsymbol{r}_i). \tag{3.20}$$

This expression can be written in a more symmetric way, because $e_i\phi_j(\boldsymbol{r}_i)$ is always equal to $e_j\phi_i(\boldsymbol{r}_j)$ for all values of i and j with $i \neq j$. This can be seen from eqn. (3.17), since it is clearly irrelevant for the amount of work whether e_2 is brought from infinity into the field of e_1 or e_1 is brought from infinity into the field of e_2. Thus we may also write:

$$W = \sum_{i=1}^{n} \sum_{j=1}^{i-1} e_j\phi_i(\boldsymbol{r}_j) = \sum_{j=1}^{n} \sum_{i=1}^{j-1} e_i\phi_j(\boldsymbol{r}_i). \tag{3.21}$$

The third member of this equation is obtained from the second one by interchanging the dummy indices i and j. Adding the expressions (3.20) and (3.21) for W and using:

$$\sum_{j=1}^{n} \sum_{i=1}^{j-1} = \sum_{i=1}^{n} \sum_{j=i+1}^{n},$$

we obtain:

$$2W = \sum_{i=1}^{n} \sum_{\substack{j=1 \\ j\neq i}}^{n} e_i\phi_j(\boldsymbol{r}_i),$$

or:

$$W = \tfrac{1}{2}\sum_{i} \sum_{j\neq i} e_i\phi_j(\boldsymbol{r}_i). \tag{3.22}$$

It is possible to rewrite this expression in a more suggestive way by defining a potential ϕ_0^i, due to all charges except the i-th charge:

$$\phi_0^i(\boldsymbol{r}) = \sum_{j\neq i} \phi_j(\boldsymbol{r}). \tag{3.23}$$

This leads to:

$$W = \tfrac{1}{2}\sum_i e_i \phi_0^i(\boldsymbol{r}_i). \tag{3.24}$$

To calculate the work of assembly of a continuous charge distribution ρ we will adopt another procedure. A continuous volume charge distribution ρ inside a volume V is formed by bringing infinitesimally small charge elements from infinity to their actual positions. At a certain stage of this process the charge distribution will have reached the value $\lambda\rho$, where λ is a parameter ranging from 0 to 1. Since the potential is a linear function of the charge distribution, it will have the value $\lambda\phi$, where ϕ is the final value of the potential. When λ is increased to a value $\lambda + \delta\lambda$ by bringing charge elements $\rho\delta\lambda\,\mathrm{d}v$ from infinity to all points inside the volume V, the total charge increases from $\iiint_V \lambda\rho\,\mathrm{d}v$ to $\iiint_V (\lambda + \delta\lambda)\rho\,\mathrm{d}v$. Since the charge element $\rho\delta\lambda\,\mathrm{d}v$ is brought into a potential $\lambda\phi$, this requires an amount of work $(\rho\delta\lambda\,\mathrm{d}v)\lambda\phi$. Therefore, the amount of work δW necessary to increase the charge distribution with an amount $\rho\delta\lambda$ everywhere is given by:

$$\delta W = \iiint_V \lambda\delta\lambda\rho\phi\,\mathrm{d}v. \tag{3.25}$$

The whole charge distribution can be built up in this way, and the total amount of work necessary can be found by integrating from $\lambda = 0$ to $\lambda = 1$:

$$W = \int_{\lambda=0}^{\lambda=1} \delta W = \tfrac{1}{2} \iiint_V \rho\phi\,\mathrm{d}v. \tag{3.26}$$

Thus, the work required to assemble a continuous charge distribution from infinitesimally small charge elements at infinity is given by an expression having the same structure as eqn. (3.24). Eqn. (3.26) is, however, not a generalization of eqn. (3.24) for the continuous case, since the expression in (3.26) is based on infinitesimally small charge elements instead of point charges. This can be shown by substituting into (3.26) the charge density appropriate to a set of point charges e_i,

$$\rho(\boldsymbol{r}) = \sum_i e_i \delta(\boldsymbol{r} - \boldsymbol{r}_i). \tag{3.27}$$

Combining (3.26) with (3.27) we have:

$$W = \tfrac{1}{2} \sum_i e_i \phi(\boldsymbol{r}_i). \tag{3.28}$$

In this case $\phi(\boldsymbol{r})$ will be given by:

$$\phi(\boldsymbol{r}) = \sum_j e_j \frac{1}{|\boldsymbol{r} - \boldsymbol{r}_j|}, \tag{3.29}$$

and comparing (3.29) with (3.23) we see that (3.28) can be rewritten as:

$$W = \tfrac{1}{2} \sum_i e_i \phi_0^i(\boldsymbol{r}_i) + \tfrac{1}{2} \sum_i \frac{e_i^2}{|\boldsymbol{r}_i - \boldsymbol{r}_i|}. \tag{3.30}$$

Thus the work of assembly starting from infinitesimally small charge elements differs from the work of assembly starting from point charges by a term that is evidently equal to the work necessary to form the point charges from infinitesimally small charge elements. This last amount of work is infinitely large. Therefore, it is not allowable to use eqn. (3.26) for point charges.

It is, however, possible to use eqn. (3.26) for another mathematical abstraction, *i.e.* a surface charge distribution, which can be built up of infinitesimally small charge elements in the same way as a volume charge distribution. If we suppose that there are conductors, each with volume V_i and enclosed by a surface S_i, given by an equation $f_i(\boldsymbol{r}) = 0$ and carrying surface charges σ_i, we can write using the δ-function (*cf.* eqn. (A2.21)):

$$\rho(\boldsymbol{r}) = \rho_0(\boldsymbol{r}) + \sum_i \sigma_i \delta\{f_i(\boldsymbol{r})\}, \tag{3.31}$$

where ρ_0 is the volume charge distribution; therefore:

$$\iiint_V \rho \, \mathrm{d}v = \iiint_{V - \sum_i V_i} \rho_0 \, \mathrm{d}v + \sum_i \oiint_{S_i} \sigma_i \, \mathrm{d}S, \tag{3.32}$$

since inside a conductor there is no volume charge distribution. Substitution of (3.31) into (3.26) and use of the fact that the potential on a conductor is a constant, leads to:

$$W = \tfrac{1}{2} \iiint_{V - \sum_i V_i} \rho_0 \phi \, \mathrm{d}v + \tfrac{1}{2} \sum_i \phi(S_i) \oiint_{S_i} \sigma_i \, \mathrm{d}S. \tag{3.33}$$

Since ρ_0 and σ_i are finite, this expression will remain finite. For $\rho_0 = 0$ everywhere we obtain the work of assembly of a set of charged conductors. We then find:

$$W = \tfrac{1}{2} \sum_i e_i \phi_i, \tag{3.34}$$

where e_i is the total charge on the i-th conductor. This is the same expression as can be derived directly with the help of electrostatic theory. Therefore, expression (3.26) may be used for surface charge distributions.

For the work of assembly of a continuous charge distribution, an expression can be derived which is more generally applicable. Again, we build up the charge distribution by assembling infinitesimally small charge elements. At a certain stage the charge distribution has attained the value ρ and the potential due to this charge distribution the value ϕ. Now we add an increment of charge density $\delta\rho$ to each volume element $\mathrm{d}v$. The amount of work necessary to do this is given by:

$$\delta W = \iiint_V \phi \delta\rho \, \mathrm{d}v. \tag{3.35}$$

In vacuo, the source equation $\operatorname{div} \boldsymbol{E} = 4\pi\rho$ (eqn. (1.9)) is valid. Therefore, we can write:

$$\delta\rho = \frac{1}{4\pi} \operatorname{div} \boldsymbol{\delta E}. \tag{3.36}$$

Substituting (3.36) into (3.35), and using the vector identity (A1.25):

$$\phi \operatorname{div} \boldsymbol{\delta E} = \operatorname{div}(\phi \boldsymbol{\delta E}) - \boldsymbol{\delta E} \cdot \operatorname{grad} \phi,$$

we find:

$$\delta W = \frac{1}{4\pi} \iiint_V \operatorname{div}(\phi \boldsymbol{\delta E}) \, \mathrm{d}v - \frac{1}{4\pi} \iiint_V \boldsymbol{\delta E} \cdot \operatorname{grad} \phi \, \mathrm{d}v. \tag{3.37}$$

Using the divergence theorem (A1.31) and expanding the volume to infinity, we see that the first integral will go to zero, since ϕ and $\boldsymbol{\delta E}$ will go to zero at least as fast as r^{-1} and r^{-2} when r goes to infinity. The second integral can be rewritten with the help of $\boldsymbol{E} = -\operatorname{grad} \phi$; thus, we find:

$$\delta W = \frac{1}{4\pi} \iiint_\infty \boldsymbol{E} \cdot \boldsymbol{\delta E} \, \mathrm{d}v. \tag{3.38}$$

This is the amount of work necessary to increase the field from $\boldsymbol{E}$ to $\boldsymbol{E} + \boldsymbol{\delta E}$ by the addition of infinitesimally small charge elements to the charge distribution which is the source of the field. Since $\boldsymbol{E} \cdot \boldsymbol{\delta E} = \tfrac{1}{2}\delta(\boldsymbol{E} \cdot \boldsymbol{E})$, integration from $\boldsymbol{E} = \boldsymbol{0}$ to $\boldsymbol{E} = \boldsymbol{E}$ will now yield the total amount of work:

$$W = \int_{\boldsymbol{E}=\boldsymbol{0}}^{\boldsymbol{E}=\boldsymbol{E}} \delta W = \frac{1}{8\pi} \iiint_{\infty} \boldsymbol{E} \cdot \boldsymbol{E}\, \mathrm{d}v. \tag{3.39}$$

This is the expression for the work of assembly of the charge distribution constituting the source of the field. It is often loosely called the energy of the electric field; this field is then considered to have an energy density $\boldsymbol{E} \cdot \boldsymbol{E}/8\pi$.*

Expression (3.39) for the work of assembly is equivalent to (3.26). This can be checked by substituting $\operatorname{div} \boldsymbol{E} = 4\pi\rho$ into (3.26) and performing the same operations as lead from (3.35) to (3.39). In fact, the linear relation between the potential ϕ and its source ρ which was used to derive (3.26), is an immediate consequence of the equations $\operatorname{div} \boldsymbol{E} = 4\pi\rho$ and $\boldsymbol{E} = -\operatorname{grad} \phi$ which were used to derive (3.39).

The advantages of the derivation of (3.39) starting from (3.35) become apparent in cases when a dielectric is present during the assembly of the charge distribution. Now we have to take $\operatorname{div} \boldsymbol{D} = 4\pi\rho$ (compare section 6) and $\boldsymbol{E} = -\operatorname{grad} \phi$, which will lead to a relation analogous to (3.38):

$$\delta W = \frac{1}{4\pi} \iiint_{\infty} \delta \boldsymbol{D} \cdot \boldsymbol{E}\, \mathrm{d}v. \tag{3.40}$$

The total amount of work can now be obtained by integrating from $\boldsymbol{D} = \boldsymbol{0}$ to $\boldsymbol{D} = \boldsymbol{D}$. To be able to derive an explicit expression, however, we must know the relation between $\boldsymbol{D}$ and $\boldsymbol{E}$ during the assembly of the charge distribution. Here, we will use the empirical fact that for many dielectrics the relation $\boldsymbol{D} = \varepsilon \boldsymbol{E}$ is valid, where ε is a function of density and of temperature (see section 8). If we now specify that no mechanical work is done on the dielectric during the assembling process (*i.e.* the volume V, and thus the density, are kept constant), and that the dielectric is enclosed in a heat bath (*i.e.* the temperature is kept constant), we may write:

$$\delta \boldsymbol{D} = \varepsilon \delta \boldsymbol{E}. \tag{3.41}$$

Substituting this into eqn. (3.40) we can perform the integration and obtain:

---

* For electromagnetic waves an equation analogous to (3.39) can be derived. In that case it makes sense to speak of the energy density of the electromagnetic field. Eqn. (3.39) may be treated as a limiting case of the general expression, so that in this sense it would indeed be possible to speak of the energy of the electric field.

$$W = \frac{1}{8\pi}\iiint_{\infty} \varepsilon \boldsymbol{E}\cdot\boldsymbol{E}\,\mathrm{d}v = \frac{1}{8\pi}\iiint_{\infty} \boldsymbol{D}\cdot\boldsymbol{E}\,\mathrm{d}v. \tag{3.42}$$

This is the work of assembly of a charge distribution when the temperature and the volume of all dielectrics present, are kept constant. When there are no dielectrics, $\varepsilon = 1$ everywhere, and (3.42) reduces to (3.39).

When the relation $\boldsymbol{D} = \varepsilon\boldsymbol{E}$ is not valid, we will generally obtain different results, even when the same thermodynamic conditions are satisfied. When the material is anisotropic the dielectric constant must be replaced by a tensor and we have:

$$\boldsymbol{D} = \boldsymbol{\varepsilon}\cdot\boldsymbol{E}. \tag{3.43}$$

This will lead to the same expression (3.42), since the relation is still linear. In section 13 we will prove that $\delta W = \mathrm{d}F$, the differential of the free energy. Thus δW is a total differential, and the argument of Appendix I, section 7 applies to prove that $\boldsymbol{\varepsilon}$ is a symmetric tensor.

When the relation between $\boldsymbol{D}$ and $\boldsymbol{E}$ is not linear, we have, for instance (compare eqn. (2.47)):

$$\boldsymbol{D} = (\varepsilon + 4\pi\xi E^2)\boldsymbol{E}. \tag{3.44}$$

This leads to:

$$\delta\boldsymbol{D} = (\varepsilon + 12\pi\xi E^2)\delta\boldsymbol{E}, \tag{3.45}$$

and thus:

$$W = \frac{1}{8\pi}\iiint_{\infty} (\varepsilon + 6\pi\xi E^2)\boldsymbol{E}\cdot\boldsymbol{E}\,\mathrm{d}v. \tag{3.46}$$

In contrast with (3.42) this cannot be written as a function of the product $\boldsymbol{D}\cdot\boldsymbol{E}$, but we have:

$$W = \frac{1}{8\pi}\iiint_{\infty} \boldsymbol{D}\cdot\boldsymbol{E}\,\mathrm{d}v + \iiint_{\infty} \tfrac{1}{4}\xi E^4\,\mathrm{d}v. \tag{3.47}$$

For an electret with fixed dipole density $\boldsymbol{P}$, there is no relation at all between $\boldsymbol{P}$ and $\boldsymbol{E}$. From this, and the general relation $\boldsymbol{D} = \boldsymbol{E} + 4\pi\boldsymbol{P}$ (*cf.* eqn. (2.41)) we derive:

$$\delta\boldsymbol{D} = \delta\boldsymbol{E}. \tag{3.48}$$

The integration should now proceed from $\boldsymbol{E}'$ to $\boldsymbol{E}$, where $\boldsymbol{E}'$ is the field due to the electret, and one obtains for the work of assembly of the charge distribution in the presence of an electret:

$$W = \frac{1}{8\pi}\iiint_{\infty} (\boldsymbol{E}\cdot\boldsymbol{E} - \boldsymbol{E}'\cdot\boldsymbol{E}')\,\mathrm{d}v. \tag{3.49}$$

§13. The work of assembly as a thermodynamic quantity

When the effect of a change in volume or temperature on the relation between $\boldsymbol{D}$ and $\boldsymbol{E}$ is known, it is possible to integrate (3.40) under the specified conditions and thus obtain an expression for the work of assembly under these conditions. If, however, the thermodynamic relations of W are known, such expressions can be calculated directly with the help of thermodynamics.

The amount of work δW necessary to increase a charge distribution by an amount $\delta\rho$, as given in (3.40), is expended in a reversible process (see section 11). In this case we may write, according to the first and second law of thermodynamics:

$$\delta W = \delta U - T\delta S, \tag{3.50}$$

where δU is the change in internal energy, T the absolute temperature and δS the change in entropy. At constant volume and temperature, the amount of work in a reversible process is also equal to the change in work content or Helmholtz free energy F:

$$\delta F = \delta(U - TS)_{T,V} = \delta U - T\delta S = \delta W. \tag{3.51}$$

Taking $\boldsymbol{D} = \varepsilon\boldsymbol{E}$, we may integrate (3.40) at constant temperature, obtaining:

$$F - F^0 = \int \delta F = \int_{\boldsymbol{D}=\boldsymbol{0}}^{\boldsymbol{D}=\boldsymbol{D}} \delta W = \frac{1}{8\pi}\iiint_\infty \boldsymbol{D}\cdot\boldsymbol{E}\,\mathrm{d}v, \tag{3.52}$$

where $F^0 = F^0(T, V)$ is the work content in the absence of the field. Thus, the work of assembly is equal to the change in the Helmholtz free energy of the system consisting of the charge distribution and the dielectric.

From the general relation:

$$\mathrm{d}F = -S\,\mathrm{d}T - p\,\mathrm{d}V + \mathrm{d}W_{\mathrm{el}}, \tag{3.53}$$

where $\mathrm{d}W_{\mathrm{el}}$ is the amount of electric work given by eqn. (3.40), V the volume of the dielectric, and p the external pressure, it follows:

$$S = -\left(\frac{\partial F}{\partial T}\right)_{V,\boldsymbol{D}}, \tag{3.54}$$

or, using eqn. (3.52):

$$S = S^0 - \frac{1}{8\pi}\iiint_\infty \boldsymbol{D}\cdot\left(\frac{\partial}{\partial T}\frac{\boldsymbol{D}}{\varepsilon}\right)_{V,\boldsymbol{D}} \mathrm{d}v$$

$$= S^0 + \frac{1}{8\pi}\iiint_\infty E^2\left(\frac{\partial\varepsilon}{\partial T}\right)_V \mathrm{d}v, \tag{3.55}$$

where $S^0 = -\left(\frac{\partial F^0}{\partial T}\right)_V$, the entropy in the absence of the field. Expressions for the pressure p in the presence of an electric field will be derived in section 45, where the subject of electrostriction is treated.

The internal energy U of the charge distribution and the dielectric will be given by:

$$\begin{aligned} U &= F + TS \\ &= U^0 + \frac{1}{8\pi} \iiint_\infty E^2 \left(\varepsilon + T \left(\frac{\partial \varepsilon}{\partial T} \right)_V \right) \mathrm{d}v \\ &= U^0 + \frac{1}{8\pi} \iiint_\infty E^2 \left(\frac{\partial(\varepsilon T)}{\partial T} \right)_V \mathrm{d}v, \end{aligned} \tag{3.56}$$

where U^0, the internal energy in the absence of the field is equal to:

$$U^0 = F^0 + TS^0. \tag{3.57}$$

From eqn. (3.55) for the entropy it is also possible to derive an expression for the heat absorbed by the dielectric during a reversible field change at constant temperature:

$$\delta Q = T \mathrm{d}S = \frac{1}{4\pi} \iiint_\infty E \, \mathrm{d}E \; T \left(\frac{\partial \varepsilon}{\partial T} \right)_V \mathrm{d}v. \tag{3.58}$$

Further relations may be derived with the help of thermodynamic methods.[1,2,3,4] The expression for the special case of a change in chemical equilibrium under the influence of an electric field will be given in section 43.

From relations (3.55) and (3.58) we can already obtain some insight into the effects of the application of an electric field. When $\left(\frac{\partial \varepsilon}{\partial T}\right)_V < 0$, which is normally the case, we have $\delta Q < 0$ and $\mathrm{d}S < 0$. Thus, heat is produced by the dielectric and given off to the heat bath. At the same time the entropy decreases, that is, the field orders the particles of the dielectric. If the dielectric were heat insulated, the temperature would rise until the ordering effect of the field was compensated for by the disorder brought about by the increased heat movements of the particles. When, on the other hand, $\left(\frac{\partial \varepsilon}{\partial T}\right)_V > 0$, heat would be absorbed by the dielectric and an existing ordering of the particles would be disrupted. This can be expected to happen, for instance, when a polar liquid consists of dimers with antiparallel dipoles. The applied field would tend to align the dipoles in the same direction, thus disrupting the dimers and lessening the previous ordering. To break up the dimers, the formation energy must be applied, *i.e.* the dielectric will absorb heat. If the dielectric were heat insulated, its temperature would decrease when an external electric field was applied, until the disorder caused by the field would compensate for the order inside the dielectric due to the lower temperature.

The production of heat by a dielectric as a consequence of a change in the applied field can be used to attain very low temperatures. This process is called adiabatic depolarization, and is exactly analogous to the cooling of a paramagnetic substance by adiabatic demagnetization. At a certain temperature an electric field is applied isothermally, and the heat produced by the dielectric is given off to the heat bath. Then the dielectric is heat insulated and the electric field is switched off. The polarization which has been built up inside the dielectric disappears adiabatically, and the temperature decreases. The theoretical efficiency of this process is given by the electrocaloric coefficient β_E:

$$\beta_E = \left(\frac{\partial T}{\partial E}\right)_S. \tag{3.59}$$

Although cooling by adiabatic depolarization has been known in principle for a long time, it has been realised experimentally only since 1956, with the help of solid $SrTiO_3$[5] or crystals of KCl doped with hydroxyl impurities.[6] In this way temperatures of 0.3°K have been reached.

§14. The energy of a dielectric in an external field

It is often useful to have an expression for that part of the free energy of a system consisting of a dielectric and a charge distribution, that refers to the dielectric. This can be obtained by comparing the amounts of work necessary to assemble a charge distribution in the presence of a dielectric and without it. The difference between these amounts of work will be equal to the non-mechanical part of the work required to bring the dielectric into the field of the charge distribution. Thus, this difference can be considered to be the change in free energy of the dielectric upon application of the field.

According to eqn. (3.42), the work required to assemble a charge distribution is given by:

$$W = \frac{1}{8\pi} \iiint_{\infty} \boldsymbol{E} \cdot \boldsymbol{D} \, \mathrm{d}v.$$

When there is no dielectric, the amount of work necessary to assemble the same charge distribution is given by (3.39):

$$W_0 = \frac{1}{8\pi} \iiint_{\infty} \boldsymbol{E}_0 \cdot \boldsymbol{E}_0 \, \mathrm{d}v.$$

The subscript 0 is added to indicate the absence of a dielectric, and $\boldsymbol{E}_0$ will be called the external field.

The difference $\Delta W = W - W_0$ between these two amounts of work will be the change in the free energy of the dielectric when the field is applied in an isothermal reversible process (*cf.* section 13):

$$\Delta W = \frac{1}{8\pi} \iiint_{\infty} \boldsymbol{E} \cdot \boldsymbol{D} \, \mathrm{d}v - \frac{1}{8\pi} \iiint_{\infty} \boldsymbol{E}_0 \cdot \boldsymbol{E}_0 \, \mathrm{d}v. \tag{3.60}$$

This expression can be simplified by noting that the charge distribution producing the field and polarizing the dielectric will be the same in the presence and in the absence of the dielectric, *i.e.*:

$$\operatorname{div} \boldsymbol{D} = \operatorname{div} \boldsymbol{E}_0 = 4\pi\rho, \tag{3.61}$$

or:

$$\operatorname{div} (\boldsymbol{D} - \boldsymbol{E}_0) = 0. \tag{3.62}$$

If we now consider a vector field $\psi(\boldsymbol{D} - \boldsymbol{E}_0)$ where ψ is the potential of $\boldsymbol{E}$ or $\boldsymbol{E}_0$, we may write for the divergence of this field, according to (A1.25) and (3.62):

$$\operatorname{div} \psi(\boldsymbol{D} - \boldsymbol{E}_0) = (\boldsymbol{D} - \boldsymbol{E}_0) \cdot \operatorname{grad} \psi. \tag{3.63}$$

When both sides of eqn. (3.63) are integrated over a volume V' with surface S', and the divergence theorem (A1.31) is applied to the lefthand side, we obtain:

$$\oiint_{S'} \psi(\boldsymbol{D} - \boldsymbol{E}_0) \cdot \boldsymbol{dS}' = \iiint_{V'} (\boldsymbol{D} - \boldsymbol{E}_0) \cdot \operatorname{grad} \psi \, \mathrm{d}v'. \tag{3.64}$$

When the boundary S' goes to infinity, the surface integral goes to zero, since ψ goes to zero as $1/r$ and $\boldsymbol{D}$ as well as $\boldsymbol{E}_0$ go to zero as $1/r^2$ or faster, when r becomes infinitely large. Thus we have:

$$\iiint_{\infty} (\boldsymbol{D} - \boldsymbol{E}_0) \cdot \operatorname{grad} \psi \, \mathrm{d}v = 0,$$

or:

$$\iiint_{\infty} \boldsymbol{D} \cdot \operatorname{grad} \psi \, \mathrm{d}v = \iiint_{\infty} \boldsymbol{E}_0 \cdot \operatorname{grad} \psi \, \mathrm{d}v. \tag{3.65}$$

Taking the cases $\operatorname{grad} \psi = -\boldsymbol{E}$ and $\operatorname{grad} \psi = -\boldsymbol{E}_0$, we derive from (3.65) the following identities:

$$\iiint_{\infty} \boldsymbol{D} \cdot \boldsymbol{E} \, \mathrm{d}v = \iiint_{\infty} \boldsymbol{E}_0 \cdot \boldsymbol{E} \, \mathrm{d}v, \tag{3.66}$$

$$\iiint_{\infty} \boldsymbol{D} \cdot \boldsymbol{E}_0 \, \mathrm{d}v = \iiint_{\infty} \boldsymbol{E}_0 \cdot \boldsymbol{E}_0 \, \mathrm{d}v. \tag{3.67}$$

When (3.66) and (3.67) are substituted into (3.60), we obtain for the energy of the dielectric in the field:

$$\Delta W = -\frac{1}{8\pi} \iiint_{\infty} (\boldsymbol{D} - \boldsymbol{E}) \cdot \boldsymbol{E}_0 \, \mathrm{d}v. \tag{3.68}$$

Using the general relation $\boldsymbol{D} - \boldsymbol{E} = 4\pi\boldsymbol{P}$ (eqn. (2.41)), and restricting the integration to the volume V of the dielectric, which is allowed since outside the dielectric $\boldsymbol{P}$ is zero, we find:

$$\Delta W = -\tfrac{1}{2} \iiint_{V} \boldsymbol{P} \cdot \boldsymbol{E}_0 \, \mathrm{d}v. \tag{3.69}$$

When the external field is uniform, we may write:

$$\Delta W = -\tfrac{1}{2}\boldsymbol{E}_0 \cdot \iiint_{V} \boldsymbol{P} \, \mathrm{d}v. \tag{3.70}$$

According to the definition of the polarizability of a macroscopic body (eqn. (2.100)), we have for the total moment $\boldsymbol{m}$:

$$m = \alpha E_0 = \iiint_V P \, dv. \tag{3.71}$$

Thus we find:

$$\Delta W = -\tfrac{1}{2}\alpha E_0^2. \tag{3.72}$$

When the dielectric is completely characterized by its dipole density P, eqn. (3.69) for the energy of the dielectric in an external field can be derived by another method, which is connected with the way in which the dielectric is polarized by the external field. The energy ΔW is split up into two parts V_P, the potential energy of the dipole density in the external field as well as in its own field, and W_{pol}, the work of polarization of the dielectric:

$$\Delta W = V_P + W_{pol}. \tag{3.73}$$

If we denote the (unknown) field due to the dipole density working on a volume element of the dipole density, by E_{dip}, we find, by integrating eqn. (3.9) over the volume of the dielectric, for the potential energy V_P:

$$V_P = -\iiint_V P \cdot E_0 \, dv - \tfrac{1}{2}\iiint_V P \cdot E_{dip} \, dv. \tag{3.74}$$

The factor $\frac{1}{2}$ in the last term of (3.74) has to be introduced to avoid counting twice the energy of two elements of the dipole density with respect to each other (compare the occurrence of the factor $\frac{1}{2}$ in eqns. (3.22) and (3.26)). The work of polarization may be found by adding increments of dipole density δP to each volume element dv against the total field $E_0 + E_{dip}$. This virtual polarizing process can be described with the help of a parameter λ ranging from 0 to 1 (compare the virtual charging process in section 12). When the total field has the value $\lambda(E_0 + E_{dip})$, the polarization will have the value λP and the increment of polarization δP will be given by $P\delta\lambda$. Thus we have:

$$\delta W_{pol} = \iiint_V \lambda \delta\lambda P \cdot (E_0 + E_{dip}) \, dv, \tag{3.75}$$

(compare (3.93)), and therefore:

$$W_{pol} = \int_{\lambda=0}^{\lambda=1} \delta W_{pol} = \tfrac{1}{2}\iiint_V P \cdot (E_0 + E_{dip}) \, dv. \tag{3.76}$$

Adding the contributions V_P and W_{pol} as given by (3.74) and (3.76), we obtain for the energy of the dielectric:

$$V_P + W_{pol} = -\tfrac{1}{2}\iiint_V P \cdot E_0 \, dv, \tag{3.77}$$

which is the same as expression (3.69) derived above.

Expression (3.69) for the energy of a dielectric in an external field can be used to derive the thermodynamic relations of the dielectric. This is possible because the work of assembly of the charge distribution in the absence of the dielectric, W_0, is independent of the thermodynamic state and the properties of the dielectric. Therefore, it may be subtracted from the work of assembly of the charge distribution in the presence of the dielectric, W, to obtain the free energy of the dielectric.

In this way we obtain for the free energy of the dielectric F_{diel} the following expression:

$$F_{diel} - F^0 = -\tfrac{1}{2} \iiint_V \boldsymbol{P} \cdot \boldsymbol{E}_0 \, dv, \tag{3.78}$$

where F^0 is the free energy in the absence of the field (*cf.* section 13). Using the Gibbs-Helmholtz relation:

$$U = F + T\left(\frac{\partial F}{\partial T}\right)_{V,\boldsymbol{E}_0}, \tag{3.79}$$

in the same way as in section 13 (*cf.* eqns. (3.54) and (3.56)), we find for the internal energy of the dielectric:

$$U_{diel} - U^0 = -\tfrac{1}{2} \iiint_V \left(\frac{\partial T\boldsymbol{P}}{\partial T}\right)_{V,\boldsymbol{E}_0} \cdot \boldsymbol{E}_0 \, dv, \tag{3.80}$$

where U^0 is the internal energy in the absence of the field.

Since we have taken $\boldsymbol{D} = \varepsilon \boldsymbol{E}$ in the derivation of (3.42), we may write $\boldsymbol{P} = \chi \boldsymbol{E}$; eqn. (3.80) now becomes:

$$U_{diel} - U^0 = -\tfrac{1}{2} \iiint_V \left(\frac{\partial T\chi}{\partial T}\right)_{V,\boldsymbol{E}_0} \boldsymbol{E} \cdot \boldsymbol{E}_0 \, dv. \tag{3.81}$$

For low values of χ we have $\boldsymbol{E} \simeq \boldsymbol{E}_0$; if the external field $\boldsymbol{E}_0$ is uniform, we may simplify eqn. (3.81) to:

$$U_{diel} - U^0 \simeq \tfrac{1}{2} \left(\frac{\partial T\chi}{\partial T}\right)_{V,\boldsymbol{E}_0} E_0^2 V. \tag{3.82}$$

When the relation $\boldsymbol{D} = \varepsilon \boldsymbol{E}$ is not assumed, the fundamental differential relation (3.40) cannot be integrated to the simple form of eqn. (3.42). It is possible, however, to derive an expression for the differential amount of work $d\,\Delta W$ required when the external field is increased by an amount $\boldsymbol{dE}_0$, by combining eqns. (3.38) and (3.40):

$$d\Delta W = dW - dW_0 = \frac{1}{4\pi} \iiint_\infty \boldsymbol{E} \cdot \boldsymbol{dD} \, dv - \frac{1}{4\pi} \iiint_\infty \boldsymbol{E}_0 \cdot \boldsymbol{dE}_0 \, dv. \tag{3.83}$$

This expression can be transformed in the same way as eqn. (3.60) by considering vector fields $\psi(\boldsymbol{dD} - \boldsymbol{dE}_0)$ and $d\psi(\boldsymbol{D} - \boldsymbol{E}_0)$. The result of this transformation is:

$$d\Delta W = -\iiint_V \boldsymbol{P} \cdot \boldsymbol{dE}_0 \, dv. \tag{3.84}$$

In the literature there is some confusion regarding the differential amount of work $d\Delta W$ upon a change in the external electric field $\boldsymbol{dE}_0$.

When there is a linear relation between $\boldsymbol{P}$ and $\boldsymbol{E}_0$ (*i.e.*, when $\boldsymbol{P}$ is a linear function of $\boldsymbol{E}$), eqn. (3.84) can be rewritten as:

$$(d\Delta W)_{\text{linear}} = -\iiint_V \boldsymbol{dP} \cdot \boldsymbol{E}_0 \, dv. \tag{3.85}$$

This is the same expression as the expression for the potential energy of an infinitesimally small dipole density $\boldsymbol{dP}$ in an external field $\boldsymbol{E}_0$ as it follows from eqn. (3.9), and is sometimes derived from this equation. Since $d\Delta W$ includes the work of polarization as well as the potential energy in the external field (compare eqn. (3.73)), the derivation of (3.85) from (3.9) is not correct; when $\boldsymbol{P}$ is not a linear function of $\boldsymbol{E}_0$ the use of eqn. (3.9) still leads to eqn. (3.85), which is then an incorrect result.

It is possible to obtain a correct expression containing $\boldsymbol{dP}$ instead of $\boldsymbol{dE}_0$, by taking a Legendre transformation of the free energy F according to:

$$\mathcal{I} = F + \iiint_V \boldsymbol{P} \cdot \boldsymbol{E}_0 \, dv. \tag{3.86}$$

Differentiation of eqn. (3.86) and substitution of expression (3.84) for dF now leads to:

$$d\mathcal{I} = \iiint_V \boldsymbol{dP} \cdot \boldsymbol{E}_0 \, dv. \tag{3.87}$$

This expression, however, is not connected with the work required to bring a dielectric into an external field. It is the electrostatic analogon of the term $V dp$ in the expression for the free enthalpy or Gibbs free energy $dG = -S\,dT + V\,dp$, which is the result of the Legendre transformation of the enthalpy with respect to the volumetric work. For this reason the quantity $\mathcal{I}$ is sometimes called the free enthalpy of the dielectric.

There is still another expression in use for the free energy of a dielectric, which is derived from eqn. (3.40). With the help of $\boldsymbol{D} = \boldsymbol{E} + 4\pi\boldsymbol{P}$, eqn. (3.40) is written as:

$$dW = \frac{1}{4\pi}\iiint_\chi \boldsymbol{dE} \cdot \boldsymbol{E} \, dv + \iiint_V \boldsymbol{dP} \cdot \boldsymbol{E} \, dv. \tag{3.88}$$

In the second term of this expression the integration is restricted to the volume of the dielectric, since $\boldsymbol{P}$ and $\boldsymbol{dP}$ are zero outside the dielectric.

The two terms on the righthand side are identified as being the work done on the vacuum and on the dielectric, so that the increment of free energy of the dielectric, which is equal to the amount of work in a reversible isothermal process, is sometimes taken to be equal to

$\iiint_V d\boldsymbol{P} \cdot \boldsymbol{E}\, dv$. This is clearly not justified, since the polarized dielectric will contribute to the field $\boldsymbol{E}$ everywhere. Thus $\boldsymbol{E}$ is not an independent variable.

When $\boldsymbol{E} \simeq \boldsymbol{E}_0$, however, the error introduced in this way is small, and the incorrect expression may be used for practical purposes. It should be noted, however, that it is an approximation of expression (3.87) for the transformed free energy $\mathcal{J}$ and not for the free energy itself.

§15. The energy of an induced dipole in an external field

In section 10 a scalar or tensorial polarizability was attributed to molecules and ions in analogy with the behaviour of macroscopic bodies in an external field. It is to be expected that this analogy can be extended to the energy of such a particle in an external field, so that it will be given by eqn. (3.72):

$$\Delta W = -\tfrac{1}{2}\alpha E_0^2.$$

A derivation of this result will be given in this section.

When a molecule or ion is brought into an electric field, the positive and negative charges (nuclei and electrons) of the particle will move with respect to each other. Thus, the dipole moment, as well as all higher multipole moments, will change. Denoting the dipole moment of the particle in the absence of an electric field, the so-called permanent dipole moment, by $\boldsymbol{\mu}$, and the total moment when the field has been applied by $\boldsymbol{m}$, we may write the induced dipole moment $\boldsymbol{p}$ as:

$$\boldsymbol{p} = \boldsymbol{m} - \boldsymbol{\mu}. \tag{3.89}$$

In the same way, the induced quadrupole and octupole moments[7] can be written as the differences of the total moments $\boldsymbol{Q}$ and $\boldsymbol{U}$ and the permanent moments $\boldsymbol{\Theta}$ and $\boldsymbol{\Omega}$. In this section we shall neglect all effects due to permanent or induced multipole moments higher than the dipole moment. This approximation will suffice as long as the gradient of the external field is small, for in this case the potential energy of the higher moments will be small (compare eqns. (3.12) and (3.13)).

The induced dipole moment $\boldsymbol{p}$ will be a function of the applied field strength $\boldsymbol{E}_0$. For small field strengths the displacements of the charges from their equilibrium positions will be a linear function of $\boldsymbol{E}_0$, so that $\boldsymbol{p}$ will also be linear in $\boldsymbol{E}_0$ (for a discussion of the induced moment at higher field strength, see section 44). For isotropic particles the induced moment will have the same direction as the applied field, so that we may write:

$$\boldsymbol{p} = \alpha \boldsymbol{E}_0, \tag{3.90}$$

where α is called the (scalar) polarizability.

When a molecule with zero net charge and permanent dipole moment $\boldsymbol{\mu}$ is placed in an external field $\boldsymbol{E}_0$, the amount of work necessary to accomplish this will not be given solely by the potential energy V_m of the total moment in the external field. A certain amount of work W_{pol}, the work of polarization, has to be expended to form the induced dipole against the internal forces of the molecule. Thus, we may write:

$$W = V_m + W_{\text{pol}}. \tag{3.91}$$

When there is equilibrium between the external and the internal forces working on the microscopic charges, the energy of the molecule will be minimal with respect to an infinitesimal change in induced moment:

$$\mathrm{d}W = 0 \quad \text{for all } \boldsymbol{dp}. \tag{3.92}$$

After applying (3.92) to (3.91) and using eqns. (3.9) and (3.89), we obtain:

$$\mathrm{d}W_{\text{pol}} = -\mathrm{d}V_m = \boldsymbol{E}_0 \cdot \boldsymbol{dp}. \tag{3.93}$$

This relation is valid for any dependence of $\boldsymbol{p}$ on $\boldsymbol{E}_0$. When we have the proportionality relation (3.90), we may also write:

$$\mathrm{d}W_{\text{pol}} = \frac{1}{\alpha}\boldsymbol{p} \cdot \boldsymbol{dp} = \frac{1}{2\alpha}\mathrm{d}(p^2). \tag{3.94}$$

If the induced moment $\boldsymbol{p}$ is formed in a reversible process, there is equilibrium between the internal and external forces at each stage, and eqn. (3.94) can be integrated. This leads to:

$$W_{\text{pol}} = \int \mathrm{d}W_{\text{pol}} = \frac{1}{2\alpha}\int_0^p \mathrm{d}(p^2) = \frac{1}{2\alpha}p^2. \tag{3.95}$$

This can be written in several equivalent forms with the help of eqn. (3.90):

$$W_{\text{pol}} = \frac{1}{2\alpha}\boldsymbol{p} \cdot \boldsymbol{p} = \tfrac{1}{2}\boldsymbol{p} \cdot \boldsymbol{E}_0 = \tfrac{1}{2}\alpha \boldsymbol{E}_0 \cdot \boldsymbol{E}_0. \tag{3.96}$$

When eqns. (3.9) and (3.96) are substituted into eqn. (3.91) for the amount of work necessary to bring the molecule into the external field, we find:

$$\begin{aligned} W &= -(\boldsymbol{\mu} + \boldsymbol{p}) \cdot \boldsymbol{E}_0 + \tfrac{1}{2}\boldsymbol{p} \cdot \boldsymbol{E}_0 \\ &= -\boldsymbol{\mu} \cdot \boldsymbol{E}_0 - \tfrac{1}{2}\alpha E_0^2. \end{aligned} \tag{3.97}$$

When the polarizability is not a scalar but a tensor $\boldsymbol{\alpha}$ (compare eqn. (2.103)), the derivation of the work of polarization proceeds in the same manner. When we introduce the inverse of $\boldsymbol{\alpha}$ by:

$$\boldsymbol{E}_0 = \boldsymbol{\alpha}^{-1} \cdot \boldsymbol{p}, \tag{3.98}$$

(compare (A1.64)), we may write instead of eqn. (3.94):

$$\mathrm{d}W_{\mathrm{pol}} = (\boldsymbol{\alpha}^{-1} \cdot \boldsymbol{p}) \cdot \boldsymbol{dp}, \tag{3.99}$$

which can be integrated to give:

$$W_{\mathrm{pol}} = \tfrac{1}{2}(\boldsymbol{\alpha}^{-1} \cdot \boldsymbol{p}) \cdot \boldsymbol{p} = \tfrac{1}{2}\boldsymbol{p} \cdot \boldsymbol{E}_0 = \tfrac{1}{2}(\boldsymbol{\alpha} \cdot \boldsymbol{E}_0) \cdot \boldsymbol{E}_0. \tag{3.100}$$

The total amount of work W will now be given by an expression similar to (3.97):

$$W = -\boldsymbol{\mu} \cdot \boldsymbol{E}_0 - \tfrac{1}{2}\boldsymbol{E}_0 \cdot \boldsymbol{\alpha} \cdot \boldsymbol{E}_0. \tag{3.101}$$

From the relation between W and the free energy we can deduce, according to the argument of Appendix I, section 7, that $\boldsymbol{\alpha}$ is a symmetric tensor.

Eqn. (3.97) or (3.101) for the energy of a polarizable particle with a permanent dipole moment shows that the energy is a function of the position and the orientation of the particle with respect to the field. Therefore expressions can be found for the forces acting on such a particle. In general, the component F_α of the force working on the particle in the direction of the α-th coordinate will be given by the negative derivative of the energy of the particle with respect to the α-th coordinate. Thus the force in the direction of the x-axis, exerted by an external field $\boldsymbol{E}_0$ on a particle with permanent dipole moment $\boldsymbol{\mu}$ and scalar polarizability α, is obtained by taking the derivative of expression (3.97):

$$F_x = -\frac{\partial W}{\partial x} = \boldsymbol{\mu} \cdot \frac{\partial \boldsymbol{E}_0}{\partial x} + \tfrac{1}{2}\alpha \frac{\partial E_0^2}{\partial x}. \tag{3.102}$$

The total translational force will be the vector sum of the components in the direction of the x-, y-, and z-axes:

$$\boldsymbol{F} = -\boldsymbol{\nabla} W = \boldsymbol{\mu} \cdot \boldsymbol{\nabla}\boldsymbol{E}_0 + \alpha \boldsymbol{E}_0 \cdot \boldsymbol{\nabla}\boldsymbol{E}_0. \tag{3.103}$$

Since the translational force is proportional to the gradient of the external field, it will be zero in a uniform field. In an inhomogeneous field the particle will move in the direction given by (3.103).

In a uniform as well as in a non-uniform field there will be a torque working on the particle, which tends to orient the dipole moment in the direction of the external field. This torque will be given by the negative derivative of the energy (eqn. (3.97)) with respect to θ, the angle between the external field and the dipole moment vector:

$$T_\theta = -\frac{\partial W}{\partial \theta} = -\mu E_0 \sin\theta. \tag{3.104}$$

Here, we have taken the external field to be uniform. When the field is not uniform, the expression for the torque will contain extra terms which are equal to the moment of the translational force with respect to the origin of the coordinate system.

Expression (3.104) can also be derived from the general expression for the torque working on a non-ideal dipole with moment $\boldsymbol{\mu}$, consisting of charges $+e$ and $-e$ at a distance $\boldsymbol{l}$. The forces on the charges are $+e\boldsymbol{E}_0$ and $-e\boldsymbol{E}_0$ and the torque is given by:

$$\boldsymbol{T} = -e\boldsymbol{E}_0 \times \boldsymbol{l} = -\boldsymbol{E}_0 \times \boldsymbol{\mu}. \tag{3.105}$$

The torque $\boldsymbol{T}$ is a vector in the direction of the axis around which the dipole rotates. Eqn. (3.105) remains valid in the limiting case of an ideal dipole.

From the fact that according to eqn. (3.105) there is a torque working on a polarizable particle with permanent dipole moment $\boldsymbol{\mu}$ in a uniform external field, we see that there is an important difference between polar compounds and non-polar compounds. A polar compound, consisting of molecules with a non-zero permanent dipole moment, will be influenced by an external field in two ways: the field will form induced dipole moments $\boldsymbol{p}$ in the direction of the field, and it will tend to align the permanent dipole moments in its own direction. For a non-polar compound we have $\mu = 0$, and the external field will have no directing influence on the molecules.

When the polarizability is anisotropic, the induced moment and the energy of this moment will depend on the orientation of the particle with respect to the external field, and the torque will not be zero even when there is no permanent dipole moment. In this case the expression for the torque is:

$$\boldsymbol{T} = -\boldsymbol{E}_0 \times \boldsymbol{p} = -\boldsymbol{E}_0 \times (\boldsymbol{\alpha} \cdot \boldsymbol{E}_0). \tag{3.106}$$

This expression will only be zero when $\boldsymbol{p} = \boldsymbol{\alpha} \cdot \boldsymbol{E}_0$ and $\boldsymbol{E}_0$ have the same direction. This will be the case when the particle is orientated in such a way that one of the principal axes of the polarizability tensor is in the direction of the external field (see Appendix I, section 7). Otherwise, the particle will rotate under the influence of the external field until it has reached its orientation of stable equilibrium, which occurs when the principal axis with the highest polarizability lies in the direction of the external field. This orientational effect depends on E_0^3 instead of on E_0. Its contribution to the polarization will become important only at high field strengths and will therefore be discussed in section 44.

From expression (3.106) for the torque working on a particle with anisotropic polarizability, a more direct argument can be derived showing that $\boldsymbol{\alpha}$ is a symmetric tensor. If $\boldsymbol{\alpha}$ were not symmetric, it could be split up into a symmetric and an antisymmetric part:

$$\boldsymbol{\alpha} = \boldsymbol{\alpha}^{\mathrm{S}} + \boldsymbol{\alpha}^{\mathrm{AS}}, \tag{3.107}$$

with:

$$(\boldsymbol{\alpha}^{\mathrm{S}})_{ij} = \tfrac{1}{2}(\alpha_{ij} + \alpha_{ji}), \tag{3.108}$$

and:

$$(\boldsymbol{\alpha}^{\mathrm{AS}})_{ij} = \tfrac{1}{2}(\alpha_{ij} - \alpha_{ji}), \tag{3.109}$$

(compare eqns. (A1.91) and (A1.92)).
Eqn. (3.106) now becomes:

$$\boldsymbol{T} = -\boldsymbol{E}_0 \times (\boldsymbol{\alpha}^{\mathrm{S}} \cdot \boldsymbol{E}_0) - \boldsymbol{E}_0 \times (\boldsymbol{\alpha}^{\mathrm{AS}} \cdot \boldsymbol{E}_0). \tag{3.110}$$

The symmetric tensor $\boldsymbol{\alpha}^{\mathrm{S}}$ can be transformed to principal axes; the antisymmetric tensor $\boldsymbol{\alpha}^{\mathrm{AS}}$, however, cannot be transformed in this way. Thus, we obtain, when $\boldsymbol{E}_0$ is in the direction of one of the principal axes of $\boldsymbol{\alpha}^{\mathrm{S}}$:

$$\boldsymbol{T} = -\boldsymbol{E}_0 \times (\boldsymbol{\alpha}^{\mathrm{AS}} \cdot \boldsymbol{E}_0). \tag{3.111}$$

This will never become zero,* and the rotational energy of the particle would increase indefinitely. This is clearly impossible, so we may conclude that $\boldsymbol{\alpha}^{\mathrm{AS}}$ should be zero, or $\boldsymbol{\alpha}$ is a symmetric tensor.

§16. The electrostatic interaction of two particles

When discussing the interaction between two particles it is often possible to neglect interactions which are not electrostatic by origin. This makes it useful to know the electrostatic interaction energy of two particles. When the permanent and induced multipole moments of the particles under consideration are known, the electrostatic interaction energy can be calculated with the help of the relations derived in this chapter.

* In the special case that the axial vector corresponding to $\boldsymbol{\alpha}^{\mathrm{AS}}$ is in the direction of the principal axis with the highest polarizability, the torque $\boldsymbol{T}$ in eqn. (3.110) becomes zero when $\boldsymbol{E}_0$ is also in this direction. It can be shown, however, that the equilibrium is unstable in this case.

The electrostatic interaction energy of two particles is defined as the work required to bring the particles from an infinite distance from each other to a given separation s, as far as this work is a result of the fact that one or both of the particles have a net charge and/or a multipole moment or multipole polarizability differing from zero. In practice, it is always assumed that the higher multipole moments and polarizabilities may be neglected. This requires the separation between the particles to be rather large, since otherwise the effects of a charge distribution should be described by the complete multipole expansion of the potential.

Moreover, when the distance between the particles becomes of the order of the radius of the particles, the charge distributions of the particles will overlap, and a quantum-mechanical calculation of the interaction energy should be preferred.[8,9] Since the exact quantum-mechanical calculations are quite unfeasible and the approximate calculations are not always very conclusive, we will in the following apply a number of results obtained for large separation between the particles to the case of real particles at a short distance. In this way the purely electrostatic contribution to the interaction energy can be estimated.

First we shall calculate the interaction energy of two unpolarizable particles characterized by their permanent multipole moments e, $\boldsymbol{\mu}$, $\boldsymbol{\Theta}$, $\boldsymbol{\Omega}$, etc. Then we introduce the dipole polarizability of the particles. The use of a point polarizability leading to an induced point dipole is an idealization, which will lead to considerable deviations when the distance becomes too small. For a special case an estimate of these deviations will be given with the help of a macroscopic model. Due to the nature of this model it will then not be necessary to take quadrupoles and higher multipoles into account.

To calculate the electrostatic interaction energy of two non-polarizable charge distributions ρ_1 and ρ_2, we shall express this energy as the energy of charge distribution ρ_1 in the field of charge distribution ρ_2 (Fig. 19).

According to eqn. (1.111), the potential due to charge distribution ρ_2 is given by:

$$\phi_2(\boldsymbol{s}') = \frac{e_2}{s'} + \boldsymbol{\mu}_2 \cdot \nabla \frac{1}{s'} + \boldsymbol{\Theta}_2 : \nabla\nabla \frac{1}{s'} + \cdots, \tag{3.112}$$

where $\boldsymbol{s}' = -\boldsymbol{s}$. The centre of mass of charge distribution ρ_2 is taken as the origin, and all differentiations are performed at this origin. If we now trans-

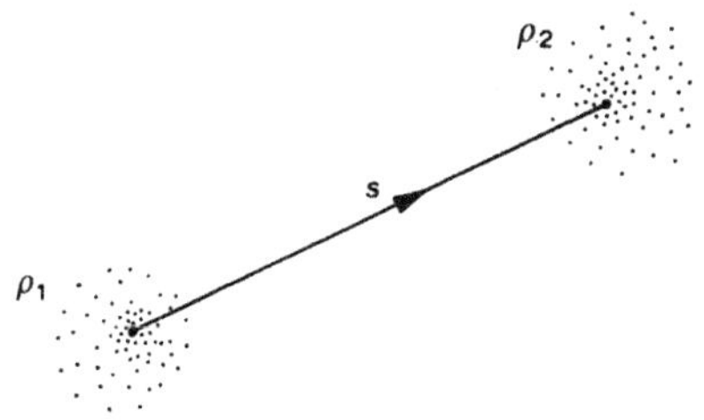

Fig. 19
The interaction of two charge distributions.

form to the coordinate system with the centre of mass of charge distribution ρ_1 as origin, we must substitute $-\boldsymbol{s}$ for $\boldsymbol{s}'$. As a result of this transformation, all gradients are now taken at the endpoint of the vectors instead of at the origin, *e.g.*:

$$\boldsymbol{\nabla}_{\text{origin}}\frac{1}{s'} = \frac{+1}{s^3}\boldsymbol{s}' = \frac{-1}{s^3}\boldsymbol{s} = \boldsymbol{\nabla}_{\text{endpoint}}\frac{1}{s}.$$

In the coordinate system with the origin at the centre of mass of ρ_1 we thus have:

$$\phi_2(\boldsymbol{0}) = \frac{e_2}{s} + \boldsymbol{\mu}_2\cdot\boldsymbol{\nabla}\frac{1}{s} + \boldsymbol{\Theta}_2:\boldsymbol{\nabla\nabla}\frac{1}{s} + \cdots. \tag{3.113}$$

In the same way, the field due to charge distribution ρ_2 is given by:

$$\boldsymbol{E}_2(\boldsymbol{0}) = \boldsymbol{\nabla}_{\text{origin}}\phi_2(\boldsymbol{s}') = \boldsymbol{\nabla}_{\text{endpoint}}\phi_2(\boldsymbol{0}). \tag{3.114}$$

The energy W_{12} of charge distribution ρ_1 in the field due to ρ_2 can now be given with the help of eqn. (3.16):

$$W_{12} = e_1\phi_2(\boldsymbol{0}) - \boldsymbol{\mu}_1\cdot\boldsymbol{\nabla}\phi_2(\boldsymbol{0}) + \boldsymbol{\Theta}_1:\boldsymbol{\nabla\nabla}\phi_2(\boldsymbol{0}) - \cdots, \tag{3.115}$$

where all differentiations are performed at the endpoint of the vector $\boldsymbol{s}$. Substitution of (3.113) into (3.115) now leads to:

$$W_{12} = \frac{e_1e_2}{s} + (e_1\boldsymbol{\mu}_2 - e_2\boldsymbol{\mu}_1)\cdot\boldsymbol{\nabla}\frac{1}{s} + (e_1\boldsymbol{\Theta}_2 - \boldsymbol{\mu}_1\boldsymbol{\mu}_2 + e_2\boldsymbol{\Theta}_1):\boldsymbol{\nabla\nabla}\frac{1}{s} +$$

$$+ (e_1\boldsymbol{\Omega}_2 - \boldsymbol{\mu}_1\boldsymbol{\Theta}_2 + \boldsymbol{\Theta}_1\boldsymbol{\mu}_2 - e_2\boldsymbol{\Omega}_1)\,\vdots\,\boldsymbol{\nabla\nabla\nabla}\frac{1}{s} + \cdots. \tag{3.116}$$

This is the general form of the electrostatic interaction energy, expressed in terms of the multipole moments of the charge distributions with respect to their own centre of mass.* The order of the multipole moments in the tensor products is not important, since they are all multiplied by completely symmetric tensors.

Since each gradient introduces an extra factor s^{-1}, it is clear that in expression (3.116) for the interaction energy, different types of interaction can be distinguished according to their dependence on the distance s. At very long distances only charge-charge interaction terms, which depend on s^{-1}, will be important. At intermediate distances the charge-dipole interaction, which depends on s^{-2}, will become important (although the charge-charge interaction will always dominate the other interactions). At smaller distances the charge-quadrupole and the dipole-dipole interactions, both depending on s^{-3}, will begin to play a part, and at still smaller distances we will find the charge-octupole and dipole-quadrupole interactions, depending on s^{-4}.

In the case of neutral particles (molecules) the leading interaction term will be the dipole-dipole interaction, followed by the dipole-quadrupole interaction and, depending on s^{-5}, by dipole-octupole and quadrupole-quadrupole interaction. Since for the theory of electric polarization the dipole-dipole interaction term will be the most important one, we will study the interaction between two permanent dipoles in somewhat more detail.

From eqn. (3.116) we see that the dipole-dipole interaction energy is given by:

$$W_{\mu_1\mu_2} = -\boldsymbol{\mu}_1\boldsymbol{\mu}_2 : \nabla\nabla\frac{1}{s}, \tag{3.117}$$

which can also be written with the help of the dipole-dipole interaction tensor $\boldsymbol{T}$ given in eqns. (1.32) and (1.34):

$$\boldsymbol{T} = -\nabla\nabla\frac{1}{s} = \frac{1}{s^3}\left(\boldsymbol{I} - \frac{3}{s^2}\boldsymbol{ss}\right),$$

as:

$$W_{\mu_1\mu_2} = \boldsymbol{\mu}_1\boldsymbol{\mu}_2 : \boldsymbol{T} = \boldsymbol{\mu}_1 \cdot \boldsymbol{T} \cdot \boldsymbol{\mu}_2. \tag{3.118}$$

This expression can also be obtained by substituting eqn. (1.31) for the field

* A different multipole development of the electrostatic interaction energy is given in ref. 10. These expressions offer certain advantages when quantum-mechanical calculations are necessary.

due to dipole $\boldsymbol{\mu}_2$ into eqn. (3.9) for the energy of dipole $\boldsymbol{\mu}_1$ in an external field.

The interaction energy $W_{\mu_1\mu_2}$ will be a function of the distance s, and the angles between the dipole vectors and the distance vector $\boldsymbol{s}$. This can be shown explicitly by taking two spherical coordinate systems with their origins in the centres of mass of the two particles, and the axes $\theta = 0$ in the direction of $\boldsymbol{s}$ (Fig. 20). The angle γ between the directions of $\boldsymbol{\mu}_1$ and $\boldsymbol{\mu}_2$ can be expressed in terms of θ_1, θ_2, and $\varphi_2 - \varphi_1$. To obtain this expression, we take $s = 0$ and use the cosine rule repeatedly, starting with the distance between the endpoints of the vectors $\boldsymbol{\mu}_1$ and $\boldsymbol{\mu}_2$. The result of this calculation is:

$$\cos\gamma = \sin\theta_1 \sin\theta_2 \cos(\varphi_2 - \varphi_1) + \cos\theta_1 \cos\theta_2. \tag{3.119}$$

When eqn. (3.118) is written out, we find:

$$W_{\mu_1\mu_2} = \frac{1}{s^3}\boldsymbol{\mu}_1 \cdot \boldsymbol{\mu}_2 - \frac{3}{s^5}(\boldsymbol{\mu}_1 \cdot \boldsymbol{s})(\boldsymbol{\mu}_2 \cdot \boldsymbol{s})$$

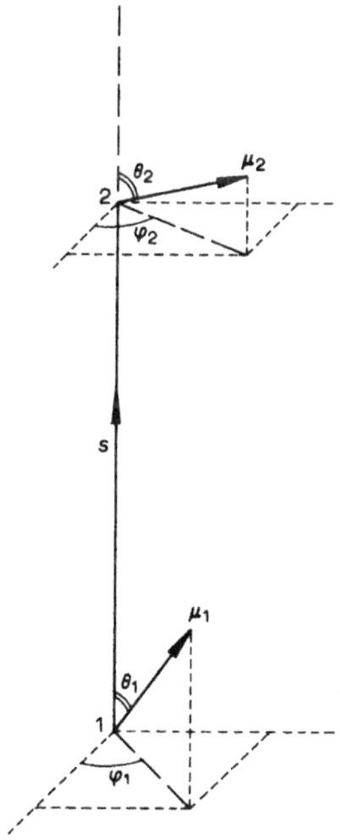

Fig. 20
The interaction of two dipoles.

$$= \frac{1}{s^3}(\mu_1\mu_2 \cos\gamma - 3\mu_1\mu_2 \cos\theta_1 \cos\theta_2)$$

$$= \frac{\mu_1\mu_2}{s^3}\{\sin\theta_1 \sin\theta_2 \cos(\varphi_2 - \varphi_1) - 2\cos\theta_1 \cos\theta_2\}. \quad (3.120)$$

Expression (3.120) is minimal when $\theta_1 = \theta_2 = 0$ or $\theta_1 = \theta_2 = \pi$; the minimum value is:

$$(W_{\mu_1\mu_2})_{\text{min}} = -2\frac{\mu_1\mu_2}{s^3}. \quad (3.121)$$

The maximum value of $W_{\mu_1\mu_2}$ is reached when $\theta_1 = 0$ and $\theta_2 = \pi$ or $\theta_1 = \pi$, $\theta_2 = 0$. It is given by:

$$(W_{\mu_1\mu_2})_{\text{max}} = +2\frac{\mu_1\mu_2}{s^3}. \quad (3.122)$$

A relatively favourable configuration is obtained when the dipoles lie anti-parallel in planes perpendicular to the distance vector **s**. In such a case we have $\varphi_1 - \varphi_2 = \pi$ and $\theta_1 = \theta_2 = \pi/2$, and the value of the energy in this position is given by:

$$(W_{\mu_1\mu_2})_{\text{antipar}} = -\frac{\mu_1\mu_2}{s^3}. \quad (3.123)$$

When the form of the particles is such that s is sufficiently smaller in an antiparallel configuration than in a parallel configuration, the energy $W_{\mu_1\mu_2}$ will be more favourable in the antiparallel configuration (see also Fig. 21).

For polar molecules with large dipole moments and small polarizabilities, the interaction between the permanent moments determines which configuration will be the most favourable. Owing to thermal movements, how-

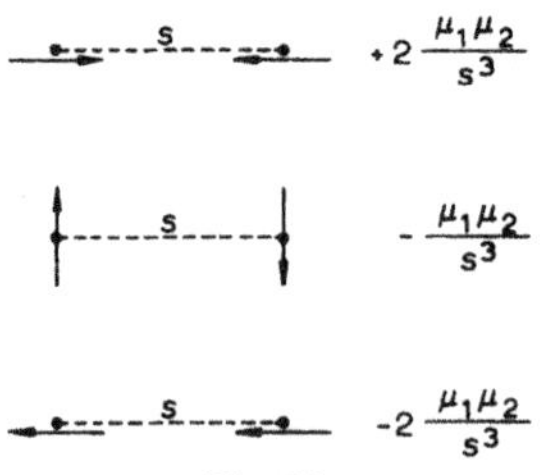

Fig. 21
Configurations of two dipoles in a plane.

ever, the less favourable configurations will also occur, although less frequently. When the attractive forces between two particles are strong, the distance between them will become small and the most favourable configuration will occur almost exclusively. In this way a stable intermolecular compound can be formed. Generally, however, forces other than dipole-dipole interaction will determine whether such an intermolecular compound will be formed. Thus, in the well-known example of the compound formed by ethoxyethane and chloroform (with dipole moments 1.15*D* and 1.01*D*)

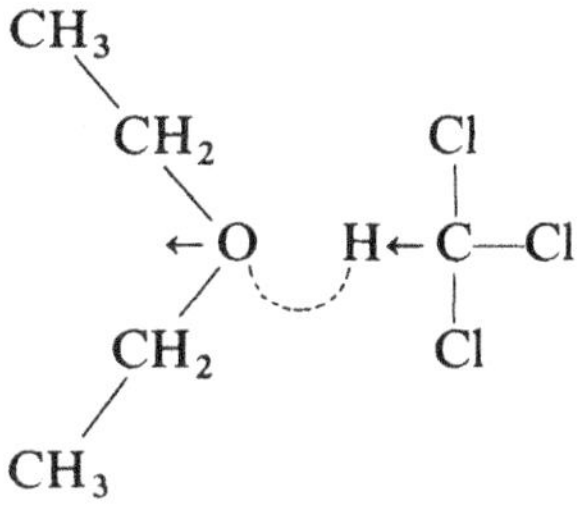

the dipole-dipole interaction energy will be 0.2–0.3 kcal/mole when the dipole moments are parallel. This is much too small to explain the stability of this compound, and here the formation of a hydrogen-bond will be the determining factor. Only when the permanent dipole moments of both molecules are high will it be possible to approximate the interaction energy with the help of the interaction between the permanent dipoles.

Generally, a description of the interaction energy has to take into account the polarizability of the particles. This can be done by adding the work of polarization to the expression in eqn. (3.116). We will limit ourselves to the case of two particles with charges e_1 and e_2, permanent dipole moments $\boldsymbol{\mu}_1$ and $\boldsymbol{\mu}_2$, and dipole polarizabilities α_1 and α_2. The induced moments will be given by:

$$\boldsymbol{p}_1 = \alpha_1 \boldsymbol{E}_2(\mathbf{0}), \tag{3.124}$$

and:

$$\boldsymbol{p}_2 = \alpha_2 \boldsymbol{E}_1(\boldsymbol{s}) \tag{3.125}$$

(compare Fig. 19). The total dipole moments will be denoted by $\boldsymbol{m}_1$ and $\boldsymbol{m}_2$. According to (3.113) and (3.114), $\boldsymbol{E}_2(\mathbf{0})$ is given by:

$$\boldsymbol{E}_2(\boldsymbol{0}) = e_2\nabla\frac{1}{s} + \boldsymbol{m}_2\cdot\nabla\nabla\frac{1}{s}. \tag{3.126}$$

The field $\boldsymbol{E}_1(\boldsymbol{s})$ is given by (*cf.* eqns. (1.5) and (1.33)):

$$\boldsymbol{E}_1(\boldsymbol{s}) = -e_1\nabla\frac{1}{s} + \boldsymbol{m}_1\cdot\nabla\nabla\frac{1}{s}. \tag{3.127}$$

The interaction energy is given by the relevant terms of eqn. (3.116) and the work of polarization as given in (3.96):

$$W_{12} = \frac{e_1e_2}{s} + (e_1\boldsymbol{m}_2 - e_2\boldsymbol{m}_1)\cdot\nabla\frac{1}{s} - \boldsymbol{m}_1\boldsymbol{m}_2:\nabla\nabla\frac{1}{s} +$$

$$+ \tfrac{1}{2}\boldsymbol{p}_1\cdot\boldsymbol{E}_2(\boldsymbol{0}) + \tfrac{1}{2}\boldsymbol{p}_2\cdot\boldsymbol{E}_1(\boldsymbol{s}). \tag{3.128}$$

By substitution of (3.126) and (3.127) into (3.128) and noting that $\boldsymbol{m}_1 = \boldsymbol{\mu}_1 + \boldsymbol{p}_1$, $\boldsymbol{m}_2 = \boldsymbol{\mu}_2 + \boldsymbol{p}_2$, the expression for the interaction energy is reduced to:

$$W_{12} = \frac{e_1e_2}{s} + (e_1\boldsymbol{\mu}_2 - e_2\boldsymbol{\mu}_1)\cdot\nabla\frac{1}{s} - \boldsymbol{\mu}_1\boldsymbol{\mu}_2:\nabla\nabla\frac{1}{s} +$$

$$+ \tfrac{1}{2}(e_1\boldsymbol{p}_2 - e_2\boldsymbol{p}_1)\cdot\nabla\frac{1}{s} - \tfrac{1}{2}(\boldsymbol{p}_1\boldsymbol{\mu}_2 + \boldsymbol{p}_2\boldsymbol{\mu}_1):\nabla\nabla\frac{1}{s}. \tag{3.129}$$

The induced moments $\boldsymbol{p}_1$ and $\boldsymbol{p}_2$ can be found by substitution of (3.126) and (3.127) into (3.124) and (3.125), respectively:

$$\boldsymbol{p}_1 = \alpha_1e_2\nabla\frac{1}{s} + \alpha_1\boldsymbol{\mu}_2\cdot\nabla\nabla\frac{1}{s} + \alpha_1\boldsymbol{p}_2\cdot\nabla\nabla\frac{1}{s}, \tag{3.130}$$

$$\boldsymbol{p}_2 = -\alpha_2e_1\nabla\frac{1}{s} + \alpha_2\boldsymbol{\mu}_1\cdot\nabla\nabla\frac{1}{s} + \alpha_2\boldsymbol{p}_1\cdot\nabla\nabla\frac{1}{s}. \tag{3.131}$$

The simultaneous vector eqns. (3.130) and (3.131) for $\boldsymbol{p}_1$ and $\boldsymbol{p}_2$ can be solved by substitution of (3.131) into (3.130); after performing the differentiations we obtain:

$$\boldsymbol{p}_1 = -\frac{\alpha_1}{s^3}e_2\boldsymbol{s} - \frac{\alpha_1}{s^3}\boldsymbol{\mu}_2 + \frac{3\alpha_1}{s^5}(\boldsymbol{\mu}_2\cdot\boldsymbol{s})\boldsymbol{s} +$$

$$+ \frac{2\alpha_1\alpha_2}{s^6}e_1\boldsymbol{s} + \frac{\alpha_1\alpha_2}{s^6}\left(\boldsymbol{\mu}_1 + \boldsymbol{p}_1 + \frac{3(\boldsymbol{\mu}_1 + \boldsymbol{p}_1)\cdot\boldsymbol{s}}{s^2}\boldsymbol{s}\right). \tag{3.132}$$

This equation can be solved for $\boldsymbol{p}_1$ when $\boldsymbol{p}_1\cdot\boldsymbol{s}$, the component of the induced moment in the direction of $\boldsymbol{s}$, is known. The expression for $\boldsymbol{p}_1\cdot\boldsymbol{s}$ can be obtained from eqn. (3.132) itself by inner multiplication of both sides by $\boldsymbol{s}$. The solution for $\boldsymbol{p}_1 \cdot \boldsymbol{s}$ is then substituted into the last term of eqn. (3.132). This leads to an equation in $\boldsymbol{p}_1$, which can be solved directly. The other induced moment $\boldsymbol{p}_2$ can be found by an analogous procedure. To give the expressions for $\boldsymbol{p}_1$ and $\boldsymbol{p}_2$ in a convenient form, we introduce abbreviations A_n:

$$A_n = 1 - n\alpha_1\alpha_2/s^6, \tag{3.133}$$

where n is an integer.

Using (3.133) we find for $\boldsymbol{p}_1$ and $\boldsymbol{p}_2$:

$$\begin{aligned} \boldsymbol{p}_1 = & -\frac{\alpha_1}{s^3}\left(\frac{e_2}{A_4}\boldsymbol{s} + \frac{1}{A_1}\boldsymbol{\mu}_2 - \frac{3A_2}{A_1A_4}\frac{\boldsymbol{\mu}_2\cdot\boldsymbol{s}}{s^2}\boldsymbol{s}\right) + \\ & + \frac{\alpha_1\alpha_2}{s^6}\left(\frac{2e_1}{A_4}\boldsymbol{s} + \frac{1}{A_1}\boldsymbol{\mu}_1 + \frac{3}{A_1A_4}\frac{\boldsymbol{\mu}_1\cdot\boldsymbol{s}}{s^2}\boldsymbol{s}\right), \end{aligned} \tag{3.134}$$

$$\begin{aligned} \boldsymbol{p}_2 = & -\frac{\alpha_2}{s^3}\left(\frac{-e_1}{A_4}\boldsymbol{s} + \frac{1}{A_1}\boldsymbol{\mu}_1 - \frac{3A_2}{A_1A_4}\frac{\boldsymbol{\mu}_1\cdot\boldsymbol{s}}{s^2}\boldsymbol{s}\right) + \\ & + \frac{\alpha_1\alpha_2}{s^6}\left(\frac{-2e_2}{A_4}\boldsymbol{s} + \frac{1}{A_1}\boldsymbol{\mu}_2 + \frac{3}{A_1A_4}\frac{\boldsymbol{\mu}_2\cdot\boldsymbol{s}}{s^2}\boldsymbol{s}\right). \end{aligned} \tag{3.135}$$

Substitution of the expressions (3.134) and (3.135) in (3.129) and working out the differentiations, leads, after some rearrangement, to:

$$\begin{aligned} W_{12} = & \frac{A_2}{A_4}\frac{e_1e_2}{s} + \frac{1}{A_4}\frac{e_2(\boldsymbol{\mu}_1\cdot\boldsymbol{s})/s}{s^2} - \frac{1}{A_4}\frac{e_1(\boldsymbol{\mu}_2\cdot\boldsymbol{s})/s}{s^2} + \\ & + \frac{1}{A_1}\frac{\boldsymbol{\mu}_1\cdot\boldsymbol{\mu}_2}{s^3} - \frac{3A_2}{A_1A_4}\frac{(\boldsymbol{\mu}_1\cdot\boldsymbol{s})(\boldsymbol{\mu}_2\cdot\boldsymbol{s})/s^2}{s^3} - \\ & - \frac{\alpha_1}{2s^3}\left\{\frac{1}{A_4}\frac{e_2^2}{s} - \frac{4}{A_4}\frac{e_2(\boldsymbol{\mu}_2\cdot\boldsymbol{s})/s}{s^2} + \frac{1}{A_1}\frac{\boldsymbol{\mu}_2\cdot\boldsymbol{\mu}_2}{s^3} + \frac{3}{A_1A_4}\frac{(\boldsymbol{\mu}_2\cdot\boldsymbol{s})^2/s^2}{s^3}\right\} - \\ & - \frac{\alpha_2}{2s^3}\left\{\frac{1}{A_4}\frac{e_1^2}{s} + \frac{4}{A_4}\frac{e_1(\boldsymbol{\mu}_1\cdot\boldsymbol{s})/s}{s^2} + \frac{1}{A_1}\frac{\boldsymbol{\mu}_1\cdot\boldsymbol{\mu}_1}{s^3} + \frac{3}{A_1A_4}\frac{(\boldsymbol{\mu}_1\cdot\boldsymbol{s})^2/s^2}{s^3}\right\}. \end{aligned} \tag{3.136}$$

Thus, the interaction energy consists of three sets of terms. First, we have the charge-charge, charge-permanent dipole and permanent dipole-permanent dipole interaction term, multiplied by factors which depend on both polarizabilities and are always higher than one: the particles polarize each other in such a way that these terms in the interaction energy are enhanced. The second and the third sets of terms denote the interaction of the charge and dipole moment of one particle with itself via the moment induced in the other particle.

By substitution of the appropriate values for e_1, e_2, $\boldsymbol{\mu}_1$, $\boldsymbol{\mu}_2$, α_1, and α_2 into eqn. (3.136) the interaction energy for several special types of electrostatic interactions can be derived. We shall discuss the cases of two charges, of a charge and a dipole, two dipoles, and of a charge or a dipole and a polarizable particle.

When we have two polarizable particles, without net charge and dipole moment, the electrostatic interaction energy will be zero. Nevertheless, there will be another interaction due to the fluctuating dipole moments that are always present in all particles. Although the average dipole moments $\bar{\boldsymbol{p}}_1$ and $\bar{\boldsymbol{p}}_2$ will be zero, the interaction energy, which depends on the averaged product $\overline{\boldsymbol{p}_1 \cdot \boldsymbol{T} \cdot \boldsymbol{p}_2}$ (eqn. (3.118)), will not be zero, due to the correlation between the instantaneous values of $\boldsymbol{p}_1$ and $\boldsymbol{p}_2$. This type of interaction energy is called the London[11]-Van der Waals[12] energy, or dispersion energy, and a quantum-mechanical calculation shows that it is proportional to $\alpha_1\alpha_2/s^6$. The London-Van der Waals interaction is always present when two particles have non-zero polarizabilities, although it may often be neglected when s is not very small and interactions falling off more slowly than $1/s^6$ are present.

When we take $\boldsymbol{\mu}_1 = \boldsymbol{\mu}_2 = \boldsymbol{0}$ in (3.136) we obtain the electrostatic interaction energy of two polarizable ions, which may be used as a rough estimate for the work required to form a molecule from the infinitely separated ions. When we also take $e_1 = -e_2 = e$, we obtain:

$$W_{12} = -\frac{e^2}{s}\left(1 + \frac{\alpha_1 + \alpha_2 + 4\alpha_1\alpha_2/s^3}{2A_4 s^3}\right). \tag{3.137}$$

If we take NaCl as an example, we calculate with $\alpha_1 = 0.2\,\text{Å}^3$, $\alpha_2 = 3.0\,\text{Å}^3$ (see Table 5, p. 88) and $s = 2.8$ Å according to Pauling's ion radii:

$$W_{12} = -\frac{e^2}{s}\left(1 + \frac{3.2 + 0.03}{43.9\,(1 - 0.005)}\right) = -\frac{e^2}{s}(1 + 0.073).$$

Thus the introduction of polarizabilities leads to a correction of 7 % in the interaction energy. From this example it is clear that in most cases second order terms such as $4\alpha_1\alpha_2/s^3$ can be neglected, and that the factors A_n may be put equal to 1.

When we take $\boldsymbol{\mu}_1 = \mathbf{0}$ and $e_2 = 0$ we obtain the interaction between an ion and a polar molecule, which is a frequently occurring phenomenon, responsible for the solution of electrolytes in a polar solvent and for the formation of hydrates and ammoniates. When we use the coordinate system of Fig. 20, we find with the help of (3.136):

$$W_{12} = \frac{-1}{A_4}\left(\frac{e_1\mu_2}{s}\cos\theta_2 + \frac{\alpha_2 e_1^2}{2s^4} + \frac{2\alpha_1\mu_2^2}{s^6}\frac{A_4 + 3\cos^2\theta_2}{4A_1}\right). \quad (3.138)$$

This expression reaches its minimum value for $\cos\theta_2 = +1$, if e_1 is positive and for $\cos\theta_2 = -1$ if e_1 is negative. This means that the state of lowest energy is that configuration where the dipole vector points from a positive ion or towards a negative ion.

In some special cases the interaction between the charge and the dipole induced by the charge will become of the same order of magnitude as the interaction between charge and permanent moment, but only when s is not too large. This explains why small and highly charged ions such as Al^{+++} and Mg^{++} form much more stable complexes with the ammonia molecule than with the water molecule.[13] The dipole moment of H_2O is larger, but NH_3 has the larger polarizability; the latter becomes decisive when e is large and s is small.

When $e_1 = e_2 = 0$ we obtain the interaction between two polarizable dipoles. It is possible to derive from eqn. (3.136) the general expression for the interaction energy. We will only give the expressions for the three configurations depicted in Fig. 21, for the special cases $\mu_1 = \mu_2 = \mu$ and $\alpha_1 = \alpha_2 = \alpha$:

$$(W_{12})_{\text{min}} = \frac{-2}{1 - 2\alpha/s^3}\,\frac{\mu^2}{s^3}, \quad (3.139)$$

$$(W_{12})_{\text{antipar}} = \frac{-1}{1 - \alpha/s^3}\,\frac{\mu^2}{s^3}, \quad (3.140)$$

$$(W_{12})_{\text{max}} = \frac{+2}{1 + 2\alpha/s^3}\,\frac{\mu^2}{s^3}. \quad (3.141)$$

Comparing these expressions with eqns. (3.121), (3.122), and (3.123) we see that all values of the interaction energy are lowered by the introduction of polarizabilities. The corrections will never be very large, however, since α is smaller than r^3 (r is the molecular radius), while s^3 is at least $(2r)^3$.

For $e_2 = 0$ and $\boldsymbol{\mu}_2 = \mathbf{0}$, eqn. (3.136) gives the interaction energy between ions or polar molecules and neutral, non-polar molecules. If we also take $\boldsymbol{\mu}_1 = \mathbf{0}$ we obtain:

$$W_{12} = -\frac{\alpha_2 e_1^2}{2s^4}\frac{1}{A_4}. \tag{3.142}$$

The electrostatic energy of an ion and a neutral molecule is probably responsible for the formation of such complex ions as:

$$J^- + J_2 \rightarrow J_3^-,$$

$$S^{--} + S \rightarrow S_2^{--}.$$

Generally, the interaction energy between ions and non-polar molecules is considerably smaller than the interaction energy between ions and polar molecules (compare (3.142) with (3.138)). Therefore, the solubility of salts, *i.e.* of ionic crystals, in non-polar solvents such as hexane should be much lower than in polar solvents such as water and the lower alcohols. Solution is only possible when the interaction energy between solvent and dissolved ions is sufficiently high to compensate for the loss of energy connected with the decomposition of the ionic lattice. Therefore, only solvents consisting of small molecules with a relatively high dipole moment will be able to dissolve salts.

When in (3.136) we take $e_1 = 0$ and again $e_2 = 0$, $\boldsymbol{\mu}_2 = \mathbf{0}$ we find the interaction energy between a polar and a non-polar molecule:

$$W_{12} = -\frac{\alpha_2 \mu_1^2}{2s^6}\frac{1}{A_1}\left(1 + \frac{3}{A_4}\cos^2\theta_1\right). \tag{3.143}$$

Here, we have used the coordinate system of Fig. 20. This type of interaction plays a role in the formation of the hydrates of the inert gases.

To obtain some insight into the nature of the approximations made when a particle of finite dimensions is replaced by a point polarizability, we can perform the following calculation.

We will use eqn. (3.69) for the energy of a dielectric to calculate the energy of a dielectric sphere of dielectric constant ε and radius a in the field of a point charge e at a distance s from the centre of the sphere (Fig. 22). The value for the energy derived in this way can be compared with the interaction energy of a point charge e and a polarizable particle with point polarizability α, as given by eqn. (3.142).

According to eqns. (1.46) and (2.99) the potentials ϕ_0 and ϕ of the external field $\boldsymbol{E}_0$ and of the Maxwell field $\boldsymbol{E}$ in the dielectric are given by:

$$\phi_0(r, \theta) = e \sum_{m=0}^{\infty} \frac{r^m}{s^{m+1}} P_m(\cos\theta), \tag{3.144}$$

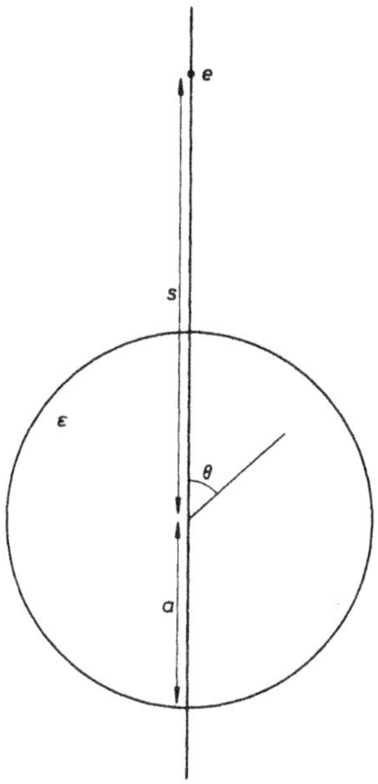

Fig. 22
A dielectric sphere in the field of a point-charge.

$$\phi(r, \theta) = e \sum_{n=0}^{\infty} \frac{2n + 1}{n\varepsilon + n + 1} \frac{r^n}{s^{n+1}} P_n(\cos \theta). \tag{3.145}$$

The fields themselves are given by $\boldsymbol{E}_0 = -\operatorname{grad} \phi_0$ and $\boldsymbol{E} = -\operatorname{grad} \phi$. Since $\boldsymbol{P} = \frac{\varepsilon - 1}{4\pi}\boldsymbol{E}$ according to (2.46), we may write for the energy of the dielectric in the field of the point charge, according to (3.69):

$$\Delta W = -\frac{\varepsilon - 1}{8\pi} \iiint_V \nabla \phi \cdot \nabla \phi_0 \, \mathrm{d}v. \tag{3.146}$$

With the help of Green's first theorem, eqn. (A1.35), expression (3.146) can be transformed to:

$$\Delta W = -\frac{\varepsilon - 1}{8\pi} \left[\oiint_S \phi \nabla \phi_0 \cdot \boldsymbol{dS} - \iiint_V \phi \Delta \phi_0 \, \mathrm{d}v \right], \tag{3.147}$$

where S is the surface of the dielectric sphere. Since there are no true charges inside the sphere, we have:

$$\Delta \phi_0 = -\operatorname{div} \boldsymbol{E}_0 = 0. \tag{3.148}$$

With the help of (A1.29) and (A1.43) $\boldsymbol{dS}$ can be written in spherical coordinates as:

$$dS = a^2 \sin\theta \, d\theta \, d\varphi \boldsymbol{e}_r. \tag{3.149}$$

Substitution of (3.148) and (3.149) into (3.147) and the use of (A1.49) now leads to:

$$\Delta W = -\frac{\varepsilon - 1}{8\pi} a^2 \int_0^{2\pi} \int_0^{\pi} \left(\phi \frac{\partial \phi_0}{\partial r} \right)_{r=a} \sin\theta \, d\theta \, d\varphi. \tag{3.150}$$

After substitution of (3.144) and (3.145) for ϕ_0 and ϕ, and integration over φ, the fact that the Legendre polynomials are orthogonal (eqn. (A3.31)) is used to obtain:

$$\begin{aligned} \Delta W &= -\frac{\varepsilon - 1}{8\pi} 2\pi a^2 e^2 \sum_{m=0}^{\infty} \sum_{n=0}^{\infty} m \frac{a^{m-1}}{s^{m+1}} \frac{2n+1}{n\varepsilon + n + 1} \frac{a^n}{s^{n+1}} \frac{2\delta_{mn}}{2n+1} \\ &= -\frac{\varepsilon - 1}{2} e^2 \sum_{n=0}^{\infty} \frac{n}{n\varepsilon + n + 1} \frac{a^{2n+1}}{s^{2n+2}}. \end{aligned} \tag{3.151}$$

Since the polarizability of the dielectric sphere is given by (2.101) as:

$$\alpha = \frac{\varepsilon - 1}{\varepsilon + 2} a^3,$$

we may also write, changing the summation-index from n to $(n + 1)$:

$$\begin{aligned} \Delta W &= -\frac{\alpha e^2}{2s^4} \sum_{n=0}^{\infty} \frac{(n+1)\varepsilon + 2n + 2}{(n+1)\varepsilon + n + 2} \left(\frac{a}{s}\right)^{2n} \\ &= -\frac{\alpha e^2}{2s^4} \left[1 + \frac{2\varepsilon + 4}{2\varepsilon + 3} \left(\frac{a}{s}\right)^2 + \frac{3\varepsilon + 6}{3\varepsilon + 4} \left(\frac{a}{s}\right)^4 + \cdots \right]. \end{aligned} \tag{3.152}$$

Comparing this with expression (3.142) for the case that the charged particle is non-polarizable, we have, when the notation is made the same as in (3.152):

$$W = -\frac{\alpha e^2}{2s^4}. \tag{3.153}$$

Thus, we see that the introduction of a homogeneously polarizable sphere instead of a point polarizability leads to correction terms in the interaction energy that will be important when the charge is close to the surface of the dielectric sphere. Although a dielectric sphere is not a realistic model for the behaviour of molecules close to a point charge, this calculation shows what type of correction terms we should expect when a description based on point polarizabilities breaks down.

References

1. E. A. Guggenheim, *Thermodynamics*, North-Holland, Amsterdam 1967.
2. G. N. Lewis and M. Randall, *Thermodynamics*, McGraw-Hill, New York 1961, Ch. 31.
3. J. G. Kirkwood and I. Oppenheim, *Chemical Thermodynamics*, McGraw-Hill, New York 1961, Ch. 14.
4. L. D. Landau and E. M. Lifshitz, *Electrodynamics of Continuous Media*, Pergamon, Oxford 1960, sections 10, 11, 12.
5. H. Gränicher, *Helv. Phys. Acta*, **29** (1956) 211.

6. I. W. Shepherd, *J. Phys. Chem. Solids*, **28** (1967) 2027.
7. A. D. Buckingham, in J. O. Hirschfelder (ed.), *Intermolecular Forces* (*Adv. Chem. Phys.* **12**), Wiley, New York 1967, p. 107.
8. H. Margenau and N. Kestner, *Theory of Intermolecular Forces*, Pergamon, London 1967.
9. J. O. Hirschfelder (ed.), *Intermolecular Forces* (*Adv. Chem. Phys.* **12**), Wiley, New York 1967.
10. J. O. Hirschfelder, C. F. Curtiss, and R. B. Bird, *Molecular Theory of Gases and Liquids*, Wiley and Chapman & Hall, New York and London 1954, p. 844.
11. F. London, *Z. Phys.* **63** (1930) 245.
12. J. D. Van der Waals, *Thesis* Leiden 1873, English translation by Threlfall and Adair, *Physical Memoirs* **1** (1890) 333.
13. A. E. van Arkel, *Molecules and Crystals in Inorganic Chemistry*, Butterworth, London 1956, p. 184.

CHAPTER IV

THE REACTION FIELD

§17. Introduction

When a molecule with a permanent dipole strength μ is surrounded by other particles, the inhomogeneous field of the permanent dipole polarizes its environment. In the surrounding particles moments proportional to the polarizability are induced, and if these particles have a permanent dipole moment their orientation is influenced. An exact calculation of the consequences of these two effects may in principle be made using the methods of statistical mechanics, but it is so difficult to take into account the interaction of the molecules and the microscopic structure of the material, that even for a gas at low pressure the statistical calculation must start from simplified models.

A fairly good approximation can be obtained, however, by using a much simpler method of calculation, first proposed by Martin.[1] Noticing that in the theory of electrolytes considerable progress had been made by considering an ion as a charged conducting sphere surrounded by a continuous dielectric, he proposed to focus attention on one molecule and to consider the environment as a continuous medium. His model for the polar molecule, two oppositely charged spheres in contact, was rather inappropriate, however. Thus a great improvement was obtained when Bell[2] introduced a better model for the dipole in a molecule: an ideal dipole in the centre of a spherical cavity.

The field of the dipole in such a cavity polarizes the surrounding matter, and the resulting inhomogeneous polarization of the environment will give rise to a field at the dipole, which is called the reaction field $\boldsymbol{R}$. For reasons of symmetry $\boldsymbol{R}$ will have the same direction as the dipole vector $\boldsymbol{\mu}$, and it is obvious that R will be proportional to μ as long as no saturation effects occur. Thus:

$$\boldsymbol{R} = f\boldsymbol{\mu}. \tag{4.1}$$

The factor f is called the factor of the reaction field. If we use not a spherical cavity but, for instance, an ellipsoidal cavity, this factor must be replaced by a reaction field tensor:

$$\boldsymbol{R} = \boldsymbol{F} \cdot \boldsymbol{\mu}. \tag{4.2}$$

The reaction field can also be defined independent of Martin and Bell's model. In each microscopic state of the sample, the electric field at the dipole of the molecule under consideration, due to the permanent and induced dipole moments of the surrounding molecules, has a certain value $\boldsymbol{E}_s$. Both the direction of the central dipole and the field strength $\boldsymbol{E}_s$ fluctuate in time. If no external field is applied ($E_0 = 0$), we find the reaction field for a certain direction of the dipole of the molecule under consideration by averaging the field $\boldsymbol{E}_s$ over all microscopic states for which the dipole of this molecule has that particular direction. We then have:

$$\boldsymbol{R} = \langle \boldsymbol{E}_s \rangle^{\mu}_{E_0 = 0}. \tag{4.3}$$

When an external field is present, we must subtract from the corresponding expression the same average field for the hypothetical case that the central molecule would have no permanent dipole moment:

$$\boldsymbol{R} = \langle \boldsymbol{E}_s \rangle^{\mu}_{E_0} - \langle \boldsymbol{E}_s \rangle^{\mu = 0}_{E_0}. \tag{4.4}$$

According to statistical mechanics in taking the averages just mentioned we must introduce a weight factor in dependence on the energy of each microscopic state. Because the energy depends on the static dielectric constant of the sample it follows that if we consider the environment as a continuum we must use for this continuum the static dielectric constant in all calculations. Some critics[3] of Martin and Bell's method have objected against this use of the static dielectric constant, because they supposed the model to ignore the dynamic effects of the field producing moment. These critics forget, however, that not only the central molecule influences the direction of the surrounding molecules, but that these molecules also influence the direction of the central molecule.

§18. The reaction field of a non-polarizable point dipole

Bell's model consists of a spherical cavity with radius a in a continuous dielectric of dielectric constant ε. Situated in the centre of the cavity is a non-polarizable* point dipole with moment $\boldsymbol{\mu}$ (Fig. 23). In this simplified picture of the interaction between a permanent dipole and its surroundings, the value chosen for a influences the results.

Bell took a about equal to the radius of the molecule but Onsager,[4] who adopted Bell's calculations in evaluating his general equations for the case that only one kind of molecule is present, used the relation:

* In section 19 the calculation is extended to a polarizable dipole.

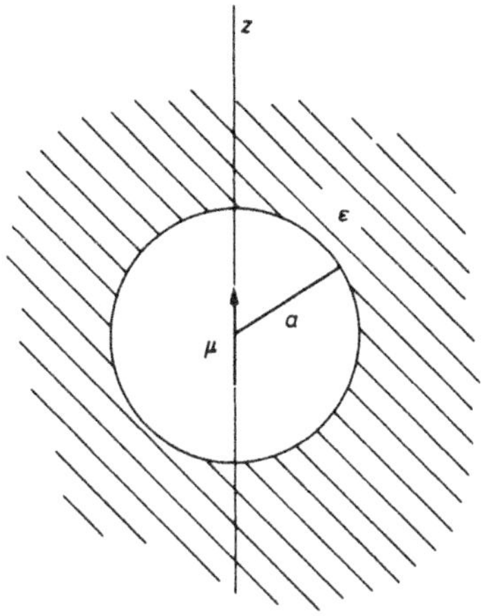

Fig. 23

$$\tfrac{4}{3}\pi N a^3 = 1, \tag{4.5}$$

where N is the number of particles per cm^3.

Eqn. (4.5) implies that the volume of the cavity is equal to the volume available to each molecule. For a gas at moderate pressure, Onsager's value for a is much larger than Bell's value. For a liquid, the opposite can occur. We will return to the question of the value of a, but it may be mentioned here that in many cases it is advisable to take for a a value approximately equal to what is generally considered to be the "molecular radius".

To calculate the reaction field, we must know the potential in the cavity due both to the dipole itself and to the interaction of the dipole with the surrounding dielectric. The method used to solve this problem is closely related to the treatment of the problems discussed in section 9. If we take the centre of the dipole as the origin of a coordinate system and choose the direction of the z-axis along the dipole vector, we have symmetry about the z-axis. As shown in Appendix II, the general solution of Laplace's equation is then given by eqn. (A2.15). As in section 9 we call the potential outside the sphere (with radius a) ϕ_1 and that inside the sphere ϕ_2, so that we may write:

$$\phi_1 = \sum_{n=0}^{\infty} \left(A_n r^n + \frac{B_n}{r^{n+1}} \right) P_n(\cos\theta),$$

$$\phi_2 = \sum_{n=0}^{\infty} \left(C_n r^n + \frac{D_n}{r^{n+1}} \right) P_n(\cos\theta).$$

The boundary conditions are in this case:

1. $$(\phi_1)_{r\to\infty} = 0, \tag{4.6}$$

2. $$(\phi_1)_{r=a} = (\phi_2)_{r=a}, \tag{4.7}$$

3. $$\varepsilon \left(\frac{\partial \phi_1}{\partial r} \right)_{r=a} = \left(\frac{\partial \phi_2}{\partial r} \right)_{r=a}. \tag{4.8}$$

The terms $\frac{D_n}{r^{n+1}} P_n(\cos\theta)$ in ϕ_2 are due to charges within the cavity (*cf.* section 4). In the present case, the only source of field lines within the cavity is the permanent dipole $\boldsymbol{\mu}$. According to (1.25), the potential due to an ideal dipole along the z-axis is given by:

$$\phi = \frac{\mu}{r^2} \cos\theta.$$

Thus all coefficients D_n are zero except D_1, which has the value $D_1 = \mu$. So we have:

$$\phi_2 = \sum_{n=0}^{\infty} C_n r^n P_n(\cos\theta) + \frac{\mu}{r^2} \cos\theta. \tag{4.9}$$

Because of (4.6) and the fact that the Legendre functions are linearly independent, we may conclude that in the expression for ϕ_1, all coefficients A_n are zero. Thus we have:

$$\phi_1 = \sum_{n=0}^{\infty} \frac{B_n}{r^{n+1}} P_n(\cos\theta). \tag{4.10}$$

Applying (4.7) and (4.8) we have for all values $n \neq 1$:

$$\frac{B_n}{a^{n+1}} = C_n a^n, \tag{4.11}$$

$$-\varepsilon(n+1) \frac{B_n}{a^{n+2}} = n C_n a^{n-1}, \tag{4.12}$$

which is only possible if for $n \neq 1$:

$$B_n = 0, \tag{4.13}$$

$$C_n = 0. \tag{4.14}$$

For $n = 1$ we find:

$$\frac{B_1}{a^2} = C_1 a + \frac{\mu}{a^2}, \tag{4.15}$$

$$-2\varepsilon\frac{B_1}{a^3} = C_1 - 2\frac{\mu}{a^3}. \tag{4.16}$$

Hence:

$$B_1 = \frac{3}{2\varepsilon + 1}\mu, \tag{4.17}$$

$$C_1 = -\frac{2(\varepsilon - 1)}{2\varepsilon + 1}\frac{\mu}{a^3}, \tag{4.18}$$

which after substitution in (4.10) and (4.9) results in:

$$\phi_1 = \frac{3}{2\varepsilon + 1}\frac{\mu}{r^2}\cos\theta, \tag{4.19}$$

$$\phi_2 = \frac{\mu}{r^2}\cos\theta - \frac{2(\varepsilon - 1)}{2\varepsilon + 1}\frac{\mu}{a^3} r\cos\theta. \tag{4.20}$$

The potentials ϕ_1' and ϕ_2', due to the apparent surface charges on the surface of the cavity, are, according to (4.19) and (4.20), given by:

$$\phi_1' = -\frac{2(\varepsilon - 1)}{2\varepsilon + 1}\frac{\mu}{r^2}\cos\theta, \tag{4.21}$$

$$\phi_2' = -\frac{2(\varepsilon - 1)}{2\varepsilon + 1}\frac{\mu}{a^3} r\cos\theta. \tag{4.22}$$

Comparing these expressions with (1.102) and (1.103), we see that the potentials ϕ_1' and ϕ_2' are due to an apparent surface charge distribution of the type given in eqn. (1.101), with a dipole moment:

$$\boldsymbol{m} = -\frac{2(\varepsilon - 1)}{2\varepsilon + 1}\boldsymbol{\mu}. \tag{4.23}$$

According to (4.20), the field in the cavity is a superposition of the dipole field *in vacuo* and a uniform field $\boldsymbol{R}$, given by:

$$R = \frac{1}{a^3} \frac{2(\varepsilon - 1)}{2\varepsilon + 1} \boldsymbol{\mu}. \tag{4.24}$$

Comparing this result with eqn. (4.1) we conclude that the factor of the reaction field is given by:

$$f = \frac{1}{a^3} \frac{2(\varepsilon - 1)}{2\varepsilon + 1}. \tag{4.25}$$

Formally, the field in the dielectric can be described as the field of a virtual dipole $\boldsymbol{\mu}_e$ at the centre of the cavity, given by:

$$\boldsymbol{\mu}_e = \frac{3\varepsilon}{2\varepsilon + 1} \boldsymbol{\mu}. \tag{4.26}$$

Onsager called $\boldsymbol{\mu}_e$ the external moment of the immersed dipole; this quantity will be used in section 39.

Bell's model involves a number of simplifications, since the dipole is assumed to be ideal and located at the centre of the molecule, which is supposed to be spherical and surrounded by a continuous dielectric. The errors introduced by these simplifications are discussed in the sections 21–23. We shall first extend our discussion to the case of polarizable point dipoles.

§19. The reaction field of a polarizable point dipole

The above calculations were made for a non-polarizable dipole. It is useful to extend Bell's derivation to the case of a polarizable permanent dipole,[5] having an average polarizability α.

In this case the reaction field $\boldsymbol{R}$ induces a dipole $\alpha\boldsymbol{R}$ and therefore satisfies the equation:

$$\boldsymbol{R} = f(\boldsymbol{\mu} + \alpha\boldsymbol{R}), \tag{4.27}$$

where $\boldsymbol{\mu}$ is the permanent dipole moment. Therefore:

$$\boldsymbol{R} = \frac{f}{1 - f\alpha} \boldsymbol{\mu}. \tag{4.28}$$

Eliminating f by using (4.25), we have:

$$\boldsymbol{R} = \frac{1}{\dfrac{2\varepsilon + 1}{2(\varepsilon - 1)} - \dfrac{\alpha}{a^3}} \frac{1}{a^3} \boldsymbol{\mu}. \tag{4.29}$$

To give an approximate calculation of the numerical values of R in the case of liquids containing one kind of particle, we use for a Onsager's approximation (4.5) and the closely related equation:

$$\frac{\alpha}{a^3} = \frac{n_D^2 - 1}{n_D^2 + 2}, \tag{4.30}$$

where n_D is the index of refraction for the Na_D-line. In section 27 it will be shown that (4.30) is a reasonable approximation when the atomic polarization can be neglected (eqn. (5.45)).

The substitution of (4.5) and (4.30) into (4.29) leads to:

$$\boldsymbol{R} = \frac{4\pi}{3} N \frac{2(\varepsilon - 1)}{2\varepsilon + n_D^2} \frac{n_D^2 + 2}{3} \boldsymbol{\mu}. \tag{4.31}$$

The number of particles per cm^3, N, can be computed from:

$$N = \frac{d}{M} N_A, \tag{4.32}$$

M being the molecular weight of the substance, d the density and N_A Avogadro's number: $N_A = 6.02\ 10^{23}$. In Table 6 we give the values of R at room temperature for some pure dipole liquids, calculated by means of eqn. (4.31). We use the name *pure dipole liquid* for a liquid containing one kind of molecule (with a permanent dipole), whereas the name *polar liquid* will be used for all liquids containing permanent dipoles, irrespective of the presence of non-polar particles.

From Table 6 we see that for the usual values of μ (0.5–5 D) the order of magnitude of the reaction field strength is 5–50 10^6 volts/cm.

Under the influence of the reaction field the dipole moment is increased considerably; the increased moment is:

$$\boldsymbol{\mu}^* = \boldsymbol{\mu} + \alpha \boldsymbol{R}. \tag{4.33}$$

TABLE 6

THE REACTION FIELD OF THE DIPOLE IN PURE DIPOLE LIQUIDS AT 20°C, CALCULATED FROM EQN. (4.31)

	M	d	ε	n_D	$\mu(D)$	R (volts/cm)
Nitrobenzene	123.1	1.203	35	1.553	4.2	43 10^6
Propanone	58.1	0.791	21.4	1.359	2.9	35 10^6
Chloroform	119.4	1.489	4.8	1.446	1.0	9 10^6

Combination of (4.33) and (4.28) gives:

$$\mu^* = \frac{\mu}{1 - f\alpha}. \tag{4.34}$$

In contrast to the quantity μ_e, introduced by Onsager (see section 18), this quantity μ^* has a direct physical meaning. To avoid misunderstandings it must be noted that in Onsager's paper the notation μ^* is used for what we have called μ_e.

Using eqn. (4.25) for f, we can replace (4.34) by:

$$\mu^* = \frac{\mu}{1 - \dfrac{\alpha}{a^3}\dfrac{2(\varepsilon - 1)}{2\varepsilon + 1}}. \tag{4.35}$$

To give an approximate calculation of this increase of the dipole moment, we again use (4.30) and obtain:

$$\frac{\mu^*}{\mu} = \frac{2\varepsilon + 1}{2\varepsilon + n_D^2}\frac{n_D^2 + 2}{3}. \tag{4.36}$$

In Table 7 the values of μ^*/μ, calculated from eqn. (4.36), are given for some pure dipole liquids at room temperature. The increase of the dipole moment, caused by its own reaction field, proves to be generally between 20 and 50%.

It must be emphasized that by "the dipole moment of a molecule" in a liquid or gas we always mean the dipole moment of the free molecule. But in considering applications, *e.g.* cohesion problems, we must not forget that these dipole moments are considerably increased by their own reaction field.

When the dipole is not surrounded by molecules of the same kind, the reaction field and the ratio μ^*/μ are changed. As is at once apparent in the derivation of eqn. (4.31), N and n_D refer in this case to the pure dipole compound, whereas ε is the dielectric constant of the mixture. Thus, the changes

TABLE 7

INCREASE OF THE DIPOLE MOMENT UNDER THE INFLUENCE OF THE REACTION FIELD IN PURE DIPOLE LIQUIDS AT 20°C, CALCULATED FROM EQN. (4.36)

	ε	n_D	μ^*/μ
Nitrobenzene	35	1.553	1.44
Propanone	21.4	1.359	1.26
Chloroform	4.8	1.446	1.24

in R and in μ^*/μ when the environment of the dipole is changed, result in a change of the factor $2(\varepsilon - 1)/(2\varepsilon + n_D^2)$ in eqn. (4.31) and a change of the factor $(2\varepsilon + 1)/(2\varepsilon + n_D^2)$ in eqn. (4.36), respectively.

As an example we may examine a very dilute solution of nitrobenzene in benzene at 20 °C. Its dielectric constant differs only slightly from the dielectric constant $\varepsilon = 2.28$ of benzene. Using the data of nitrobenzene given in Table 6 we can calculate that the factor $2(\varepsilon - 1)/(2\varepsilon + n_D^2)$ is reduced from 0.939 to 0.367, whereas the factor $(2\varepsilon + 1)/(2\varepsilon + n_D^2)$ is decreased from 0.971 to 0.798. Consequently, the reaction field of the nitrobenzene dipole is reduced from $42 \; 10^6$ volts/cm (Table 6) to $16 \; 10^6$ volts/cm, when the surrounding nitrobenzene molecules are replaced by benzene molecules. The increase of the dipole moment caused by its own reaction field is decreased from 44% (Table 7) to 17%.

So far it has been assumed that the dipole is situated at the centre of the molecule and that the molecule has a spherical shape. In many molecules with a permanent dipole moment, however, the dipole is far from being localized at the centre of the molecule, and generally the molecule is not spherical. The incorporation of these refinements in our model will be investigated in the following sections.

§20. The reaction field in an ellipsoidal cavity

In the case of an ellipsoidal cavity several possibilities for the calculation of the reaction field must be distinguished. Scholte,[6,7] who was the first to derive an expression for the reaction field in an ellipsoidal cavity, considered a homogeneous dipole density filling up the cavity. In this case the reaction field can be calculated from the polarization of a homogeneous ellipsoid with dielectric constant ε_2, immersed in a dielectric with dielectric constant ε_1.

A homogeneous external field $\boldsymbol{E}$ is applied in the direction of the a-axis of the ellipsoid. The field inside the ellipsoid is then also homogeneous, and is given by eqn. (2.79):

$$\boldsymbol{E}_2 = \frac{\varepsilon_1}{\varepsilon_1 + (\varepsilon_2 - \varepsilon_1)A_a}\boldsymbol{E}. \tag{4.37}$$

The homogeneous polarization induced by this field is given by:

$$\boldsymbol{P}_2 = \frac{\varepsilon_2 - 1}{4\pi}\boldsymbol{E}_2 = \frac{1}{4\pi}\frac{\varepsilon_1(\varepsilon_2 - 1)}{\varepsilon_1 + (\varepsilon_2 - \varepsilon_1)A_a}\boldsymbol{E}. \tag{4.38}$$

Hence, the dipole moment of the ellipsoid amounts to:

$$\boldsymbol{m} = \frac{4\pi}{3}abc\boldsymbol{P}_2 = abc\frac{\varepsilon_1(\varepsilon_2 - 1)}{3\{\varepsilon_1 + (\varepsilon_2 - \varepsilon_1)A_a\}}\boldsymbol{E}. \tag{4.39}$$

It is also possible to calculate this moment from α_a, the polarizability of the ellipsoid along the a-axis. We obtain the induced moment $\boldsymbol{m}$ by multiplying this polarizability with the total polarizing field. This total field consists of the cavity field $\boldsymbol{E}_c$, which would occur in the empty cavity, and the reaction field $\boldsymbol{R}$ of the homogeneous dipole density $\boldsymbol{P}_2$ with total dipole moment $\boldsymbol{m}$. Therefore:

$$\boldsymbol{m} = \alpha_a(\boldsymbol{E}_c + \boldsymbol{R}) = \alpha_a(\boldsymbol{E}_c + f_a\boldsymbol{m}). \tag{4.40}$$

Since $\boldsymbol{m}$ and $\boldsymbol{E}_c$ are in the same direction, we get:

$$f_a = \frac{1}{\alpha_a} - \frac{E_c}{m}. \tag{4.41}$$

We now use (2.105), (2.84) and (4.39):

$$\alpha_a = \frac{\varepsilon_2 - 1}{3\{1 + (\varepsilon_2 - 1)A_a\}}abc, \tag{4.42}$$

$$\boldsymbol{E}_c = \frac{\varepsilon_1}{\varepsilon_1 + (1 - \varepsilon_1)A_a}\boldsymbol{E}, \tag{4.43}$$

$$\boldsymbol{m} = abc\frac{\varepsilon_1(\varepsilon_2 - 1)}{3\{\varepsilon_1 + (\varepsilon_2 - \varepsilon_1)A_a\}}\boldsymbol{E}, \tag{4.44}$$

and find:

$$\begin{aligned} f_a &= \frac{3\{1 + (\varepsilon_2 - 1)A_a\}}{abc(\varepsilon_2 - 1)} - \frac{\varepsilon_1 E}{\varepsilon_1 + (1 - \varepsilon_1)A_a}\,\frac{3\{\varepsilon_1 + (\varepsilon_2 - \varepsilon_1)A_a\}}{abc\varepsilon_1(\varepsilon_2 - 1)E} \\ &= \frac{3}{abc}\frac{A_a(1 - A_a)(\varepsilon_1 - 1)}{\varepsilon_1 + (1 - \varepsilon_1)A_a}. \end{aligned} \tag{4.45}$$

In the case of a homogeneous dipole density with its direction not along one of the major axes of the ellipsoidal cavity, we must consider the polarization to be built up from components along the major axes. The components of the reaction field are then given by multiplying the components of the dipole moment of the dipole density with the reaction field factor for that direction. So if we use (4.2):

$$\boldsymbol{R} = \boldsymbol{F} \cdot \boldsymbol{\mu},$$

we find that $\boldsymbol{F}$ is a tensor with the axes of the ellipsoidal cavity as principal axes. In a coordinate system along these axes the components of $\boldsymbol{F}$ are given by:

$$\boldsymbol{F} = \begin{pmatrix} \frac{3}{abc}\frac{A_a(1-A_a)(\varepsilon-1)}{\varepsilon+(1-\varepsilon)A_a} & 0 & 0 \\ 0 & \frac{3}{abc}\frac{A_b(1-A_b)(\varepsilon-1)}{\varepsilon+(1-\varepsilon)A_b} & 0 \\ 0 & 0 & \frac{3}{abc}\frac{A_c(1-A_c)(\varepsilon-1)}{\varepsilon+(1-\varepsilon)A_c} \end{pmatrix}. \quad (4.46)$$

In the case of a sphere, we have $a = b = c$, $A_a = A_b = A_c = \frac{1}{3}$, and we find:

$$\boldsymbol{R} = f\boldsymbol{m}, \quad (4.47)$$

with:

$$f = \frac{1}{a^3}\frac{2(\varepsilon - 1)}{2\varepsilon + 1}. \quad (4.48)$$

So in the case of a spherical cavity we find for a homogeneous dipole density, filling up the cavity, the same reaction field as for an ideal dipole in the centre of the sphere (eqn. (4.25)). This was to be expected, because we found in section 9c that the field of a homogeneously polarized sphere is, outside the sphere, identical with the field of an ideal dipole with the same moment at the centre of that sphere. This does not hold for an ellipsoid, however, and therefore one does not expect (4.46) to be valid for an ideal dipole in the centre of an ellipsoidal cavity. The latter case must be discussed separately.

The calculation of the reaction field of an ideal dipole in an ellipsoidal cavity is very intricate. It was given by Abbott and Bolton[8] for the more manageable case of a prolate spheroidal cavity, with the dipole placed between the foci and directed along the axis. In this case the reaction field is not homogeneous even when the dipole is placed in the centre of the spheroid. It is obvious that this conclusion holds also in the case of a general ellipsoid.

Abbot and Bolton give for the reaction field potential:[8]

$$\phi^{R} = -\mu\frac{\varepsilon-1}{d^2}\sum_{n=1}^{\infty}\frac{(2n+1)P_n(v)P_n(w)P_n'(w_s)}{\varepsilon\dfrac{P_n(v_0)}{Q_n(v_0)}-\dfrac{P_n'(v_0)}{Q_n'(v_0)}}, \tag{4.49}$$

where $2d$ is the distance between the foci. The coordinates v and w are functions of the distances r_1 and r_2 to the foci of the spheroid:

$$v = \frac{r_1+r_2}{2d}, \tag{4.50}$$

$$w = \frac{r_1-r_2}{2d}. \tag{4.51}$$

Without indices, v and w refer to the point in which ϕ^{R} is calculated. The index s refers to the point at which the dipole is located and the index 0 to the boundary of the cavity. The functions P_n are the Legendre polynomials, P_n' their derivatives, Q_n the Legendre functions of the second kind,* and Q_n' their derivatives.

From (4.49) the reaction field can be calculated with:

$$\boldsymbol{R} = -\operatorname{grad}\phi^{R}. \tag{4.52}$$

Using $P_1(x) = x$ and $Q_1(x) = \frac{x}{2}\ln\frac{x+1}{x-1} - 1$ for $x > 1$, we find for ϕ_1^{R}, the first term of (4.49):

$$\phi_1^{R} = \frac{-3\mu(\varepsilon-1)vw\left\{\frac{1}{2}\ln\left(\frac{v_0+1}{v_0-1}\right)-\frac{1}{v_0}\right\}\left\{\frac{1}{2}\ln\left(\frac{v_0+1}{v_0-1}\right)-\frac{v_0}{v_0^2-1}\right\}}{d^2\left[\varepsilon\left\{\frac{1}{2}\ln\left(\frac{v_0+1}{v_0-1}\right)-\frac{v_0}{v_0^2-1}\right\}-\left\{\frac{1}{2}\ln\left(\frac{v_0+1}{v_0-1}\right)-\frac{1}{v_0}\right\}\right]}. \tag{4.53}$$

We can transform this expression to Cartesian coordinates by using:

$$x = -vwd. \tag{4.54}$$

The a-axis of the spheroid is then the x-axis and the centre of the spheroid is the origin of the new coordinate system. In this way we find that the first term of the reaction field is homogeneous and directed along the major axis of the spheroid:

$$\boldsymbol{R}_1 = \frac{3\mu(\varepsilon-1)\left\{\frac{1}{2}\ln\left(\frac{v_0+1}{v_0-1}\right)-\frac{1}{v_0}\right\}\left\{\frac{1}{2}\ln\left(\frac{v_0+1}{v_0-1}\right)-\frac{v_0}{v_0^2-1}\right\}}{d^3\left[\varepsilon\left\{\frac{1}{2}\ln\left(\frac{v_0+1}{v_0-1}\right)-\frac{v_0}{v_0^2-1}\right\}-\left\{\frac{1}{2}\ln\left(\frac{v_0+1}{v_0-1}\right)-\frac{1}{v_0}\right\}\right]}. \tag{4.55}$$

This can be written in a more manageable way by using:

$$d^2 = a^2 - b^2, \tag{4.56}$$

and:

$$\frac{b^2}{a^2} = \frac{v_0^2-1}{v_0^2}. \tag{4.57}$$

* See for the Legendre functions of the second kind Appendix III, section 3, and Hobson.[9]

With the help of (4.57) we can calculate A_a in the case of a prolate spheroid:

$$A_a = \frac{ab^2}{2}\int_0^\infty \frac{ds}{(s+a^2)^{3/2}(s+b^2)}$$

$$= \frac{v_0^2-1}{2v_0^2}\int_0^\infty \frac{ds}{(s+1)^{3/2}\left(s+1-\frac{1}{v_0^2}\right)}$$

$$= \frac{v_0^2-1}{2v_0^2}\int_1^\infty \frac{ds}{s^{3/2}\left(s-\frac{1}{v_0^2}\right)}$$

$$= 1 - v_0^2 + \tfrac{1}{2}v_0(v_0^2-1)\ln\frac{v_0+1}{v_0-1}. \tag{4.58}$$

Using A_a we can write (4.55) in the form:

$$\boldsymbol{R}_1 = \frac{3\boldsymbol{\mu}}{ab^2}\,\frac{(\varepsilon-1)(1-A_a)A_a}{\varepsilon+(1-\varepsilon)A_a}. \tag{4.59}$$

Hence, the first term of the reaction field derived from (4.49) is equal to the reaction field of a homogeneous dipole density filling up the cavity, with the same dipole moment (eqn. (4.45)). The higher terms have non-zero values even when the dipole is situated in the centre of the cavity, so that there results a difference with the case of eqn. (4.45).

§21. Reaction field and energy

To investigate the energy of a charge distribution in its reaction field, we need an expression for the work of assembly of a set of point charges in a cavity in a dielectric. Eqn. (3.24) for the work of assembly of a set of point charges:

$$W = \tfrac{1}{2}\sum_i e_i\phi_0^i(\boldsymbol{r}_i),$$

cannot be used for this purpose, because in the derivation of this equation it was assumed that no dielectrics were present. It is possible, however, to obtain an expression of the work of assembly of a set of point charges in a cavity in a dielectric by using the procedure followed in section 12 to derive eqn. (3.24).

The work W_1, necessary to bring a charge e_1 from infinity to its actual position $\boldsymbol{r}_1$, when no other charges are present, is zero in the absence of dielectrics. For points in a cavity in a dielectric, however, W_1 is not zero: the dielectric is polarized by the point charge, and this polarization leads to a non-zero potential in $\boldsymbol{r}_1$. W_1 can be found from a charging process: we let increase the point charge with infinitesimally small steps from zero to its actual value e_1. At a certain stage of this process the charge will have reached the value λe_1; the potential at the point charge due to the dielectric is then $\lambda\phi_1^R(\boldsymbol{r}_1)$. Here, ϕ_1^R is the potential due to the polarization induced by the charge e_1. Its gradient is the reaction field of the charge e_1. We take the reaction field potentials due to the different charge elements to be superposable; *i.e.* we assume no field dependence of the permittivity of the dielectric.

The energy necessary to bring a charge element $e_1\delta\lambda$ at $\boldsymbol{r}_1$ against the reaction field of the charge λe_1, is given by:

$$\delta W_1 = \lambda\phi_1^R(\boldsymbol{r}_1)e_1\delta\lambda. \tag{4.60}$$

Therefore the energy W_1 is given by:

$$W_1 = \int_0^1 \lambda\phi_1^R(\boldsymbol{r}_1)e_1\delta\lambda = \tfrac{1}{2}e_1\phi_1^R(\boldsymbol{r}_1). \tag{4.61}$$

When a charge e_2 is now brought to $\boldsymbol{r}_2$, the energy required for this will consist of three parts: the energy necessary to move it against the field due to the charge e_1, given by $e_2\phi_1(\boldsymbol{r}_2)$, the energy necessary to move it against the reaction field of e_1, given by $e_2\phi_1^R(\boldsymbol{r}_2)$, and the energy necessary to move it against its own reaction field, given by $\frac{1}{2}e_2\phi_2^R(\boldsymbol{r}_2)$. Therefore:

$$W_2 = e_2\{\phi_1(\boldsymbol{r}_2) + \phi_1^R(\boldsymbol{r}_2) + \tfrac{1}{2}\phi_2^R(\boldsymbol{r}_2)\}. \tag{4.62}$$

In this way we find for the j-th charge:

$$W_j = e_j\left\{\sum_{i=1}^{j-1}\phi_i(\boldsymbol{r}_j) + \sum_{i=1}^{j-1}\phi_i^R(\boldsymbol{r}_j) + \tfrac{1}{2}\phi_j^R(\boldsymbol{r}_j)\right\}. \tag{4.63}$$

So we find for the energy necessary to assemble the whole set of n point charges:

$$W = \sum_{j=1}^{n} W_j = \sum_{j=1}^{n} e_j\left\{\sum_{i=1}^{j-1}\phi_i(\boldsymbol{r}_j) + \sum_{i=1}^{j-1}\phi_i^R(\boldsymbol{r}_j) + \tfrac{1}{2}\phi_j^R(\boldsymbol{r}_j)\right\}. \tag{4.64}$$

To assemble the set of n point charges in the reversed sequence, the same amount of energy would be necessary:

$$W = \sum_{j=1}^{n} e_j\left\{\sum_{i=j+1}^{n}\phi_i(\boldsymbol{r}_j) + \sum_{i=j+1}^{n}\phi_i^R(\boldsymbol{r}_j) + \tfrac{1}{2}\phi_j^R(\boldsymbol{r}_j)\right\}. \tag{4.65}$$

Adding (4.64) and (4.65), we find:

$$2W = \sum_{j=1}^{n} e_j \left\{ \sum_{i \neq j} \phi_i(\boldsymbol{r}_j) + \sum_{i \neq j} \phi_i^{\mathrm{R}}(\boldsymbol{r}_j) + \phi_j^{\mathrm{R}}(\boldsymbol{r}_j) \right\}, \tag{4.66}$$

or:

$$W = \tfrac{1}{2} \sum_{j=1}^{n} e_j \{ \phi_0^j(\boldsymbol{r}_j) + \phi^{\mathrm{R}}(\boldsymbol{r}_j) \}, \tag{4.67}$$

in which ϕ_0^j is given by (3.23):

$$\phi_0^j(\boldsymbol{r}_j) = \sum_{j \neq i} \phi_i(\boldsymbol{r}_j),$$

and ϕ^{R} is defined by:

$$\phi^{\mathrm{R}}(\boldsymbol{r}_j) = \sum_{i=1}^{n} \phi_i^{\mathrm{R}}(\boldsymbol{r}_j). \tag{4.68}$$

ϕ^{R} is the potential of the reaction field of the whole set of point charges.

In eqn. (4.67), $\frac{1}{2} \sum_{j=1}^{n} e_j \phi_0^j(\boldsymbol{r}_j)$ is the energy necessary to assemble the charge distribution in the absence of dielectrics, as given by (3.24). Therefore, the part $\frac{1}{2} \sum_{j=1}^{n} e_j \phi^{\mathrm{R}}(\boldsymbol{r}_j)$ is the energy necessary to bring the set of point charges with fixed mutual distances into the cavity. This is called the energy of the charge distribution in the reaction field:

$$W_{\mathrm{R}} = \tfrac{1}{2} \sum_{j=1}^{n} e_j \phi^{\mathrm{R}}(\boldsymbol{r}_j). \tag{4.69}$$

Because every charge distribution can be approximated, with any desired degree of accuracy, by a distribution of point charges, eqn. (4.69) holds for an arbitrary charge distribution in a cavity in a dielectric as long as the permittivity of the dielectric is a constant during the charging process.

Returning to eqns. (4.61) and (4.62), we find for the energy necessary to bring two charges into the cavity:

$$W = W_1 + W_2 = \tfrac{1}{2} e_1 \phi_1^{\mathrm{R}}(\boldsymbol{r}_1) + e_2 \phi_1(\boldsymbol{r}_2) + e_2 \phi_1^{\mathrm{R}}(\boldsymbol{r}_2) + \tfrac{1}{2} e_2 \phi_2^{\mathrm{R}}(\boldsymbol{r}_2). \tag{4.70}$$

Since the energy does not depend on the sequence in which the charges are brought into the cavity, we have also:

$$W = \tfrac{1}{2}e_2\phi_2^R(\boldsymbol{r}_2) + e_1\phi_2(\boldsymbol{r}_1) + e_1\phi_2^R(\boldsymbol{r}_1) + \tfrac{1}{2}e_1\phi_1^R(\boldsymbol{r}_1). \tag{4.71}$$

By equating (4.70) and (4.71), we find:

$$e_2\phi_1(\boldsymbol{r}_2) + e_2\phi_1^R(\boldsymbol{r}_2) = e_1\phi_2(\boldsymbol{r}_1) + e_1\phi_2^R(\boldsymbol{r}_1). \tag{4.72}$$

In section 12 it was demonstrated that:

$$e_2\phi_1(\boldsymbol{r}_2) = e_1\phi_2(\boldsymbol{r}_1). \tag{4.73}$$

Hence, we find:

$$e_2\phi_1^R(\boldsymbol{r}_2) = e_1\phi_2^R(\boldsymbol{r}_1). \tag{4.74}$$

Thus, the energy of a point charge in the reaction field of another charge is equal to the energy of the second charge in the reaction field of the first one.

It is obvious that this conclusion holds not only for point charges but also for arbitrary charge distributions: the energy of a charge distribution ρ_1 in the reaction field of a second charge distribution ρ_2, is equal to the energy of the charge distribution ρ_2 in the reaction field of ρ_1. This will be used in section 22 to calculate the average reaction field of an arbitrary charge distribution in an ellipsoidal cavity.

We shall now use eqn. (4.69) to calculate the energy of an ideal dipole in its own reaction field. For a non-ideal dipole (compare section 2b), eqn. (4.69) gives:

$$W_R = \tfrac{1}{2}e\{\phi^R(\boldsymbol{r}_+) - \phi^R(\boldsymbol{r}_-)\}. \tag{4.75}$$

For an ideal dipole, we let $\boldsymbol{r}_+$ approach $\boldsymbol{r}_-$ and let e go to infinity, keeping $e(\boldsymbol{r}_+ - \boldsymbol{r}_-) = \boldsymbol{\mu}$, which leads to:

$$W_R = \tfrac{1}{2}\boldsymbol{\mu}\cdot\operatorname{grad}\phi^R = -\tfrac{1}{2}\boldsymbol{\mu}\cdot\boldsymbol{R}. \tag{4.76}$$

This equation holds independent of the position of the dipole in the cavity and of the particular shape of the cavity. Generally, the value of W_R will depend on the direction of $\boldsymbol{\mu}$ and on the position of the dipole in the cavity. This does not mean, however, that the dipole is displaced or rotated by the reaction field, because in the model used the dipole cannot move with respect to the boundary of the cavity.

Since according to (4.2) we have for a cavity of arbitrary shape:

$$\boldsymbol{R} = \boldsymbol{F}\cdot\boldsymbol{\mu},$$

W_R will be a quadratic form in $\boldsymbol{\mu}$. If the dipole is situated in the centre of a spherical cavity, $\boldsymbol{R}$ is given by (4.24):

$$\boldsymbol{R} = f\boldsymbol{\mu} = \frac{1}{a^3}\frac{2(\varepsilon - 1)}{2\varepsilon + 1}\boldsymbol{\mu},$$

and we have:

$$W_R = -\tfrac{1}{2}f\mu^2 = -\frac{\mu^2}{a^3}\frac{(\varepsilon - 1)}{2\varepsilon + 1}. \tag{4.77}$$

For a polarizable dipole, W_R consists of two parts: the work of polarization, W_{pol}, necessary to enlarge the dipole moment from $\boldsymbol{\mu}$ to $\boldsymbol{\mu}^*$ (compare eqn. (4.33)), and the potential energy V_R of the enlarged dipole in its own reaction field. According to (3.96), we have for the work of polarization of an induced moment $(\boldsymbol{\mu}^* - \boldsymbol{\mu})$ in a field $\boldsymbol{R}$:

$$W_{pol} = \tfrac{1}{2}(\boldsymbol{\mu}^* - \boldsymbol{\mu})\cdot\boldsymbol{R}. \tag{4.78}$$

The potential energy of $\boldsymbol{\mu}^*$ in its own reaction field $\boldsymbol{R}$ will be given by (4.76):

$$V_R = -\tfrac{1}{2}\boldsymbol{\mu}^*\cdot\boldsymbol{R}. \tag{4.79}$$

Hence:

$$W_R = W_{pol} + V_R = -\tfrac{1}{2}\boldsymbol{\mu}\cdot\boldsymbol{R}. \tag{4.80}$$

This expression has the same form as eqn. (4.76) but here $\boldsymbol{R}$ is the reaction field of a polarizable dipole.

If the polarizable point dipole is situated in the centre of a spherical cavity, $\boldsymbol{R}$ is given by eqn. (4.28):

$$\boldsymbol{R} = \frac{f}{1 - f\alpha}\boldsymbol{\mu},$$

in which f is given by (4.25):

$$f = \frac{1}{a^3}\frac{2(\varepsilon - 1)}{2\varepsilon + 1},$$

and we have:

$$W_R = -\tfrac{1}{2}\frac{f\mu^2}{1 - f\alpha}. \tag{4.81}$$

Comparing eqn. (4.81) with (4.77) we see that the introduction of a polarizability leads to a more favourable energy W_R, since $(1 - f\alpha)$ is smaller than 1.

§22. The average reaction field of an arbitrary charge distribution in an ellipsoidal cavity

In the preceding section we showed that for two arbitrary charge distributions ρ_1 and ρ_2 in an arbitrary cavity, the energy of ρ_1 in the reaction field of ρ_2 will be equal to the energy of ρ_2 in the reaction field of ρ_1. We shall now apply this to the case of an ellipsoidal cavity, with axes $2a$, $2b$ and $2c$, a point charge within it, and a homogeneous dipole density filling up the cavity.[10]

In section 20 the reaction field of a homogeneous dipole density $\boldsymbol{P}$, filling up an ellipsoidal cavity was found to be homogeneous. It is given by:

$$\boldsymbol{R} = \boldsymbol{F} \cdot \boldsymbol{m}, \tag{4.82}$$

in which $\boldsymbol{F}$ is the symmetric reaction field tensor for this case, and $\boldsymbol{m}$ is the total moment of the homogeneous dipole density:

$$\boldsymbol{m} = \frac{4\pi}{3} abc\boldsymbol{P}. \tag{4.83}$$

When a coordinate system along the axes of the ellipsoidal cavity is used, $\boldsymbol{F}$ is given by eqn. (4.46).

Because the reaction field is homogeneous, the reaction field potential in a point within the cavity with radius vector $\boldsymbol{r}$ will be given by:

$$\phi^{\mathrm{R}}(\boldsymbol{r}) = \phi^{\mathrm{R}}(\boldsymbol{0}) - \boldsymbol{r} \cdot \boldsymbol{R}. \tag{4.84}$$

For reasons of symmetry, the reaction field potential will be zero in the centre of the cavity. Hence, we choose this centre as the origin, so that:

$$\phi^{\mathrm{R}}(\boldsymbol{0}) = 0, \tag{4.85}$$

and obtain:

$$\phi^{\mathrm{R}}(\boldsymbol{r}) = -\boldsymbol{r} \cdot \boldsymbol{R}. \tag{4.86}$$

Therefore, the energy of a point charge e in the point with radius vector $\boldsymbol{r}'$ will, according to (3.6), be given by:

$$W = -e\boldsymbol{r}' \cdot \boldsymbol{R}. \tag{4.87}$$

Using (4.82) and (4.83), we find:

$$W = -\frac{4\pi}{3} abc\boldsymbol{m}' \cdot \boldsymbol{F} \cdot \boldsymbol{P}, \tag{4.88}$$

where $\boldsymbol{m}'$ is the moment of the point charge with respect to the origin:

$$\boldsymbol{m}' = e\boldsymbol{r}'. \tag{4.89}$$

The energy of the point charge in the reaction field of the homogeneous dipole density must be equal to the energy of the homogeneous dipole density in the reaction field $\boldsymbol{R}'$ of the point charge. The latter field will not be homogeneous, *i.e.* it should be written as a function of the position coordinates. The energy of the homogeneous dipole density in this reaction field is found by integrating (3.9) over the cavity:

$$W = \iiint\limits_{\text{cavity}} - \boldsymbol{P} \cdot \boldsymbol{R}'(\boldsymbol{r})\, \mathrm{d}v. \tag{4.90}$$

Since $\boldsymbol{P}$ is homogeneous, this is equal to:

$$W = -\boldsymbol{P} \cdot \iiint\limits_{\text{cavity}} \boldsymbol{R}'(\boldsymbol{r})\, \mathrm{d}v = -\frac{4\pi}{3} abc\boldsymbol{P} \cdot \overline{\boldsymbol{R}'} = -\frac{4\pi}{3} abc\overline{\boldsymbol{R}'} \cdot \boldsymbol{P}, \tag{4.91}$$

in which $\overline{\boldsymbol{R}'}$ is the value of the reaction field of the point charge averaged over the cavity:

$$\overline{\boldsymbol{R}'} = \frac{3}{4\pi abc} \iiint\limits_{\text{cavity}} \boldsymbol{R}'(\boldsymbol{r})\, \mathrm{d}v. \tag{4.92}$$

Equating (4.88) and (4.91), we have:

$$-\frac{4\pi}{3} abc\boldsymbol{m}' \cdot \boldsymbol{F} \cdot \boldsymbol{P} = -\frac{4\pi}{3} abc\overline{\boldsymbol{R}'} \cdot \boldsymbol{P}. \tag{4.93}$$

This equation holds for arbitrary direction of $\boldsymbol{P}$, and therefore:

$$\overline{\boldsymbol{R}'} = \boldsymbol{m}' \cdot \boldsymbol{F} = \boldsymbol{F} \cdot \boldsymbol{m}', \tag{4.94}$$

since $\boldsymbol{F}$ is a symmetric tensor.

Thus, the average reaction field of a point charge in an ellipsoidal cavity is obtained from the moment of the charge with respect to the centre of the cavity and the reaction field tensor appearing in the expression for the homogeneous reaction field of a homogeneous dipole density filling up the cavity.

Because of the superposition principle, this holds also for a set of point charges, and, since every charge distribution can be approximated to any degree of accuracy by a set of point charges, (4.94) holds for an arbitrary charge distribution.

In cases in which an expression for $\boldsymbol{R}$ is known over the whole cavity, the average reaction field can be calculated in a direct way. For an ideal dipole situated on the axis of a prolate spheroid between the foci and directed along this axis, Abbot and Bolton[8] demonstrated that the average of the field derived from every term with $n > 1$ in eqn. (4.49) is zero. Therefore, only the first term of (4.49) contributes to the average reaction field, and the field corresponding to this term is equal to the reaction field of a homogeneous dipole density filling up the cavity, with the same dipole moment (eqn. (4.59)).

For a spherical cavity, we may write the scalar reaction field factor (eqn. (4.25)) in tensor form as:

$$\boldsymbol{F} = \frac{1}{a^3}\frac{2(\varepsilon - 1)}{2\varepsilon + 1}\boldsymbol{I}, \tag{4.95}$$

which leads, with the help of (4.94) for an arbitrary charge distribution to:

$$\overline{\boldsymbol{R}} = \frac{\boldsymbol{m}}{a^3}\frac{2(\varepsilon - 1)}{2\varepsilon + 1}, \tag{4.96}$$

where $\boldsymbol{m}$ is the moment of the charge distribution with respect to the centre of the sphere. Eqn. (4.96) was derived by Scholte[7] and by Frood and Dekker;[11] for the case of an ideal dipole, not in the centre of the spherical cavity and directed along the radius, it was derived earlier by De Bruyn and Dekker.[12,13]

Eqn. (4.94) is of great importance, since the shape of almost all molecules can be approximated by an ellipsoid. When the induced dipole moment of a molecule is written as the product of a polarizability and a polarizing field, it is preferable to use the volume average of the field instead of the value of the field at the centre of the molecule, so as to take into account the fact that all parts of the molecule contribute to the polarizability. Proceeding in this way we find for the dipole of an arbitrary molecule enlarged by the moment induced by the reaction field:

$$\boldsymbol{\mu}^* = \boldsymbol{\mu} + \boldsymbol{\alpha}\cdot\overline{\boldsymbol{R}} = \boldsymbol{\mu} + \boldsymbol{\alpha}\cdot\boldsymbol{F}\cdot\boldsymbol{\mu}^*, \tag{4.97}$$

or:

$$\boldsymbol{\mu}^* = (\boldsymbol{I} - \boldsymbol{\alpha}\cdot\boldsymbol{F})^{-1}\cdot\boldsymbol{\mu}. \tag{4.98}$$

If $\boldsymbol{\mu}$ lies along the a-axis of the ellipsoid, and the a-axis is a principal axis of the polarizability tensor, (4.98) is reduced to:

$$\boldsymbol{\mu}^* = \frac{\boldsymbol{\mu}}{1 - f_a\alpha_a}, \tag{4.99}$$

where α_a is the polarizability along the a-axis and f_a is given by (4.45):

$$f_a = \frac{3}{abc} \frac{A_a(1 - A_a)(\varepsilon - 1)}{\varepsilon + (1 - \varepsilon)A_a}. \tag{4.100}$$

If the polarizability is equal to the polarizability of a homogeneous ellipsoid with the same dimensions as the molecular ellipsoid, α_a is given by (2.105):

$$\alpha_a = \frac{\varepsilon_i - 1}{3\{1 + (\varepsilon_i - 1)A_a\}} abc, \tag{4.101}$$

in which ε_i is the dielectric constant of the homogeneous ellipsoid.

Substituting (4.100) and (4.101) into (4.99), we have:

$$\frac{\mu^*}{\mu} = \frac{1}{1 - f_a\alpha_a} = \frac{1}{1 - \dfrac{A_a(1 - A_a)(\varepsilon - 1)}{\varepsilon + (1 - \varepsilon)A_a} \dfrac{\varepsilon_i - 1}{1 + (\varepsilon_i - 1)A_a}}$$

$$= \frac{\{\varepsilon + (1 - \varepsilon)A_a\}\{1 + (\varepsilon_i - 1)A_a\}}{\varepsilon + (\varepsilon_i - \varepsilon)A_a}. \tag{4.102}$$

In Table 8 we have computed values of μ^*/μ, for oblate and prolate spheroids in which $\boldsymbol{\mu}$ is directed along the rotation axis. For ε_i the value 2 is used.

TABLE 8

VALUES OF μ^*/μ FOR SPHEROIDAL PARTICLES, $\varepsilon_i = 2$, CALCULATED FROM EQN. (4.102)

$a/b = a/c$	A_a	$\varepsilon = 4$	$\varepsilon = 6$	$\varepsilon = 10$	$\varepsilon = 20$
0.4	0.583	1.26	1.33	1.41	1.49
0.6	0.464	1.24	1.30	1.36	1.41
0.8	0.394	1.22	1.27	1.31	1.35
1.0	0.333	1.20	1.24	1.27	1.30
1.25	0.276	1.17	1.20	1.23	1.25
1.5	0.233	1.15	1.18	1.20	1.21
2.0	0.174	1.12	1.14	1.15	1.16

It follows from Table 8 that the influence of the shape of the cavity on the increase of the dipole moment due to its own reaction field is considerable, especially for dielectrics with a high value of ε.

§23. The reaction field of an eccentric dipole in a spherical cavity

The average reaction field, which was calculated in the preceding section, is a useful quantity if we want to know the dipole moment induced in a molecule by its own reaction field. Another application of the reaction field is in the calculation of the energy of an ideal dipole in its own reaction field according to eqn. (4.76):

$$W_R = -\tfrac{1}{2}\boldsymbol{\mu}\cdot\boldsymbol{R}.$$

For this application, however, we must use the reaction field at the position of the dipole, and not the average reaction field. In most cases the computation of the reaction field at the position of the dipole is very intricate. A system for which this calculation can be performed without difficulty, is an eccentric dipole in a spherical cavity, directed along the radius vector. This calculation was first given by Dekker.[12,13]

The origin is chosen at the centre of the cavity and the positive z-axis is taken through the dipole, so that there is symmetry about the z-axis. Both in the dielectric and in the spherical shell between the dipole and the boundary of the cavity, we may use the general solution of Laplace's equation (eqn. (A2.15)). Therefore:

$$\phi_1 = \sum_{n=0}^{\infty}\left(A_n r^n + \frac{B_n}{r^{n+1}}\right)P_n(\cos\theta) \quad (r > a).$$

$$\phi_2 = \sum_{n=0}^{\infty}\left(C_n r^n + \frac{D_n}{r^{n+1}}\right)P_n(\cos\theta) \quad (s < r < a).$$

Here, a is the radius of the sphere and s is the distance from the dipole to the origin. As in section 18, the boundary conditions are given by (4.6), (4.7), and (4.8):

1. $$(\phi_1)_{r\to\infty} = 0,$$

2. $$(\phi_1)_{r=a} = (\phi_2)_{r=a},$$

3. $$\varepsilon\left(\frac{\partial\phi_1}{\partial r}\right)_{r=a} = \left(\frac{\partial\phi_2}{\partial r}\right)_{r=a}.$$

The terms $\frac{D_n}{r^{n+1}}P_n(\cos\theta)$ in ϕ_2 are due to the charge distribution within the cavity (*cf*. section 4). Because the only field source within the cavity is the permanent dipole, these terms are obtained from the potential due to the permanent dipole in the absence of the dielectric. This potential is given by eqn. (1.59):

$$\phi = \mu \sum_{n=0}^{\infty} n \frac{s^{n-1}}{r^{n+1}} P_n(\cos\theta).$$

Hence, we have:

$$D_n = \mu n s^{n-1}, \tag{4.103}$$

so that:

$$\phi_2 = \sum_{n=0}^{\infty} \left(C_n r^n + \frac{\mu n s^{n-1}}{r^{n+1}} \right) P_n(\cos\theta). \tag{4.104}$$

The first boundary condition results in:

$$A_n = 0, \tag{4.105}$$

for every n, since the Legendre polynomials are linearly independent. From the second and the third boundary condition and the linear independency of the Legendre polynomials, we find:

$$\frac{B_n}{a^{n+1}} = C_n a^n + \frac{\mu n s^{n-1}}{a^{n+1}}, \tag{4.106}$$

$$-\varepsilon(n+1)\frac{B_n}{a^{n+2}} = nC_n a^{n-1} - \mu n(n+1)\frac{s^{n-1}}{a^{n+2}}. \tag{4.107}$$

Combining (4.106) and (4.107), we obtain:

$$C_n = \frac{-\mu(\varepsilon-1)n(n+1)s^{n-1}}{(n+n\varepsilon+\varepsilon)a^{2n+1}}. \tag{4.108}$$

In ϕ_2 (eqn. (4.104)) the terms with C_n represent the contribution of the apparent surface charges, which is equal to the reaction field potential ϕ^R (*cf.* section 18). Because the apparent surface charges are the only sources of the reaction field potential, the expression for this potential will be valid in the whole region enclosed by the apparent surface charges, thus also in the region where $r < s$. Hence, we find for the whole cavity:

$$\phi^R = \sum_{n=0}^{\infty} C_n r^n P_n(\cos\theta) = -\mu(\varepsilon-1)\sum_{n=0}^{\infty} \frac{n(n+1)s^{n-1}r^n}{(n+n\varepsilon+\varepsilon)a^{2n+1}} P_n(\cos\theta). \tag{4.109}$$

From (4.109) we can derive the reaction field with:

$$\boldsymbol{R} = -\text{grad}\,\phi^{\mathrm{R}}.$$

On the z-axis, the components of $\boldsymbol{R}$ in the x- and y-direction will be zero due to the symmetry. Thus, we have:

$$\begin{aligned}\boldsymbol{R}(z) &= -\frac{\mathrm{d}}{\mathrm{d}z}\left\{\mu(\varepsilon - 1)\sum_{n=0}^{\infty}\frac{n(n+1)s^{n-1}z^{n}}{(n+n\varepsilon+\varepsilon)a^{2n+1}}P_n(1)\right\}\boldsymbol{k}\\ &= \boldsymbol{\mu}(\varepsilon - 1)\sum_{n=0}^{\infty}\frac{n^2(n+1)s^{n-1}z^{n-1}}{(n+n\varepsilon+\varepsilon)a^{2n+1}}.\end{aligned} \tag{4.110}$$

Hence, at the position of the dipole the reaction field is given by:

$$\boldsymbol{R}(s) = \boldsymbol{\mu}(\varepsilon - 1)\sum_{n=0}^{\infty}\frac{n^2(n+1)s^{2n-2}}{(n+n\varepsilon+\varepsilon)a^{2n+1}}. \tag{4.111}$$

The first non-zero term of the series for $\boldsymbol{R}(s)$ is identical with the reaction field of the same dipole in the centre of the cavity as given by eqn. (4.24):

$$\boldsymbol{R}_1(s) = \frac{\boldsymbol{\mu}}{a^3}\frac{2(\varepsilon - 1)}{2\varepsilon + 1}. \tag{4.112}$$

In Table 9 we give the ratio between $R(s)$ and $R_1(s)$ for some values of s/a and ε. It appears from Table 9 that even for $s/a = 0.25$, the influence of the higher terms is already considerable. It must be noted that the influence of ε on the ratio $R(s)/R_1(s)$ is small.

From (4.110) one can calculate the reaction field in the origin due to the eccentric dipole:

$$\boldsymbol{R}(0) = \frac{\boldsymbol{\mu}}{a^3}\frac{2(\varepsilon - 1)}{2\varepsilon + 1}. \tag{4.113}$$

Hence, the reaction field in the origin due to an eccentric dipole is independent of the position of the dipole.

TABLE 9

THE RATIO $R(s)/R_1(s)$, CALCULATED BY MEANS OF EQNS. (4.111) AND (4.112)

s/a	$\varepsilon = 2$	$\varepsilon = 6$	$\varepsilon = 10$	$\varepsilon = 20$
0	1	1	1	1
0.1	1.04	1.04	1.04	1.04
0.25	1.27	1.28	1.28	1.29
0.50	2.80	2.90	2.92	2.96

For an arbitrary charge distribution an expression of the same form as (4.113) can be derived for the reaction field in the origin. The energy of a point charge e with radius vector $\boldsymbol{r}$ in the reaction field of a central dipole $\boldsymbol{\mu}$ in a spherical cavity is found from (4.87) and (4.24) to be:

$$W = -e\boldsymbol{r} \cdot \frac{\boldsymbol{\mu}}{a^3} \frac{2(\varepsilon - 1)}{2\varepsilon + 1}. \tag{4.114}$$

The energy of the dipole in the reaction field $\boldsymbol{R}$ of the point charge is, according to (3.9), given by:

$$W' = -\boldsymbol{\mu} \cdot \boldsymbol{R}(0). \tag{4.115}$$

In section 21 we derived:

$$W = W'. \tag{4.116}$$

From this it follows that for a point charge, and therefore also for an arbitrary charge distribution:

$$\boldsymbol{R}(0) = \frac{\boldsymbol{m}}{a^3} \frac{2(\varepsilon - 1)}{2\varepsilon + 1}, \tag{4.117}$$

in which $\boldsymbol{m}$ is the moment with respect to the origin. This relation was proved in another way by Frood and Dekker.[11]

§24. The contribution of the permanent dipoles to the cohesion energy of a liquid

The reaction field can be used to calculate the contribution of the permanent dipoles to the cohesion energy of a liquid. According to eqn. (3.9), the work required to bring a dipole $\boldsymbol{\mu}$ into a field $\boldsymbol{E}$ amounts to $-\boldsymbol{\mu} \cdot \boldsymbol{E}$. In the absence of an external electric field, the average field at the permanent dipole $\boldsymbol{\mu}$ of a molecule is the reaction field $\boldsymbol{R}$. Thus, for one mole of a system of identical molecules the part of the free energy of the system due to the permanent dipoles of the particles is given by:

$$W = -\tfrac{1}{2} N_A \boldsymbol{\mu} \cdot \boldsymbol{R}, \tag{4.118}$$

where the factor $\frac{1}{2}$ is applied to avoid counting the interaction energy between two particular molecules twice, because each term $-\boldsymbol{\mu} \cdot \boldsymbol{R}$ indicates the interaction energy of one dipole with all other molecules.

Another way of deriving (4.118) starts from eqn. (4.76). The work required to surround a permanent dipole $\boldsymbol{\mu}$ by a dielectric is given by:

$$W = -\tfrac{1}{2}\boldsymbol{\mu}\cdot\boldsymbol{R}. \tag{4.119}$$

If we start with N_A molecules, infinitely separated from each other, and we condense the system to a liquid, the result will be that every dipole is surrounded by a dielectric. Thus, the total amount of work is given by:

$$W = N_A(-\tfrac{1}{2}\boldsymbol{\mu}\cdot\boldsymbol{R}) = -\tfrac{1}{2}N_A\boldsymbol{\mu}\cdot\boldsymbol{R}, \tag{4.120}$$

which agrees with (4.118).

The value of the contribution of the permanent dipoles to the cohesion energy as given by expression (4.118) depends on the shape of the molecule and the position of the dipole in the molecule. As a first approximation, we consider the molecules to be spherical and the dipole situated at the centre. For polarizable particles the reaction field is then given by eqn. (4.31):

$$\boldsymbol{R} = \frac{4\pi}{3}N\frac{2(\varepsilon - 1)}{2\varepsilon + n_D^2}\frac{n_D^2 + 2}{3}\boldsymbol{\mu}.$$

After substitution of this relation into eqn. (4.118), one has:

$$W = -\frac{4\pi}{3}\frac{d}{M}N_A^{\ 2}\frac{\varepsilon - 1}{2\varepsilon + n_D^2}\frac{n_D^2 + 2}{3}\mu^2, \tag{4.121}$$

where M is the molecular weight, and d, ε and n_D are the density, the dielectric constant, and the index of refraction for the Na_D-line respectively. In Table 10 we give the values of W, calculated for some pure dipole liquids at room temperature. The values in Table 10 give only the order of magnitude of W, since the model used for the derivation of eqn. (4.31) is that of a spherical molecule with the dipole at its centre. In section 23 we showed that the value of the energy of a dipole in its reaction field is considerably changed when the dipole is not central, and this is the case in many real molecules.

It must be emphasized that eqn. (4.121) may only be used if there is no

TABLE 10

DIPOLE CONTRIBUTION TO THE COHESION ENERGY AT 20°C, CALCULATED FROM EQN. (4.121)

Compound	ε	n_D	d	M	$\mu(D)$	$-W$(kcal/mole)
Nitrobenzene	35	1.553	1.203	123.1	4.2	4.3
Propanone	21.4	1.359	0.791	58.1	2.9	2.4
Aniline	7.0	1.586	1.022	93.1	1.5	0.49
Chloroform	4.8	1.446	1.489	119.4	1.0	0.20

TABLE 11

DIPOLE CONTRIBUTION TO THE COHESION ENERGY AT THE BOILING POINT T_b AT A PRESSURE OF 1 ATM IN COMPARISON WITH THE HEAT OF VAPORIZATION λ

Compound	T_b(°K)	d	M	ε	n_D^2	μ	$-W$	λ
Chloroform	334	1.41	119.4	4.23	2.03	1.0	0.18	6.98
Iodomethane	315	2.21	141.9	6.48	2.30	1.6	0.74	6.55
Bromoethane	311	1.41	109.0	8.81	1.98	1.8	0.80	6.39
Chlorobenzene	405	0.98	112.6	4.20	2.12	1.7	0.38	8.29
Aniline	458	0.87	93.1	4.54	2.24	1.5	0.34	9.70
Pyridine	389	0.88	79.1	9.38	2.12	2.2	1.18	8.01
Quinoline	511	0.92	129.2	5.05	2.33	2.3	0.64	10.60
Cyanomethane	355	0.72	41.1	26.2	1.76	3.9	5.6	7.12
Propanone	329	0.75	58.1	17.68	1.76	2.9	2.2	7.27
Butanone	353	0.73	72.1	14.48	1.83	2.8	1.6	7.45
Acetophenone	475	0.85	120.1	8.64	2.06	3.0	1.2	9.26
Nitromethane	374	1.03	61.0	27.75	1.80	3.5	4.4	7.00
Nitrobenzene	484	1.01	123.1	15.61	2.12	4.2	3.2	10.05
o-Nitrotoluene	495	0.96	137.1	11.82	2.10	3.6	1.9	10.30

association. With associating liquids, such as water and the alcohols, large deviations occur.

It is interesting to compare the contribution of the permanent dipoles to the cohesion energy with the heat of vaporization. To this end Table 11 gives computed values of W for a number of pure dipole liquids at their boiling points at the pressure of 1 atm, using values of ε determined by Grimm and Patrick.[14] The values of W are compared with values of the heat of vaporization λ determined by the same authors. The heat of vaporization is equal to the change of the enthalpy:

$$\lambda = \Delta U + p\Delta V. \tag{4.122}$$

If we neglect the volume of the liquid relative to the volume of the gas and apply Boyle's law to the gas, we find the change in the internal energy from:

$$\Delta U = \lambda - RT_b. \tag{4.123}$$

Part of the increase of the internal energy results from the presence of permanent dipoles, the remaining part being due to London-Van der Waals interaction. For the compounds in Table 11, where the dipole interaction makes an important contribution to the change of the internal energy, we have calculated the London-Van der Waals interaction energy which is given

TABLE 12

COMPARISON OF THE INCREASES OF THE INTERNAL ENERGY IN THE CASE OF VAPORIZATION FOR SOME POLAR COMPOUNDS WITH AN ISOMORPHIC NON-POLAR COMPOUND

Compound	λ	ΔU	$\Delta U + W$	Isomorphic compound	ΔU
Cyanomethane	7.12	6.41	0.8	Propyne	5.1
Propanone	7.27	6.61	4.4	2-methylpropene	4.8
Butanone	7.45	6.74	5.1	2-methylbutene-1	5.5
Nitromethane	7.00	6.25	1.9	Ethane	3.1
Nitrobenzene	10.05	9.08	5.9	Toluene	7.2
o-Nitrotoluene	10.30	9.31	7.4	o-Xylene	8.0

by the difference between ΔU and $-W$. In Table 12 we compare these values with the increase of the internal energy upon vaporization for isomorphic compounds without permanent dipole moment. All values of ΔU are given for pressures of 1 atm. Actually the values of ΔU for the non-polar compounds should be taken at pressures where the free volume ($(v - v_0)/v_0$, v_0 being the specific volume of the liquid at $T = 0°K$) has the same value as for the corresponding polar compound. Other sources of deviations are the approximations used in the calculations of W and the fact that the London-Van der Waals forces between various kinds of molecules are never exactly the same. Nevertheless, Table 12 indicates that an interpretation of the heats of vaporization of various polar liquids with the help of the reaction field is useful.

It must be mentioned, that it does not hold for every case in which the total permanent dipole moment of the molecule is zero that all interaction energy is due to London-Van der Waals forces. If the molecule has an important quadrupole moment, the energy of the quadrupole in the inhomogeneous reaction field must also be taken into account. For p-dichlorobenzene the quadrupole contribution to the cohesion energy is of the same order as W for the other di-chlorobenzenes. This is evident from the boiling points given in Table 13.

TABLE 13

BOILING POINTS OF THE DICHLOROBENZENES

Compound	$\mu(D)$	T_b(°K)
o-dichlorobenzene	2.50	455
m-dichlorobenzene	1.72	445
p-dichlorobenzene	0	447

References

1. A. Martin, *Phil. Mag.* **8** (1929) 550.
2. R. P. Bell, *Trans. Faraday Soc.* **27** (1931) 797.
3. B. Lindner, in J. Hirschfelder (ed.), *Intermolecular Forces* (*Adv. Chem. Phys.* **12**), Wiley, New York 1967, p. 225.
4. L. Onsager, *J. Am. Chem. Soc.* **58** (1936) 1486.
5. C. J. F. Böttcher, *Physica* **5** (1938) 635.
6. Th. G. Scholte, *Physica* **15** (1949) 437.
7. Th. G. Scholte, *Thesis* Leiden 1950.
8. J. A. Abbott and H. C. Bolton, *Trans. Faraday Soc.* **48** (1952) 422.
9. E. W. Hobson, *The Theory of Spherical and Ellipsoidal Harmonics*, University Press, Cambridge 1955.
10. P. Bordewijk, *Physica* **47** (1970) 596.
11. D. G. Frood and A. J. Dekker, *J. Chem. Phys.* **20** (1952) 1030.
12. A. J. Dekker, *Thesis* Amsterdam 1945.
13. A. J. Dekker, *Physica* **12** (1946) 209.
14. F. V. Grimm and W. A. Patrick, *J. Am. Chem. Soc.* **45** (1923) 2794.

CHAPTER V

THE DIELECTRIC CONSTANT IN THE CONTINUUM APPROACH TO THE ENVIRONMENT OF THE MOLECULE

§25. Introduction

In section 8 it has already been mentioned that in many cases the polarization is proportional to the field strength. The relation between the dielectric displacement $\boldsymbol{D}$ and the electric field strength $\boldsymbol{E}$ is then given by:

$$\boldsymbol{D} = \varepsilon \boldsymbol{E}, \tag{5.1}$$

in which ε is called the permittivity or the dielectric constant. Deviations of (5.1) occur for high field strengths, non-isotropic systems, and rapidly changing fields. Therefore, (5.1) is valid for liquids and gases in static or low-frequency fields of moderate intensity. The aim of this chapter is to investigate the dependence of ε on molecular quantities. To this end we shall investigate the dependence of the polarization on the electric field strength given, according to eqn. (2.46), by:

$$\boldsymbol{P} = \frac{\varepsilon - 1}{4\pi}\boldsymbol{E}. \tag{5.2}$$

At moderate intensity, the electric field gives rise to a dipole density by three different effects:

1. Translation (deformation) effects.
   1a: The electrons are shifted relative to the positive charges (electronic polarization).
   1b: Atoms or atom groups are displaced relative to each other (atomic polarization).
2. Rotation (orientation) effect: the electric field tends to direct the permanent dipoles.

The rotation effect is counteracted by the thermal movement of the molecules. Therefore, it is strongly dependent on the temperature, whereas the translation effects are only slightly dependent on the temperature, because they are intramolecular phenomena.

At higher field intensities there are two effects which can be neglected at moderate field intensities:

1. The field tends to direct an anisotropic particle to an orientation such that its axis of highest polarizability coincides with the direction of the external field.
2. Chemical equilibria between components with different permanent dipole moments are shifted by the electric field in favour of the component with a high permanent dipole moment.

These effects will be dealt with in Chapter VII.

To investigate the dependence of the polarization on molecular quantities it is convenient to assume the polarization $\boldsymbol{P}$ to be divided into two parts: the induced polarization $\boldsymbol{P}_\alpha$, caused by the translation effects, and the dipole polarization $\boldsymbol{P}_\mu$, caused by the orientation of the permanent dipoles. Hence, we write eqn. (5.2) as:

$$\frac{\varepsilon - 1}{4\pi}\boldsymbol{E} = \boldsymbol{P}_\alpha + \boldsymbol{P}_\mu. \tag{5.3}$$

In general, there are two methods to calculate both parts of the polarization from molecular parameters. The first method starts from the electric field acting on a single molecule. For the calculation of this field the environment of the molecule is considered to be a continuum with the macroscopic properties of the dielectric. The dipole density is then computed from the mean dipole moment of the molecule and the density. We shall denote this method as the continuum approach to the environment of the molecule; it will be treated in this chapter. It must be admitted that one can object against this method that the assumption of a continuous environment of the molecule is rather crude and that it does not permit to account for specific molecular interactions.

These interactions can be accounted for by the use of statistical mechanics. This is the second method for calculation of the polarization from molecular parameters; it will be dealt with in Chapter VI. Although in principle this method permits an exact evaluation, the calculations can only be executed if important simplifications are made, and these simplifications often have to be as crude as the simplifications made in the continuum approach to the environment of the molecule.

§26. The dependence of the polarization on the internal and the directing field

We shall now investigate the dependence of the polarization on the electric fields working on a single molecule. For the induced polarization $\boldsymbol{P}_\alpha$ we write:

$$\boldsymbol{P}_\alpha = \sum_k N_k \alpha_k (\boldsymbol{E}_\mathrm{i})_k, \tag{5.4}$$

where N is the number of particles per cm^3, α the scalar polarizability of a particle and $\boldsymbol{E}_\mathrm{i}$ the average field strength acting upon that particle. The index k refers to the k-th kind of particle.

The field $\boldsymbol{E}_\mathrm{i}$ is called the internal field. It is defined as the total electric field at the position of the particle minus the field due to the particle itself. The calculation of $\boldsymbol{E}_\mathrm{i}$ is one of the important problems associated with the theory of electric polarization. This calculation can be executed both in the continuum approach for the environment of the molecule and with the help of statistical mechanics.

The orientation polarization can be written as:

$$\boldsymbol{P}_\mu = \sum_k N_k \overline{\boldsymbol{\mu}_k}, \tag{5.5}$$

where $\overline{\boldsymbol{\mu}_k}$ is the value of the permanent dipole vector averaged over all orientations. The value $\overline{\boldsymbol{\mu}_k}$ can be computed from the energy of the permanent dipole in the electric field. This energy is dependent on that part of the electric field tending to direct the permanent dipoles. This part of the field is called the directing field $\boldsymbol{E}_\mathrm{d}$.

In the older theories $\boldsymbol{E}_\mathrm{d}$ was taken equal to $\boldsymbol{E}_\mathrm{i}$. Onsager showed, however, that this is only satisfactory for gases at low pressures, since in general only a part of the internal field $\boldsymbol{E}_\mathrm{i}$ influences the direction of the permanent dipoles. The difference between $\boldsymbol{E}_\mathrm{i}$ and $\boldsymbol{E}_\mathrm{d}$ will be discussed in section 28.

The dependence of $\overline{\boldsymbol{\mu}_k}$ on $\boldsymbol{E}_\mathrm{d}$ is computed from the energy of a dipole in an electric field as given by eqn. (3.9):

$$W = -\boldsymbol{\mu} \cdot \boldsymbol{E}_\mathrm{d} = -\mu E_\mathrm{d} \cos\theta, \tag{5.6}$$

where θ is the angle between the directions of $\boldsymbol{E}_\mathrm{d}$ and $\boldsymbol{\mu}$. Since W is the only part of the energy which depends on the orientation of the dipole, the relative probabilities of the various orientations of the dipole depend on this energy W according to Boltzmann's distribution law.

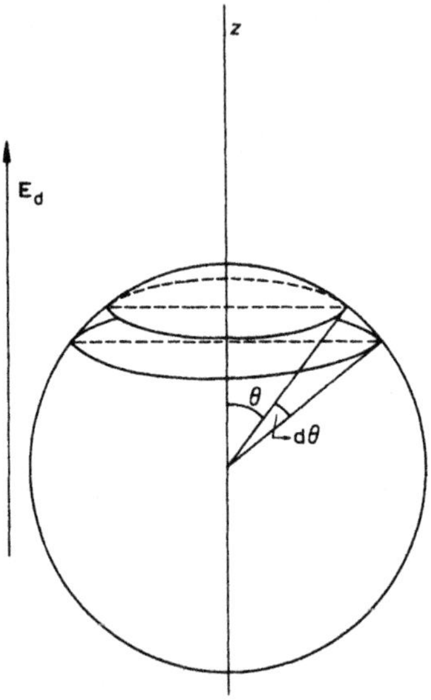

Fig. 24

If there is no directing field, we have $W = 0$ for every θ, and all directions of the dipole vector have the same probability. Therefore the probability $p(\theta)\,\mathrm{d}\theta$ of the dipole having an inclination of the dipole axis to the direction of the z-axis between θ and $\theta + \mathrm{d}\theta$ (Fig. 24), is equal to:

$$p(\theta)\,\mathrm{d}\theta = \frac{2\pi r \sin\theta\, r\,\mathrm{d}\theta}{4\pi r^2} = \tfrac{1}{2}\sin\theta\,\mathrm{d}\theta. \tag{5.7}$$

When a directing field is present a weight factor $e^{-W/kT}$ must be introduced according to Boltzmann's law, and thus (5.7) is changed* into:

$$p(\theta)\,\mathrm{d}\theta = A e^{\mu E_d \cos\theta/kT}\,\tfrac{1}{2}\sin\theta\,\mathrm{d}\theta, \tag{5.8}$$

where A is a proportionality constant amounting to:

$$A = \left[\int_0^\pi e^{\mu E_d \cos\theta/kT}\,\tfrac{1}{2}\sin\theta\,\mathrm{d}\theta\right]^{-1}, \tag{5.9}$$

* According to the principles of quantum mechanics, not all directions of the dipole in the external field are allowed, but only a definite number of discrete directions. However, the strict quantum-mechanical calculation leads only to very small deviations from the final results. In the corresponding magnetic case, however, the restriction of the allowed energy values has much more influence.

since $\int_0^\pi p(\theta)\,\mathrm{d}\theta = 1$.

To determine the extent to which the dipoles are directed by the field $\boldsymbol{E}_\mathrm{d}$, we calculate the average value $\overline{\cos\theta}$ of $\cos\theta$. For a random distribution of the dipoles we have $\overline{\cos\theta} = 0$, whereas $\overline{\cos\theta} = 1$ if all the dipoles have the same direction as $\boldsymbol{E}_\mathrm{d}$. From eqns. (5.8) and (5.9) we obtain:

$$\overline{\cos\theta} = \frac{\int\limits_0^\pi \cos\theta e^{\mu E_\mathrm{d}\cos\theta/kT}\frac{1}{2}\sin\theta\,\mathrm{d}\theta}{\int\limits_0^\pi e^{\mu E_\mathrm{d}\cos\theta/kT}\frac{1}{2}\sin\theta\,\mathrm{d}\theta}. \tag{5.10}$$

Using the abbreviations $\dfrac{\mu E_\mathrm{d}\cos\theta}{kT} = x$ and $\dfrac{\mu E_\mathrm{d}}{kT} = a$, we obtain:

$$\overline{\cos\theta} = \frac{1}{a}\,\frac{\int\limits_{-a}^{a} e^x x\,\mathrm{d}x}{\int\limits_{-a}^{a} e^x\,\mathrm{d}x} = \frac{1}{a}\,\frac{[xe^x - e^x]_{-a}^{+a}}{[e^x]_{-a}^{+a}}$$

$$= \frac{e^a + e^{-a}}{e^a - e^{-a}} - \frac{1}{a} = \operatorname{cotanh} a - \frac{1}{a} = L(a). \tag{5.11}$$

$L(a)$ is called the Langevin function, since this expression was first derived by Langevin (1905) for the case of magnetic moments in a directing magnetic field.[1,2]

In Fig. 25 the Langevin function $L(a)$ is plotted against a. $L(a)$ has a limiting value 1, which was to be expected since this is the maximum of $\cos\theta$. The exponentials in the expression:

$$\overline{\cos\theta} = \frac{1}{a}\,\frac{[xe^x - e^x]_{-a}^{+a}}{[e^x]_{-a}^{+a}}$$

can be developed in a series:

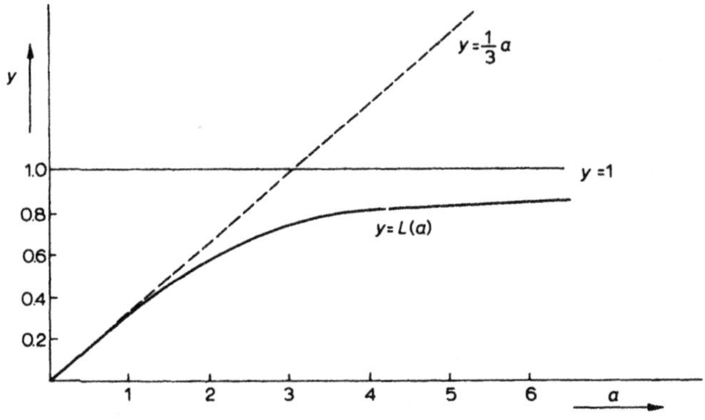

Fig. 25
The Langevin function.

$$\overline{\cos\theta} = \frac{\displaystyle\sum_{n=0,2,4\ldots}\frac{na^n}{(n+1)!}}{\displaystyle\sum_{n=1,3,5\ldots}\frac{a^n}{n!}} = a\frac{\frac{1}{3}+\frac{1}{30}a^2+\cdots}{1+\frac{1}{6}a^2+\cdots}. \tag{5.12}$$

Thus for small values of a, $\overline{\cos\theta}$ is linear in E_d:

$$\overline{\cos\theta} = \tfrac{1}{3}a = \frac{\mu E_d}{3kT} \quad \text{if} \quad 0 \leqslant a \ll 1. \tag{5.13}$$

For larger values of a we must also consider other terms in the series development for $L(a)$:

$$L(a) = \tfrac{1}{3}a - \tfrac{1}{45}a^3 + \tfrac{2}{945}a^5 - \tfrac{2}{9450}a^7 + \cdots.$$

In Table 14 the function $L(a)$ is compared with $a/3$ for some values of a. We read from this Table that for $a = 0.2$ the deviation between $L(a)$ and $a/3$

TABLE 14
THE FUNCTION $L(a)$ COMPARED WITH $a/3$

a	$L(a)$	$a/3$
0.1	0.0333	0.0333
0.2	0.0665	0.0667
0.5	0.1640	0.1667
1.0	0.3130	0.3333
2.0	0.5373	0.6667

is only 0.3%. For $a = 0.5$ the deviation is already 1.6% and for $a = 1.0$ it is more than 6%. We conclude, that the approximation of eqn. (5.13) may be used, as long as $a = \frac{\mu E_d}{kT} < 0.1$ or $E_d < \frac{0.1kT}{\mu}$. At room temperature ($T = 300\,°K$) this gives for a dipole of $4D$:

$$E_d < \frac{0.1kT}{\mu} = 3\ 10^5 \text{ volts/cm}.$$

For a value of μ smaller than the large value of $4D$, the value calculated for E_d is even larger. In usual dielectric measurements, E_d is much smaller than 10^5 volts/cm and the use of (5.13) is allowed.

From eqn. (5.13) it follows that:

$$\bar{\boldsymbol{\mu}} = \frac{\mu^2}{3kT}\boldsymbol{E}_d. \tag{5.14}$$

Substituting this into (5.5) we get:

$$\boldsymbol{P}_\mu = \sum_k N_k \frac{\mu_k^2}{3kT}(\boldsymbol{E}_d)_k. \tag{5.15}$$

We now substitute (5.4) and (5.15) into (5.3), and find:

$$\frac{\varepsilon - 1}{4\pi}\boldsymbol{E} = \sum_k N_k \left[\alpha_k(\boldsymbol{E}_i)_k + \frac{\mu_k^2}{3kT}(\boldsymbol{E}_d)_k \right]. \tag{5.16}$$

This fundamental equation will be the starting point of the following sections, in which we shall express $\boldsymbol{E}_i$ and $\boldsymbol{E}_d$ as functions of the Maxwell field $\boldsymbol{E}$ and the dielectric constant ε.

If the condition $\frac{\mu E_d}{kT} < 0.1$ is not fulfilled a saturation term proportional to $E_d^2 \boldsymbol{E}_d$ must be added; this case will be discussed in Chapter VII.

Generally, $\overline{\cos\theta}$ is much smaller than 0.01. This means that if we could make a picture of the dipole vectors of a number of molecules at a certain moment, it would be impossible to see in that picture that the dipoles prefer a certain direction. Only accurate measurement of all the angles θ would show that $\overline{\cos\theta}$ differs slightly from zero. Thus, the polarizing influence of the electric field appears only in the average of $\cos\theta$ and does not appear as an appreciable change in the direction of the individual dipole moments.

§27. The internal field for non-polar dielectrics

For a non-polar system the fundamental equation for the dielectric constant, derived in the last section (eqn. (5.16)):

$$\frac{\varepsilon - 1}{4\pi}\boldsymbol{E} = \sum_k N_k\left[\alpha_k(\boldsymbol{E}_\mathrm{i})_k + \frac{\mu_k^2}{3kT}(\boldsymbol{E}_\mathrm{d})_k\right],$$

is simplified to:

$$\frac{\varepsilon - 1}{4\pi}\boldsymbol{E} = \sum_k N_k\alpha_k(\boldsymbol{E}_\mathrm{i})_k. \tag{5.17}$$

In this case only the relation between the internal field and the Maxwell field has to be determined. The internal field has been the subject of many discussions, and the computation of this field has always been one of the major problems in the theory of electric polarization.

For non-polar dielectrics, good results are obtained when a formula derived by Lorentz[3] is used for the internal field. Lorentz calculated the internal field in homogeneously polarized matter as the field in a virtual spherical cavity. The field in such a cavity differs from the field in a real cavity, given by eqn. (2.70), because in the latter case the polarization adjusts itself to the presence of the cavity. Therefore the polarization in the environment of a real cavity is not homogeneous (see Fig. 18a), whereas the polarization in the environment of a virtual cavity remains homogeneous.

The field in a virtual spherical cavity, which we will call the Lorentz field $\boldsymbol{E}_\mathrm{L}$, is the sum of:

1. the Maxwell field $\boldsymbol{E}$ caused by the external charges and by the apparent charges on the outer surface of the dielectric, and
2. the field $\boldsymbol{E}_\mathrm{sph}$ induced by the apparent charges on the boundary of the cavity (see Fig. 26).

The field $\boldsymbol{E}_\mathrm{sph}$ is calculated by subdividing the boundary in infinitesimally small rings perpendicular to the field direction. The apparent surface charge density on these rings is $-P\cos\theta$, their surface is $2\pi r^2 \sin\theta\, \mathrm{d}\theta$, so that the total charge on each ring amounts to:

$$de = -2\pi r^2 \sin\theta\, \mathrm{d}\theta\, P\cos\theta. \tag{5.18}$$

According to Coulomb's law, a charge element de on the boundary of the cavity contributes an amount $\mathrm{d}E$ to the field component in the direction of

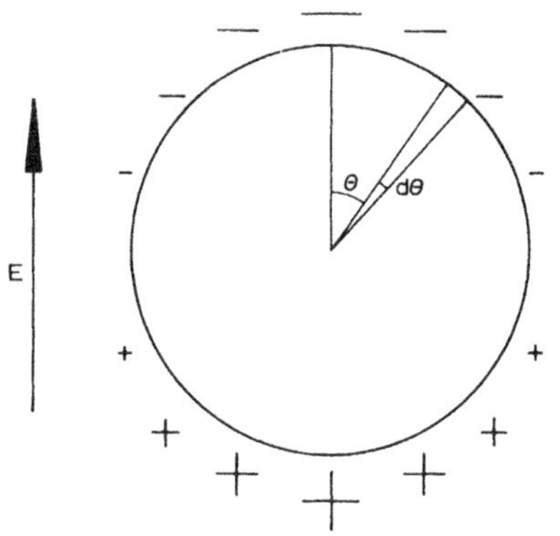

Fig. 26
The Lorentz field.

the external field, given by:

$$dE = \frac{de}{r^2}\cos\theta. \tag{5.19}$$

Combining (5.18) and (5.19), we find for the component of $\boldsymbol{E}_{\text{sph}}$ in the direction of the external field:

$$E_{\text{sph}} = P\int_0^{\pi} 2\pi \sin\theta \cos^2\theta \, d\theta = \frac{4\pi}{3}P. \tag{5.20}$$

For reasons of symmetry, the other components of $\boldsymbol{E}_{\text{sph}}$ are zero, and we have:

$$\boldsymbol{E}_{\text{sph}} = \frac{4\pi}{3}\boldsymbol{P}, \tag{5.21}$$

and, with $\boldsymbol{E}_{\text{L}} = \boldsymbol{E} + \boldsymbol{E}_{\text{sph}}$:

$$\boldsymbol{E}_{\text{L}} = \frac{\varepsilon + 2}{3}\boldsymbol{E}. \tag{5.22}$$

This is Lorentz's equation for the internal field.

Lorentz demonstrated that for a cubical arrangement of identical particles, the internal field is exactly given by his formula. To prove this the internal field is considered as the sum of two parts: the field in a virtual spherical cavity of macroscopic diameter and with its centre at the particle at which the internal field has to be calculated, and the field due to the other particles within the sphere. The first field is given exactly by eqn. (5.22), because the cavity has macroscopic dimensions in this case. Due to the symmetry, all particles within the cavity have the same induced moment, $\boldsymbol{p}$. If we take the centre of the sphere as the origin, and the axes of the coordinate system along the symmetry axes of the cubic arrangement, we have for the electric field due to a particle at the point (x_i, y_i, z_i), at a distance r_i from the centre, according to eqns. (1.31) and (1.32):

$$\boldsymbol{E} = -\boldsymbol{T}_i \cdot \boldsymbol{p} = -\frac{1}{r_i^3}\left(\boldsymbol{I} - \frac{3\boldsymbol{r}_i\boldsymbol{r}_i}{r_i^2}\right) \cdot \boldsymbol{p}. \tag{5.23}$$

For the x-component this gives:

$$E_x = \frac{3x_i^2 - r_i^2}{r_i^5} p_x + \frac{3x_i y_i}{r_i^5} p_y + \frac{3x_i z_i}{r_i^5} p_z. \tag{5.24}$$

In a cubic arrangement, for each particle with coordinates (x_i, y_i, z_i) there is a particle with coordinates $(-x_i, y_i, z_i)$. For these two particles the second and the third terms of (5.24) annihilate each other. Therefore, the x-component of the field in the origin due to all the particles within the sphere is given by:

$$E_x = p_x \sum_i \frac{3x_i^2 - r_i^2}{r_i^5}. \tag{5.25}$$

For a cubic arrangement, the three axes are equivalent. Hence:

$$\sum_i \frac{3x_i^2 - r_i^2}{r_i^5} = \sum_i \frac{3y_i^2 - r_i^2}{r_i^5} = \sum_i \frac{3z_i^2 - r_i^2}{r_i^5}. \tag{5.26}$$

But we also know:

$$\frac{3x_i^2 - r_i^2}{r_i^5} + \frac{3y_i^2 - r_i^2}{r_i^5} + \frac{3z_i^2 - r_i^2}{r_i^5} = 0. \tag{5.27}$$

Therefore:

$$\sum_i \frac{3x_i^2 - r_i^2}{r_i^5} = 0. \tag{5.28}$$

Thus, in a cubic arrangement of identical particles the field contributions due to all dipoles within the sphere compensate each other, and the internal field is given exactly by the Lorentz field (eqn. (5.22)). The same exact agreement is found for the idealized case that the distribution of the surrounding molecules is completely random. Only for diluted gases is such a distribution realized approximately (see section 37, p. 223).

Substituting (5.22) into (5.17), we find:

$$\frac{\varepsilon - 1}{\varepsilon + 2} = \frac{4\pi}{3} \sum_k N_k \alpha_k. \tag{5.29}$$

This relation is generally called the Clausius-Mossotti equation. For a pure compound it is reduced to:

$$\frac{\varepsilon - 1}{\varepsilon + 2} = \frac{4\pi}{3} N\alpha. \tag{5.30}$$

Eqn. (5.29) was first derived in 1847 by Mossotti,[4] whereas Clausius[5] gave a similar derivation independently. Both authors used the same model for the dielectric: a system of conducting spheres, representing the molecules. They did not derive an explicit expression for the internal field, since they used a general potential function. Implicitly, however, their derivation involved the notion of the internal field, which was introduced later, particularly by Lorentz.

The Clausius-Mossotti equation can also be derived by writing the internal field as the sum of the cavity field and the reaction field of the induced dipole:

$$\boldsymbol{E}_\mathrm{i} = \boldsymbol{E}_\mathrm{c} + \boldsymbol{R}. \tag{5.31}$$

According to eqn. (2.70) $\boldsymbol{E}_\mathrm{c}$ is given by:

$$\boldsymbol{E}_\mathrm{c} = \frac{3\varepsilon}{2\varepsilon + 1}\boldsymbol{E}. \tag{5.32}$$

The reaction field $\boldsymbol{R}$ is given by:

$$\boldsymbol{R} = f\boldsymbol{p} = f\alpha(\boldsymbol{E}_\mathrm{c} + \boldsymbol{R}). \tag{5.33}$$

From (5.33) it follows that:

$$\boldsymbol{R} = \boldsymbol{E}_\mathrm{c}\frac{f\alpha}{1 - f\alpha}. \tag{5.34}$$

If we substitute this expression in (5.31), we find:

$$\boldsymbol{E}_\mathrm{i} = \boldsymbol{E}_\mathrm{c}\frac{1}{1 - f\alpha}. \tag{5.35}$$

According to eqn. (4.25) for a spherical cavity the reaction field factor is determined by:

$$f = \frac{1}{a^3}\frac{2(\varepsilon - 1)}{2\varepsilon + 1}.$$

Hence, we find for the internal field:

$$\boldsymbol{E}_\mathrm{i} = \frac{3\varepsilon}{2\varepsilon + 1}\frac{1}{1 - \dfrac{\alpha}{a^3}\dfrac{2(\varepsilon - 1)}{2\varepsilon + 1}}\boldsymbol{E} = \frac{3\varepsilon}{2\varepsilon + 1 - 2\dfrac{\alpha}{a^3}(\varepsilon - 1)}\boldsymbol{E}. \tag{5.36}$$

If we now take for a^3 Onsager's approximation given in eqn. (4.5):

$$\frac{4\pi}{3}Na^3 = 1,$$

we find:

$$E_i = \frac{3\varepsilon}{2\varepsilon + 1 - \frac{8\pi}{3} N\alpha(\varepsilon - 1)}. \tag{5.37}$$

Substituting this expression for E_i in (5.17) for a pure compound:

$$\frac{\varepsilon - 1}{4\pi} E = N\alpha E_i, \tag{5.38}$$

we find:

$$\frac{\varepsilon - 1}{4\pi} = \frac{3\varepsilon N\alpha}{2\varepsilon + 1 - \frac{8\pi}{3} N\alpha(\varepsilon - 1)}, \tag{5.39}$$

which after rearrangement leads to the Clausius-Mossotti equation (eqn. (5.30)):

$$\frac{\varepsilon - 1}{\varepsilon + 2} = \frac{4\pi}{3} N\alpha.$$

The advantage of the latter derivation of the Clausius-Mossotti equation is that it reveals the connexion between the Clausius-Mossotti equation and Onsager's approximation for the radius of the cavity. In section 32 it will be shown that it is often possible to explain deviations from the Clausius-Mossotti equation as a consequence of the inapplicability of Onsager's approximation.

From the Clausius-Mossotti equation for a pure compound (eqn. (5.30)) it follows that it is useful to define a molar polarization $[P]$:

$$[P] = \frac{\varepsilon - 1}{\varepsilon + 2} \frac{M}{d}, \tag{5.40}$$

where d is the density and M the molecular weight. When the Clausius-Mossotti equation is valid $[P]$ is a constant for a given substance:

$$[P] = \frac{4\pi}{3} N_A \alpha. \tag{5.41}$$

According to (5.40) and (5.41), the ratio $(\varepsilon - 1)/(\varepsilon + 2)$ must be proportional to the density of the dielectric, as far as the polarizability α is

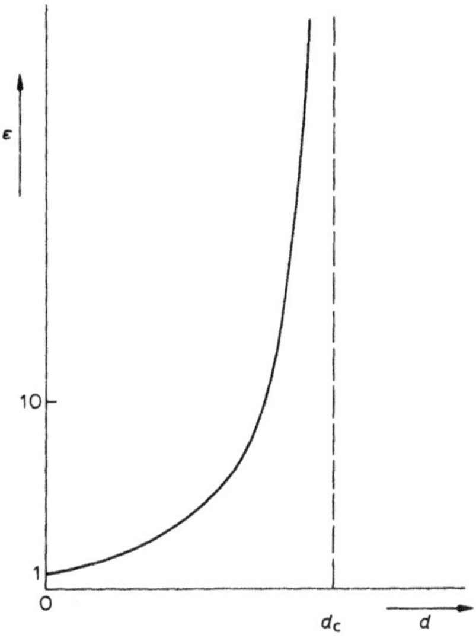

Fig. 27
The dielectric constant of a non-polar dielectric as a function of the density, according to the Clausius-Mossotti equation.

independent of the density. Thus, with increasing density the dielectric constant increases much stronger than linearly (see Fig. 27).

For a density d_c, corresponding to a value N_c of N, given by:

$$\frac{4\pi}{3}N_c\alpha = 1, \tag{5.42}$$

the dielectric constant would be infinite, and for $d > d_c$ it would be negative, according to (5.40). This is called the Clausius-Mossotti catastrophe and has been considered as an objection to the Clausius-Mossotti equation.

Such a critical value N_c, however, is never reached. The volume available to a particle at the critical density d_c is equal to $1/N_c$. Therefore, the condition (5.42) can only be fulfilled if $(4\pi/3)\alpha$ is larger than the volume of a particle itself. For a conducting spherical body we calculated that $\alpha = r^3$ (eqn. (2.102)), and moreover the polarizability of a non-conducting homogeneous spherical body is smaller than that of a conducting body of the same radius (see section 10). Atoms and molecules, however, are not homogeneous, but the experimental values of $(4\pi/3)\alpha$ are in all cases considerably smaller than

TABLE 15

THE POLARIZATION OF HYDROGEN GAS

MEASUREMENTS OF MICHELS, SANDERS AND SCHIPPER[6] AT 99.93°C

p (atm)	d (10^{-3} g/cm^3)	ε	$\frac{\varepsilon - 1}{\varepsilon + 2}\frac{1}{d}$
13.52	0.439	1.00266	2.018
46.46	1.482	1.00898	2.014
88.06	2.751	1.01670	2.012
141.47	4.305	1.02628	2.017
221.39	6.484	1.03966	2.012
478.76	12.496	1.07750	2.015
926.10	20.374	1.12840	2.014
1229.25	24.504	1.15620	2.019
1425.36	26.833	1.17232	2.025

the corresponding values of the volume of the particles. This explains why the critical value N_c is never reached.

The validity of the Clausius-Mossotti equation can be examined by measuring the dielectric constant as a function of the density, for instance in the case of a gas as a function of the pressure at constant temperature. For gases at moderate pressures the Clausius-Mossotti equation shows good agreement with experiment. An excellent example is hydrogen. In Table 15 we give the measurements of Michels, Sanders and Schipper[6] at pressures up to 1425 atm. In this case, where ε varies from 1.002 to 1.17 and $10^3\, d$ from 0.4 to 26.8, the molar polarization is a constant within the experimental error, not only at moderate pressures but also at high pressures.

In section 32 it will be shown that such perfect agreement between theory and experiment at pressures of more than 1000 atm, is obtained only for molecules with a small polarizability such as the hydrogen molecule. Otherwise deviations occur at high pressures.

The Clausius-Mossotti equation can also be used for polar systems in high-frequency alternating fields:

$$\frac{\varepsilon_\infty - 1}{\varepsilon_\infty + 2} = \frac{4\pi}{3}\sum_k N_k \alpha_k, \tag{5.43}$$

in which ε_∞ is the dielectric constant at a frequency at which the permanent dipoles (*i.e.* the orientation polarization) can no longer follow the changes of the field but where the atomic and the electronic polarization are still the same as in static fields. Therefore, ε_∞ is the dielectric constant characteristic for the induced polarization.

Often the Clausius-Mossotti equation is used for still higher frequencies, where the atomic polarization too cannot follow the changes of the field. With the help of the Maxwell relation for the dielectric constant and the refractive index:

$$\varepsilon = n^2, \tag{5.44}$$

it is possible to write:

$$\frac{n^2 - 1}{n^2 + 2} = \frac{4\pi}{3} \sum_k N_k \alpha_k^e, \tag{5.45}$$

where α_k^e denotes the part of the polarizability related with the electronic polarization. Eqn. (5.45) is called the Lorenz-Lorentz equation.[7,8] This equation and also the Maxwell relation will be discussed in more detail in Volume II.

The Lorenz-Lorentz equation has already been used in section 19 to derive eqns. (4.31) and (4.36). The advantage of (5.45) as compared to (5.43) is, that in most cases n can be determined much more accurately than ε_∞, the dielectric constant at frequencies where only the orientation polarization cannot follow the changes of the field. In practice, the atomic polarization can often be neglected, and n^2 is used instead of ε_∞ in the calculation of α.

§28. The internal and the directing fields for polar dielectrics

In the case of polar dielectrics, the molecules have a permanent dipole moment $\boldsymbol{\mu}$, and both parts of the fundamental equation (eqn. (5.16)):

$$\frac{\varepsilon - 1}{4\pi} \boldsymbol{E} = \sum_k N_k \left[\alpha_k (\boldsymbol{E}_\mathrm{i})_k + \frac{\mu_k^2}{3kT} (\boldsymbol{E}_\mathrm{d})_k \right]$$

must be taken into account. In the older theories the Lorentz field (eqn. (5.22)) was used for both the internal and the directing fields. Onsager[9] showed, however, that this is not permissible because of the way in which the Lorentz field is built up.

In section 27 we demonstrated that in the case of non-polar liquids the internal field can be considered as the sum of two parts; one being the cavity field and the other the reaction field of the dipole induced in the molecule (eqn. (5.31)). For polar molecules the internal field can also be built up from the cavity field and the reaction field, in this case the reaction field of the total dipole moment of the molecule. The angle between the reaction field of the permanent part of the dipole moment and the permanent dipole moment itself will be constant during the movements of the molecule (*cf.* section 21, p. 144). In fact in a spherical cavity the permanent dipole moment and the reaction field caused by it will have the same direction. Therefore, this reaction field $\boldsymbol{R}$ does not influence the direction of the dipole moment of the molecule under consideration, and does not contribute to the directing field $\boldsymbol{E}_d$. On the other hand, the reaction field does contribute to the internal field $\boldsymbol{E}_i$, because it polarizes the molecule. As a result, we find a difference between the internal field $\boldsymbol{E}_i$ and the directing field $\boldsymbol{E}_d$. Since the reaction field $\boldsymbol{R}$ belongs to one particular orientation of the dipole moment, the difference between $\boldsymbol{E}_i$ and $\boldsymbol{E}_d$ will be given by the value of the reaction field averaged over all orientations of the polar molecule:

$$\boldsymbol{E}_i - \boldsymbol{E}_d = \bar{\boldsymbol{R}}. \tag{5.46}$$

If we use eqns. (3.9) and (4.4), we see that the average energy of a molecule with dipole moment $\boldsymbol{\mu}$, in the field $\boldsymbol{E}_0 + \boldsymbol{E}_s$, is given by:

$$W = -\langle \boldsymbol{\mu} \cdot (\boldsymbol{E}_0 + \boldsymbol{E}_s) \rangle_{\boldsymbol{E}_0}^{\boldsymbol{\mu}} = -\boldsymbol{\mu} \cdot \boldsymbol{E}_0 - \boldsymbol{\mu} \cdot \langle \boldsymbol{E}_s \rangle_{\boldsymbol{E}_0}^{\boldsymbol{\mu}}$$

$$= -\boldsymbol{\mu} \cdot \boldsymbol{E}_0 - \boldsymbol{\mu} \cdot \boldsymbol{R} - \boldsymbol{\mu} \cdot \langle \boldsymbol{E}_s \rangle_{\boldsymbol{E}_0}^{\boldsymbol{\mu}=0}. \tag{5.47}$$

In this expression the term $-\boldsymbol{\mu} \cdot \boldsymbol{R} = -\mu R$ is independent of the direction of $\boldsymbol{\mu}$ and does not influence the average orientation of $\boldsymbol{\mu}$. Therefore the orientational distribution of $\boldsymbol{\mu}$ depends only on the orientational energy W_d:

$$W_d = -\boldsymbol{\mu} \cdot \langle \boldsymbol{E}_0 + \boldsymbol{E}_s \rangle_{\boldsymbol{E}_0}^{\boldsymbol{\mu}=0}. \tag{5.48}$$

This energy can be written as the product of $\boldsymbol{\mu}$ and a directing field $\boldsymbol{E}_d$:

$$\boldsymbol{E}_d = \langle \boldsymbol{E}_0 + \boldsymbol{E}_s \rangle_{\boldsymbol{E}_0}^{\boldsymbol{\mu}=0}. \tag{5.49}$$

The directing field $\boldsymbol{E}_d$ can be obtained by the following procedure: remove the permanent dipole of a molecule without changing its polarizability; let the surrounding dielectric adapt itself to the new situation; then fix the charge distribution of the surroundings and remove the central molecule. The average field in the cavity so obtained is equal to the value of $\boldsymbol{E}_d$ that is to be calculated, since we have eliminated the contribution of $\boldsymbol{R}$ to $\boldsymbol{E}_i$ by removing the permanent dipole of the molecule. With the aid of this procedure it is possible to calculate $\boldsymbol{E}_{\mathrm{d}}$ in a simple way. If in the above procedure we had also removed the polarizability of the molecule before fixing the charge distribution of the surroundings, we would have formed a "physical" cavity in the dielectric, *i.e.* a cavity where the surroundings are allowed to adapt themselves to the new situation. For a spherical cavity, the homogeneous cavity field $\boldsymbol{E}_{\mathrm{c}}$ is given by eqn. (2.70):

$$\boldsymbol{E}_{\mathrm{c}} = \frac{3\varepsilon}{2\varepsilon + 1}\boldsymbol{E}.$$

To calculate $\boldsymbol{E}_{\mathrm{d}}$, the polarizability of the molecule must now be taken into account. The field $\boldsymbol{E}_{\mathrm{d}}$ causes a dipole $\alpha\boldsymbol{E}_{\mathrm{d}}$ with reaction field $f\alpha\boldsymbol{E}_{\mathrm{d}}$, where f is the reaction field factor given by (4.25). Thus, $\boldsymbol{E}_{\mathrm{d}}$ will be given by the equation:

$$\boldsymbol{E}_{\mathrm{d}} = \boldsymbol{E}_{\mathrm{c}} + f\alpha\boldsymbol{E}_{\mathrm{d}}, \tag{5.50}$$

or:

$$\boldsymbol{E}_{\mathrm{d}} = \frac{1}{1 - f\alpha}\boldsymbol{E}_{\mathrm{c}}. \tag{5.51}$$

Combining (2.70) and (5.51), we have:

$$\boldsymbol{E}_{\mathrm{d}} = \frac{1}{1 - f\alpha}\,\frac{3\varepsilon}{2\varepsilon + 1}\boldsymbol{E}. \tag{5.52}$$

If the dielectric consists of different kinds of molecules, we must calculate the directing field for each kind of molecule separately. We then have for the k-th kind of molecule:

$$(\boldsymbol{E}_{\mathrm{d}})_k = \frac{1}{1 - f_k\alpha_k}\,\frac{3\varepsilon}{2\varepsilon + 1}\boldsymbol{E}, \tag{5.53}$$

with:

$$f_k = \frac{1}{a_k^3}\,\frac{2(\varepsilon - 1)}{2\varepsilon + 1}, \tag{5.54}$$

where ε is the dielectric constant of the mixture and a_k the radius of the cavity belonging to the particle of the k-th kind.

In the ratio between E_{c} and E_{d} we again meet the factor $(1 - f\alpha)$:

$$\frac{\mu^*}{\mu} = \frac{E_{\mathrm{d}}}{E_{\mathrm{c}}} = \frac{1}{1 - f\alpha}. \tag{5.55}$$

As mentioned in section 19, this ratio generally lies between 1.1 and 1.5 in liquids, and in gases it is almost equal to 1.

The internal field is now found from eqn. (5.46):

$$\boldsymbol{E}_{\mathrm{i}} = \boldsymbol{E}_{\mathrm{d}} + \bar{\boldsymbol{R}}, \tag{5.56}$$

where $\boldsymbol{R}$ is the total reaction field connected with the permanent part of the dipole moment. This will be given by the expression for the reaction field of a polarizable point dipole, eqn. (4.28):

$$\boldsymbol{R} = \frac{f}{1 - f\alpha}\boldsymbol{\mu}.$$

Therefore, $\bar{\boldsymbol{R}}$ is given by:

$$\bar{\boldsymbol{R}} = \frac{f}{1 - f\alpha}\bar{\boldsymbol{\mu}}, \tag{5.57}$$

in which $\bar{\boldsymbol{\mu}}$ is the value of $\boldsymbol{\mu}$ averaged over all orientations. $\bar{\boldsymbol{\mu}}$ is given in eqn. (5.14):

$$\bar{\boldsymbol{\mu}} = \frac{\mu^2}{3kT}\boldsymbol{E}_{\mathrm{d}}.$$

Hence:

$$\begin{aligned}\boldsymbol{E}_{\mathrm{i}} &= \boldsymbol{E}_{\mathrm{d}} + \frac{f}{1 - f\alpha}\frac{\mu^2}{3kT}\boldsymbol{E}_{\mathrm{d}} \\ &= \left(1 + \frac{f}{1 - f\alpha}\frac{\mu^2}{3kT}\right)\frac{1}{1 - f\alpha}\frac{3\varepsilon}{2\varepsilon + 1}\boldsymbol{E}. \end{aligned} \tag{5.58}$$

For a mixture of different kinds of molecules we find for the internal field at the k-th kind of molecule:

$$(\boldsymbol{E}_{\mathrm{i}})_k = \left(1 + \frac{f_k}{1 - f_k\alpha_k}\frac{\mu_k^2}{3kT}\right)\frac{1}{1 - f_k\alpha_k}\frac{3\varepsilon}{2\varepsilon + 1}\boldsymbol{E}. \tag{5.59}$$

It must be stressed, that eqns. (5.52), (5.53), (5.58) and (5.59) hold only for spherical particles. It is not necessary that the dipole be situated in the centre of the sphere, however, if the polarizability of these particles is considered to be a point polarizability in the centre of the molecule or homogeneously distributed over the whole sphere. This follows from the fact, that for an arbitrary charge distribution in a spherical cavity both the reaction field in the centre and the average reaction field are equal to the reaction field of an ideal dipole with the same moment in the centre of the cavity (see eqns. (4.117) and (4.96)).

§29. The Onsager equation

We can now substitute eqns. (5.53) and (5.59) for the directing and the internal field, respectively, into the fundamental equation, eqn. (5.16):

$$\frac{\varepsilon - 1}{4\pi}\boldsymbol{E} = \sum_k N_k \left[\alpha_k(\boldsymbol{E}_\mathrm{i})_k + \frac{\mu_k^2}{3kT}(\boldsymbol{E}_\mathrm{d})_k\right].$$

We then find:

$$\frac{(\varepsilon - 1)(2\varepsilon + 1)}{12\pi\varepsilon} = \sum_k N_k \frac{1}{1 - f_k\alpha_k}\left[\alpha_k + \frac{1}{3kT}\frac{\mu_k^2}{1 - f_k\alpha_k}\right], \tag{5.60}$$

in which f_k is given by (5.54):

$$f_k = \frac{1}{a_k^3}\frac{2(\varepsilon - 1)}{2\varepsilon + 1}.$$

Using μ_k^* and introducing a quantity α_k^* given by:

$$\frac{\alpha_k^*}{\alpha_k} = \frac{\mu_k^*}{\mu_k} = \frac{1}{1 - f_k\alpha_k}, \tag{5.61}$$

we can write (5.60) in the more simple form:

$$\frac{(\varepsilon - 1)(2\varepsilon + 1)}{12\pi\varepsilon} = \sum_k N_k\left[\alpha_k^* + \frac{(\mu_k^*)^2}{3kT}\right]. \tag{5.62}$$

Taking for the radius of the cavity and the polarizability the values of these quantities as found for each pure compound from respectively Onsager's approximation, eqn. (4.5), and eqn. (5.43):

$$\frac{4\pi}{3}Na_k^3 = 1,$$

$$\frac{(\varepsilon_\infty)_k - 1}{(\varepsilon_\infty)_k + 2} = \frac{4\pi}{3}N\alpha_k,$$

we obtain for the ratio α_k/a_k^3:

$$\frac{\alpha_k}{a_k^3} = \frac{(\varepsilon_\infty)_k - 1}{(\varepsilon_\infty)_k + 2}. \tag{5.63}$$

From this we find:

$$\frac{1}{1 - f_k\alpha_k} = \frac{((\varepsilon_\infty)_k + 2)(2\varepsilon + 1)}{3(2\varepsilon + (\varepsilon_\infty)_k)}. \tag{5.64}$$

Hence, (5.60) can be written in the form:

$$\frac{\varepsilon - 1}{4\pi\varepsilon} = \sum_k N_k \frac{(\varepsilon_\infty)_k + 2}{2\varepsilon + (\varepsilon_\infty)_k}\left[\alpha_k + \frac{((\varepsilon_\infty)_k + 2)(2\varepsilon + 1)}{3(2\varepsilon + (\varepsilon_\infty)_k)}\frac{\mu_k^2}{3kT}\right]. \tag{5.65}$$

For pure dipole liquids it is useful to substitute α, as found from eqn. (5.43), into (5.65):

$$\frac{\varepsilon - 1}{4\pi} = \frac{3\varepsilon}{4\pi}\frac{\varepsilon_\infty - 1}{2\varepsilon + \varepsilon_\infty} + \frac{(\varepsilon_\infty + 2)^2(2\varepsilon + 1)\varepsilon}{(2\varepsilon + \varepsilon_\infty)^2}\frac{N\mu^2}{9kT}. \tag{5.66}$$

Hence:

$$\mu^2 = \frac{9kT}{4\pi N}\frac{(\varepsilon - \varepsilon_\infty)(2\varepsilon + \varepsilon_\infty)}{\varepsilon(\varepsilon_\infty + 2)^2}. \tag{5.67}$$

Eqn. (5.67) is generally called the Onsager equation. It makes possible the computation of the permanent dipole moment of a molecule from the dielectric constant of the pure dipole liquid if the density and ε_∞ are known. For the derivation of this equation, it is supposed that the particles are spherical and that no specific molecular interactions between the particles occur (*cf.* sections 30 and 39).

In Table 16 dipole moments calculated with (5.67) are given for a number of non-associating organic liquids and compared with the actual values of the dipole moments as measured in the gas phase. The main problem in the calculation is the value we must choose for ε_∞, the dielectric constant at frequencies so high, that no orientation polarization occurs any longer but nevertheless low enough to include the contributions of the electronic and the atomic polarization. The contribution of the electronic polarization to ε_∞ is calculated from the refractive index. The contribution of the atomic polarization can only be estimated, however. From data on non-polar liquids, where no orientation polarization occurs and therefore $\varepsilon_\infty = \varepsilon$, we

TABLE 16

ONSAGER MOMENTS FOR A NUMBER OF NON-ASSOCIATING ORGANIC COMPOUNDS

Compound	t(°C)	ε	d	ε_∞	$\mu_{ons}(D)$	$\mu_{gas}(D)$
Chloromethane	−20	12.6	1.005[a]	1.97[b]	1.74	1.87
Bromomethane	0	9.82	1.730[a]	2.17[b]	1.55	1.81
Iodomethane	20	7.100	2.279	2.46	1.27	1.62
Dichloromethane	15	9.28	1.335	2.13[c]	1.68	1.60
Dibromomethane	15	7.41	2.510	2.49[c]	1.37	1.43
Chloroform	20	4.777	1.489	2.20	1.18	1.01
Bromoform	20	4.39	2.891	2.70	0.92	0.99
Nitromethane	30	35.9	1.125	1.99[c]	3.44	3.46
Cyanomethane	25	34.58	0.777	1.89	3.39	3.92
Ethanethiol	25	6.667	0.833	2.14	1.50	1.58
Bromoethane	20	9.39	1.461	2.13	1.84	2.03
Iodoethane	25	7.64	1.925	2.39	1.58	1.91
1,1,1-Trichloroethane	20	5.66	1.338	2.17	1.52	1.78
Ethanal	21	21.1	0.816	2.08[c]	2.50	2.69
Cyanoethane	20	27.2	0.782	1.96	3.36	4.02
2,2-Dichloropropane	20	11.37	1.091	2.10[d]	2.45	2.27
Propanone	25	20.7	0.785	1.93	3.03	2.88
Trimethylamine	0	2.57	0.658[e]	1.95[b]	0.66	0.61
1-Chlorobutane	20	7.39	0.887	2.06	1.92	2.05
1-Bromobutane	20	7.07	1.276	2.18	1.83	2.08
1-Iodobutane	20	6.22	1.615	2.36	1.63	2.12
2-Chloro-2-methylpropane	20	9.90	0.841	2.02	2.39	2.00
Ethoxyethane	20	4.376	0.714	1.92	1.40	1.15
Ethylthioethane	20	6.004	0.836	2.18[d]	1.64	1.54
Pyridine	25	12.3	0.798	2.38	2.12	2.19
Triethylamine	25	2.42	0.723	2.05	0.66	0.66
Chlorobenzene	20	5.641	1.107	2.43	1.40	1.69
Bromobenzene	25	5.40	1.488	2.55	1.33	1.70
Iodobenzene	20	4.485	1.831	2.73[d]	1.05	1.70
Aniline	20	6.89	1.022	2.64	1.44	1.53
Nitrobenzene	25	34.89	1.198	2.52	4.06	4.22
Cyanobenzene	15	26.0	1.009	2.44[d]	3.48	4.18
Acetophenone	25	17.39	1.024	2.46	3.04	3.02
Quinoline	25	9.00	1.090	2.77	1.91	2.29

[a] From Morgan and Lowry.[13]

[b] Based on values of n_D from Grosse, Wackher and Linn.[14]

[c] Based on the refractive index for the He_r-line.

[d] Based on the refractive index for the H_α-line.

[e] From Grosse, Wackher and Linn.[14]

find that for organic compounds ε_∞ is given approximately by $\varepsilon_\infty = 1.05n^2$, in which n is the refractive index. This relation will be used here to estimate ε_∞ for pure dipole liquids.

In Table 16 where no references are given, the static dielectric constants and densities were taken from Timmermans,[10] or from Maryott and Smith.[11] If not otherwise indicated the refractive indices were taken from the values given by Timmermans for the Na_D-line. The gas values for the dipole moments were taken from Nelson, Lide and Maryott.[12] It is evident from Table 16, that for non-associating compounds the difference between the value of μ, calculated with (5.67), and the value measured in the gas phase is generally of the order of 10%, with a largest deviation of 38% for iodobenzene.

In Table 17 dipole moments, calculated from Onsager's equation, are given for some associating compounds and compared with the value in the gaseous state, or if this value is not known with a value determined in dilute solution, As a rule, the differences between these moments and the Onsager moments are much larger than for non-associating compounds.

As mentioned above, the inadequacies of the Onsager equation result from the assumption of a spherical shape for the particles; when strong specific interactions are present, the Onsager equation appears to be inapplicable. Furthermore, the uncertainty in the atomic polarization influences the

TABLE 17

ONSAGER MOMENTS FOR SOME ASSOCIATING ORGANIC COMPOUNDS

Compound	t(°C)	ε	d	ε_∞	$\mu_{ons}(D)$	$\mu_{gas}(D)$
Hydrogen cyanide	20	114.9	0.687	1.67	5.66	2.98
Methanol	20	33.6	0.791	1.85	2.95	1.70
Formamide	20	111.5	1.133	2.20	4.91	3.73
Propionic acid	20	3.34	0.993	2.02	0.87	1.75
Heptanol-1	20	11.8[a]	0.823[a]	2.13[b]	2.90	1.66[c]
3-Ethylpentanol-3	25	3.24	0.837[d]	2.14[e]	1.08	1.61[f]

[a] From Middelhoek.[15]
[b] Based on the value of n_D from this author.[15]
[c] Measured in cyclohexane by this author.[15]
[d] From Owen, Quayle and Beavers.[16]
[e] From Doering, Cortes and Knox.[17]
[f] From unpublished measurements in cyclohexane.

results and thus hampers the comparison between the Onsager moment and the gas moment.

The specific interactions between the particles, which are responsible for the large deviations found for the associating compounds, can only be taken into account by statistical mechanical calculation. It is possible, however, in the continuum approach to the environment of the molecule to investigate the effects of deviations from the spherical shape. This will be done in the next section.

§30. The generalization of the Onsager equation for ellipsoidal molecules

We shall now calculate the polarization in an electric field for a dielectric consisting of polar molecules with an ellipsoidal shape. We restrict ourselves to the case that the principal axes of the polarizability tensor of the molecule lie in the same direction as the axes of the ellipsoid and that the dipole moment of the molecule lies in the direction of one of these axes, which will be denoted as the a-axis.

According to eqn. (5.3) the polarization generally consists of two parts:

$$\frac{\varepsilon - 1}{4\pi}\boldsymbol{E} = \boldsymbol{P}_\alpha + \boldsymbol{P}_\mu,$$

in which the dipole polarization $\boldsymbol{P}_\mu$ is given by eqn. (5.5):

$$\boldsymbol{P}_\mu = \sum_k N_k \bar{\boldsymbol{\mu}}_k.$$

The value of the permanent dipole moment averaged over all orientations can be found from the directing field. As in the case of spherical particles, for ellipsoidal particles this directing field consists of the cavity field and of the reaction field of the dipole induced by the cavity field and this reaction field itself. In the case of ellipsoidal molecules, however, the cavity field depends on the orientation of the molecule with respect to the electric field and therefore eqn. (5.14):

$$\bar{\boldsymbol{\mu}} = \frac{\mu^2}{3kT}\boldsymbol{E}_\mathrm{d},$$

cannot be used. Instead we must start again from the dependence of the energy of the permanent dipole in the directing field on the angle θ_a between the permanent dipole (lying in the direction of the a-axis) and the electric field.

The energy of the dipole in the directing field is given by eqn. (3.9):

$$W = -\boldsymbol{\mu} \cdot \mathbf{E}_{\mathrm{d}} = -\mu(E_{\mathrm{d}})_a, \tag{5.68}$$

in which $(E_{\mathrm{d}})_a$ is the component of the directing field in the direction of the permanent dipole. Due to the coincidence of the a-axis of the ellipsoid with a principal axis of the polarizability tensor, $(E_{\mathrm{d}})_a$ can be found from:

$$(E_{\mathrm{d}})_a = (E_{\mathrm{c}})_a + f_a \alpha_a (E_{\mathrm{d}})_a, \tag{5.69}$$

so that:

$$(E_{\mathrm{d}})_a = \frac{1}{1 - f_a \alpha_a} (E_{\mathrm{c}})_a. \tag{5.70}$$

Since a is an axis of the ellipsoid, we find $(E_{\mathrm{c}})_a$ from eqn. (2.84):

$$(E_{\mathrm{c}})_a = \frac{\varepsilon}{\varepsilon + (1 - \varepsilon)A_a} E_a = \frac{\varepsilon}{\varepsilon + (1 - \varepsilon)A_a} E \cos \theta_a. \tag{5.71}$$

Hence, we find for $(E_{\mathrm{d}})_a$ from (5.70) and (5.71):

$$(E_{\mathrm{d}})_a = \frac{\varepsilon E \cos \theta_a}{(1 - f_a \alpha_a)\{\varepsilon + (1 - \varepsilon)A_a\}}. \tag{5.72}$$

Because the cavity field is homogeneous and all parts of the molecule contribute to the polarizability, the directing field can also be considered to be homogeneous. We now substitute (5.72) into (5.68):

$$W = \frac{-\mu \varepsilon E \cos \theta_a}{(1 - f_a \alpha_a)\{\varepsilon + (1 - \varepsilon)A_a\}}. \tag{5.73}$$

From this expression we find for $(\bar{\boldsymbol{\mu}})_E$, the component of $\bar{\boldsymbol{\mu}}$ in the direction of the external field (*cf.* the derivation of (5.13) from (5.6)):

$$(\bar{\boldsymbol{\mu}})_E = \mu \overline{\cos \theta_a} = \mu \frac{\int_0^\pi 2\pi \sin \theta_a \cos \theta_a e^{-W/kT} \, \mathrm{d}\theta_a}{\int_0^\pi 2\pi \sin \theta_a e^{-W/kT} \, \mathrm{d}\theta_a}$$

$$= \frac{\mu^2}{3kT} \frac{\varepsilon E}{(1 - f_a \alpha_a)\{\varepsilon + (1 - \varepsilon)A_a\}}. \tag{5.74}$$

Since the dielectric is isotropic, the other components of $\bar{\mu}$ will be zero, and we have:

$$P_\mu = \sum_k N_k \bar{\mu}_k = \sum_k \frac{N_k \mu_k^2}{3kT} \frac{\varepsilon}{\{1 - (f_a)_k(\alpha_a)_k\}\{\varepsilon + (1 - \varepsilon)(A_a)_k\}} E. \quad (5.75)$$

We shall now calculate the induced polarization P_α as the sum of the induced moments of the particles. These induced moments are given as the products of polarizability and internal field. According to eqn. (5.46), the internal field is built up of two parts, the directing field and the reaction field of the permanent dipole averaged over all orientations:

$$E_i = E_d + \bar{R}.$$

We will consider the contributions of both parts of the internal field to the polarization separately.

The reaction field induces in the molecule a moment p_R, which can be found from the enlarged dipole moment given by eqn. (4.99):

$$\mu^* = \frac{\mu}{1 - f_a \alpha_a}.$$

Hence:

$$p_R = \mu^* - \mu = \frac{f_a \alpha_a}{1 - f_a \alpha_a} \mu, \quad (5.76)$$

so that:

$$\bar{p}_R = \frac{f_a \alpha_a}{1 - f_a \alpha_a} \bar{\mu}. \quad (5.77)$$

The moment p_d induced by the directing field is given by its components in the direction of the axes of the ellipsoid. Denoting the angle between the λ-axis ($\lambda = a, b, c$) and the electric field by θ_λ, one has:

$$(p_d)_\lambda = \alpha_\lambda (E_d)_\lambda = \frac{\alpha_\lambda}{1 - f_\lambda \alpha_\lambda} (E_c)_\lambda$$

$$= \frac{\alpha_\lambda}{1 - f_\lambda \alpha_\lambda} \frac{\varepsilon}{\varepsilon + (1 - \varepsilon) A_\lambda} E \cos \theta_\lambda. \quad (5.78)$$

The component in the direction of the external field is calculated as the inner product of p_d and the unit vector in the direction of the field and given by:

$$(p_d)_E = \sum_{\lambda=a,b,c} (p_d)_\lambda \cos\theta_\lambda = \sum_{\lambda=a,b,c} \frac{\alpha_\lambda}{1-f_\lambda\alpha_\lambda} \frac{\varepsilon}{\varepsilon+(1-\varepsilon)A_\lambda} E\cos^2\theta_\lambda. \quad (5.79)$$

In averaging (5.79) we may consider all orientations equally probable, which is equivalent to neglecting all terms with powers of E higher than the first. Using:

$$\overline{\cos^2\theta_\lambda} = \tfrac{1}{2}\int_0^\pi \cos^2\theta \sin\theta \, d\theta = \tfrac{1}{3}, \quad (5.80)$$

and the fact that the averages of the components perpendicular to $\boldsymbol{E}$ are zero, we find:

$$\bar{\boldsymbol{p}}_d = \tfrac{1}{3}\varepsilon\boldsymbol{E} \sum_{\lambda=a,b,c} \frac{\alpha_\lambda}{(1-f_\lambda\alpha_\lambda)\{\varepsilon+(1-\varepsilon)A_\lambda\}}$$

$$= \varepsilon\boldsymbol{E}\, \overline{\alpha_\lambda/[(1-f_\lambda\alpha_\lambda)\{\varepsilon+(1-\varepsilon)A_\lambda\}]}, \quad (5.81)$$

where the bar denotes an average over the three axes.

For ellipsoids whose eccentricity is not too large, the value of this expression will differ only very slightly from the corresponding expression for spherical particles, *i.e.* we may take:

$$\overline{\alpha_\lambda/[(1-f_\lambda\alpha_\lambda)\{\varepsilon+(1-\varepsilon)A_\lambda\}]} \cong \frac{\alpha}{(1-f\alpha)\{\varepsilon+(1-\varepsilon)\frac{1}{3}\}}, \quad (5.82)$$

where α and f denote the polarizability and the reaction field factor for a spherical particle with the same volume as the ellipsoidal particle under consideration. Using this approximation, we find:

$$\boldsymbol{P}_\alpha = \sum_k N_k\{(\bar{\boldsymbol{p}}_R)_k + (\bar{\boldsymbol{p}}_d)_k\}$$

$$= \sum_k N_k \left[\frac{(\alpha_a)_k(f_a)_k}{1-(\alpha_a)_k(f_a)_k} \frac{\mu_k^2}{3kT} \frac{\varepsilon\boldsymbol{E}}{\{1-(\alpha_a)_k(f_a)_k\}\{\varepsilon+(1-\varepsilon)(A_a)_k\}} + \right.$$

$$\left. + \frac{3\varepsilon\boldsymbol{E}\alpha_k}{(1-f_k\alpha_k)(2\varepsilon+1)} \right]. \quad (5.83)$$

Substitution of (5.75) and (5.83) into (5.3) now gives:

$$\frac{\varepsilon-1}{4\pi} = \sum_k N_k \left[\frac{3\varepsilon}{2\varepsilon+1} \frac{\alpha_k}{1-f_k\alpha_k} + \frac{\varepsilon}{\varepsilon+(1-\varepsilon)(A_a)_k} \frac{\mu_k^2}{3kT\{1-(f_a)_k(\alpha_a)_k\}^2} \right]. \quad (5.84)$$

This equation was first derived by Scholte;[18] it is valid for ellipsoidal particles, where the dipole moment is directed along one of the axes and the principal axes of the polarizability tensor coincide with the axes of the ellipsoid, and is independent of the place of the dipole in the molecule (*cf.* section 22). If the particles are spherical, eqn. (5.84) reduces to eqn. (5.60):

$$\frac{(\varepsilon - 1)(2\varepsilon + 1)}{12\pi\varepsilon} = \sum_k \frac{N_k}{1 - f_k\alpha_k}\left[\alpha_k + \frac{1}{3kT}\frac{\mu_k^2}{1 - f_k\alpha_k}\right].$$

To obtain an expression analogous to the Onsager equation, we apply eqn. (5.84) to the case of a pure dipole liquid:

$$\frac{\varepsilon - 1}{4\pi} = N\left[\frac{3\varepsilon}{2\varepsilon + 1}\frac{\alpha}{1 - f\alpha} + \frac{\varepsilon}{\varepsilon + (1 - \varepsilon)A_a}\frac{\mu^2}{3kT(1 - f_a\alpha_a)^2}\right], \tag{5.85}$$

and express f, α, f_a and α_a in the macroscopic properties of the liquid. To this end we follow Weaver and Parry,[19] who considered the volume of the cavity to be equal to the volume which is available to each molecule:

$$\frac{4\pi}{3}Nabc = 1. \tag{5.86}$$

This approximation is equivalent to eqn. (4.5), which was used to derive the Onsager equation and which follows from (5.86) for spherical molecules:

$$\frac{4\pi}{3}Na^3 = 1.$$

We further follow Weaver and Parry in considering the polarizability of the molecule to be uniformly distributed throughout the cavity. From this assumption it follows that E_d indeed can be considered to be homogeneous (compare the remark after eqn. (5.72)). Hence, we find the polarizability α_a from (2.105):

$$\alpha_a = \frac{\varepsilon_i - 1}{3\{1 + (\varepsilon_i - 1)A_a\}}abc = \frac{\varepsilon_i - 1}{4\pi N\{1 + (\varepsilon_i - 1)A_a\}}, \tag{5.87}$$

whereas the expression $1/(1 - f_a\alpha_a)$ is given by eqn. (4.102):

$$\frac{1}{1 - f_a\alpha_a} = \frac{\{\varepsilon + (1 - \varepsilon)A_a\}\{1 + (\varepsilon_i - 1)A_a\}}{\varepsilon + (\varepsilon_i - \varepsilon)A_a}.$$

The values of α and $1/(1 - f\alpha)$ are given by the corresponding expressions for spherical particles:

$$\alpha = \frac{3(\varepsilon_i - 1)}{4\pi N(\varepsilon_i + 2)}, \tag{5.88}$$

and

$$\frac{1}{1 - f\alpha} = \frac{(2\varepsilon + 1)(\varepsilon_i + 2)}{3(2\varepsilon + \varepsilon_i)}. \tag{5.89}$$

We must now express the internal dielectric constant ε_i in the macroscopic properties of the liquid. To this end we note that if we substitute $\mu = 0$ in eqn. (5.85), it reduces to the corresponding equation for ε_∞, the dielectric constant at a frequency high enough to exclude the orientational polarization, while the induced polarization is not yet affected. Substituting eqns. (5.88) and (5.89) in (5.85) with $\mu = 0$ and replacing ε by ε_∞, we obtain:

$$\frac{\varepsilon_\infty - 1}{4\pi} = N \frac{3\varepsilon_\infty}{2\varepsilon_\infty + 1} \frac{3(\varepsilon_i - 1)}{4\pi N(\varepsilon_i + 2)} \frac{(2\varepsilon_\infty + 1)(\varepsilon_i + 2)}{3(2\varepsilon_\infty + \varepsilon_i)}$$

$$= \frac{3\varepsilon_\infty(\varepsilon_i - 1)}{4\pi(2\varepsilon_\infty + \varepsilon_i)}, \tag{5.90}$$

and from this it follows that:

$$\varepsilon_i = \varepsilon_\infty. \tag{5.91}$$

Using eqns. (5.88), (5.89), (4.102), and the identity (5.91), we can write (5.85) in the form:

$$\frac{\varepsilon - 1}{4\pi} = \frac{3\varepsilon}{2\varepsilon + 1} \frac{3}{4\pi} \frac{\varepsilon_\infty - 1}{\varepsilon_\infty + 2} \frac{(2\varepsilon + 1)(\varepsilon_\infty + 2)}{3(2\varepsilon + \varepsilon_\infty)} +$$

$$+ \frac{N\mu^2}{3kT} \frac{\varepsilon}{\varepsilon + (1 - \varepsilon)A_a} \frac{\{\varepsilon + (1 - \varepsilon)A_a\}^2 \{1 + (\varepsilon_\infty - 1)A_a\}^2}{\{\varepsilon + (\varepsilon_\infty - \varepsilon)A_a\}^2}, \tag{5.92}$$

or:

$$\mu^2 = \frac{3kT}{4\pi N} \frac{\{\varepsilon + (\varepsilon_\infty - \varepsilon)A_a\}^2}{\{\varepsilon + (1 - \varepsilon)A_a\}\{1 + (\varepsilon_\infty - 1)A_a\}^2} \left[\frac{\varepsilon - 1}{\varepsilon} - \frac{3(\varepsilon_\infty - 1)}{2\varepsilon + \varepsilon_\infty}\right]$$

$$= \frac{3kT}{4\pi N} \frac{(\varepsilon - \varepsilon_\infty)(2\varepsilon + 1)}{\varepsilon(2\varepsilon + \varepsilon_\infty)} \frac{\{\varepsilon + (\varepsilon_\infty - \varepsilon)A_a\}^2}{\{\varepsilon + (1 - \varepsilon)A_a\}\{1 + (\varepsilon_\infty - 1)A_a\}^2}. \tag{5.93}$$

This is the generalization of Onsager's equation for ellipsoidal molecules. For spherical molecules, one has $A_a = \frac{1}{3}$, and (5.93) reduces to (5.67):

$$\mu^2 = \frac{9kT}{4\pi N} \frac{(\varepsilon - \varepsilon_\infty)(2\varepsilon + \varepsilon_\infty)}{\varepsilon(\varepsilon_\infty + 2)^2}.$$

Table 18 gives dipole moments calculated with eqn. (5.93) for those compounds of Table 16 and Table 17, for which the dipole is directed along one of the axes of the ellipsoid. The estimated ratios of the axes are based on the interatomic distances and the atomic radii; these estimates are taken from the literature. It must be mentioned that the methods by which these ratios are calculated by different authors are not the same, leading to slight differences in the values of b/a and c/a. In Table 18 μ_{ons} denotes the dipole moment calculated with (5.67), and μ_{ell} the dipole moment calculated with

TABLE 18

DIPOLE MOMENTS CALCULATED FROM EQN. (5.93)

Compound	b/a	c/a	ref.	μ_{ons}	μ_{ell}	μ_{gas}
Chloromethane	0.8	0.8	a, b, c	1.74	1.86	1.87
Bromomethane	0.7	0.7	a, b, c	1.55	1.86	1.81
Iodomethane	0.8	0.8	b, c	1.27	1.43	1.62
Chloroform	1.4	1.4	a	1.18	1.07	1.01
Bromoform	1.6	1.6	d	0.92	0.82	0.99
Nitromethane	0.9	0.9	a	3.44	3.56	3.46
Cyanomethane	0.7	0.7	a, c	3.39	3.83	3.92
2,2-Dichloropropane	1.2	1.2	a	2.45	2.31	2.27
Propanone	1.3	1.3	a	3.03	2.78	2.88
Trimethylamine	1.8	1.8	b	0.66	0.61	0.61
2-Chloro-2-methylpropane	1.1	1.1	b	2.39	2.31	2.13
Ethoxyethane	2.0	2.0	a, b	1.40	1.19	1.15
Ethylthioethane	1.8	1.8	a	1.64	1.38	1.54
Chlorobenzene	0.8	0.4	b, d	1.40	1.67	1.67
Bromobenzene	0.8	0.4	e	1.33	1.59	1.70
Iodobenzene	0.7	0.4	e	1.05	1.25	1.70
Nitrobenzene	0.8	0.4	b, d	4.06	5.44	4.22
Triethylamine	1.7	1.7	a	0.66	0.62	0.66
Cyanobenzene	0.8	0.4	b	3.48	4.44	4.18
Hydrogen cyanide	0.6	0.6	b	5.66	6.59	2.98

[a] Estimate of Buckley and Maryott.[20]
[b] Estimate of Weaver and Parry.[19]
[c] Estimate of Abbott and Bolton.[21]
[d] Estimate of Scholte.[18]
[e] For b/a the estimate of Buckley and Maryott is used, for c/a the same value is taken as for chlorobenzene.

(5.93). For the calculation of μ_{ell} the same values of ε, d, and ε_∞ are used as for the calculation of μ_{ons}.

It appears that for $a > b$ and $a > c$, μ_{ell} is larger than μ_{ons} and that for $a < b$ and $a < c$, μ_{ell} is smaller than μ_{ons}. As a rule, the values of μ_{ell} give a better agreement with μ_{gas} than the values of μ_{ons}. A number of cases in which the agreement between μ_{ell} and μ_{gas} is still unsatisfactory must, however, be noted.

1. For hydrogen cyanide, both μ_{ons} and μ_{ell} differ greatly from μ_{gas}. This must be explained from the association between the molecules. On account of the formation of hydrogen bonds, the environment of a molecule cannot be approximated by a continuum.
2. For nitrobenzene, the agreement between μ_{ons} and μ_{gas} is much better than the agreement between μ_{ell} and μ_{gas}. The poor agreement between μ_{ell} and μ_{gas} is mostly explained by supposing association between the nitrobenzene molecules to occur, which makes the continuum approach for the molecular environment inapplicable. The good agreement between μ_{ons} and μ_{gas} is then supposed to come from the circumstance that the errors made by considering the molecule as spherical and by considering the molecular environment as a continuum, compensate each other (see also section 43, p. 298).
3. For the iodo-compounds iodomethane and iodobenzene, the agreement between μ_{ell} and μ_{gas} is better than μ_{ons} and μ_{gas}, but is still rather poor. In this case our model of a polarizability uniformly distributed over the whole cavity is not very accurate. The polarizability of these molecules is rather high, as appears from the high values of ε_∞; since this high polarizability is due to the iodium atom, the polarizability cannot be homogeneous.
4. Remarkable deviations between μ_{ell} and μ_{gas} also occur for triethylamine, ethylthioethane, and bromoform. For triethylamine, the factor $(\varepsilon - \varepsilon_\infty)$ in (5.93) is very small, and therefore the error in the atomic polarization has a strong influence. In the case of ethylthioethane, the model used for the calculation of b/a and c/a may be inappropriate, because it considers the aliphatic chains to be extended as far as possible; nevertheless this model gives good results for ethoxyethane. No explanation is put forward for the deviations in the case of bromoform.

Sometimes better results are obtained by special assumptions. For the iodo-compounds, one might prefer to consider the polarizability as isotropic, instead of equal to the polarizability of a homogeneous ellipsoid with the dimensions of the cavity. In that case one has:

$$\alpha_a = \alpha. \tag{5.94}$$

If we assume eqn. (5.85) to be valid for this case, and then substitute (5.88), (5.89), (5.94), and (4.45):

$$f_a = \frac{3}{abc} \frac{A_a(1 - A_a)(\varepsilon - 1)}{\varepsilon + (1 - \varepsilon)A_a},$$

and further use (5.91), we have:

$$\frac{\varepsilon - 1}{4\pi} = \frac{3\varepsilon}{2\varepsilon + 1} \frac{3}{4\pi} \frac{\varepsilon_\infty - 1}{\varepsilon_\infty + 2} \frac{(2\varepsilon + 1)(\varepsilon_\infty + 2)}{3(2\varepsilon + \varepsilon_\infty)} + \frac{N\mu^2}{3kT} \frac{\varepsilon}{\varepsilon + (1 - \varepsilon)A_a} \cdot$$
$$\cdot \left[\frac{(\varepsilon_\infty + 2)\{\varepsilon + (1 - \varepsilon)A_a\}}{(\varepsilon_\infty + 2)\{\varepsilon + (1 - \varepsilon)A_a\} - (\varepsilon_\infty - 1)A_a(1 - A_a)(\varepsilon - 1)} \right]^2, \tag{5.95}$$

or:

$$\mu^2 = \frac{3kT}{4\pi N} \frac{(2\varepsilon + 1)(\varepsilon - \varepsilon_\infty)}{\varepsilon(2\varepsilon + \varepsilon_\infty)(\varepsilon_\infty + 2)^2} \cdot$$
$$\cdot \frac{[(\varepsilon_\infty + 2)\{\varepsilon + (1 - \varepsilon)A_a\} - (\varepsilon_\infty - 1)A_a(1 - A_a)(\varepsilon - 1)]^2}{\varepsilon + (1 - \varepsilon)A_a}. \tag{5.96}$$

In Table 19 values of μ calculated from (5.96) are compared with values calculated from (5.93) and (5.67) and the gas value. For iodomethane, (5.96) gives a good result, but for iodobenzene it is still 15% too low.

TABLE 19

VALUES OF μ FOR TWO IODO-COMPOUNDS, CALCULATED WITH VARIOUS EQUATIONS

Compound	(5.67)	(5.93)	(5.96)	μ_{gas}
Iodomethane	1.27	1.43	1.64	1.62
Iodobenzene	1.05	1.25	1.44	1.70

§31. The Debye equation

We have already mentioned that before Onsager's publication of 1936, the internal field $\boldsymbol{E}_\mathrm{i}$ and the directing field $\boldsymbol{E}_\mathrm{d}$ were not distinguished from each other. Following Debye,[22] one used for both fields the Lorentz field, given by eqn. (5.22):

$$\boldsymbol{E}_{\mathrm{L}} = \frac{\varepsilon + 2}{3}\boldsymbol{E}.$$

If one substitutes this expression in the fundamental equation, eqn. (5.16), for both $\boldsymbol{E}_{\mathrm{i}}$ and $\boldsymbol{E}_{\mathrm{d}}$, one finds:

$$\frac{\varepsilon - 1}{\varepsilon + 2} = \frac{4\pi}{3}\sum_k N_k\left(\alpha_k + \frac{\mu_k^2}{3kT}\right). \tag{5.97}$$

This is generally called the Debye equation.

Using the molar polarization $[P]$, defined by (5.40):

$$[P] = \frac{\varepsilon - 1}{\varepsilon + 2}\frac{M}{d},$$

we can write the Debye equation for a pure compound as:

$$[P] = \frac{4\pi}{3}N_{\mathrm{A}}\left(\alpha + \frac{\mu^2}{3kT}\right). \tag{5.98}$$

Thus, according to the Debye equation, the molar polarization of a compound at a given temperature is a constant. For example, it must be independent of the pressure and have the same value in the gaseous and the liquid states. In many cases, however, the Debye equation is in considerable disagreement with the experiment; for instance in the case of polar liquids it is of doubtful value to use the concept of the molar polarization, because we know from Onsager's theory that for polar dielectrics it is not permissible to use the Lorentz field.

Nevertheless, for gases at normal pressures the Debye equation gives a fairly good approximation. In this case one has $\varepsilon - 1 \ll 1$ (*cf.* Table 4), and eqn. (5.97) can be written as:

$$\varepsilon - 1 = 4\pi\sum_k N_k\left(\alpha_k + \frac{\mu_k^2}{3kT}\right). \tag{5.99}$$

In this form, the Debye equation is still used. For gases at normal pressures eqn. (5.60), which is based on Onsager's theory, gives the same result. In fact, for gases at normal pressures, the differences between the fields $\boldsymbol{E}$, $\boldsymbol{E}_{\mathrm{i}}$, $\boldsymbol{E}_{\mathrm{d}}$, and $\boldsymbol{E}_{\mathrm{L}}$ can be neglected.

For pure compounds, eqn. (5.99) changes into:

$$\varepsilon - 1 = 4\pi N\left(\alpha + \frac{\mu^2}{3kT}\right). \tag{5.100}$$

By measuring at different temperatures, one can determine α and μ. The values of μ found in this way are not dependent on the model used for the calculation of $\boldsymbol{E}_i$ and $\boldsymbol{E}_d$, because we can use for both fields in a good approximation the Maxwell field $\boldsymbol{E}$. Therefore values of μ determined in the gas phase were used in the preceding sections to check the results of the eqns. (5.67), (5.93) and (5.96). The values of μ determined with (5.100) also do not depend on the values chosen for the atomic polarization. Because α includes both parts of the induced polarization it is possible to compare α with α^e, as found from the refractive index. With this method values of $\alpha^a = \alpha - \alpha^e$ can be calculated. However, the accuracy of the experimental results proves to be too low to give good results for the rather small differences between α and α^e.

Eqn. (5.97) gives also a reasonable approximation for diluted solutions of polar molecules in non-polar solvents. In this way the dipole moments of non-volatile compounds can be determined. This will be treated extensively in Volume II.

Before Onsager's 1936 publication, many other attempts were made to explain the deviations of the Debye equation in the case of polar liquids. Many of these authors assumed that the apparent dipole moment was lowered by association. Van Arkel and Snoek[23] suggested the following modification of the Debye equation:

$$\frac{\varepsilon - 1}{\varepsilon + 2} = \frac{4\pi}{3} N\left(\alpha + \frac{\mu^2}{3kT + cN\mu^2}\right), \tag{5.101}$$

where c is a constant. This equation was based on the idea that the mutual dipole interaction would counteract the directing action of the external field in a manner similar to the effects caused by thermal movements. The mutual dipole interaction energy is proportional to μ^2/r^3, and thus to $N\mu^2$. Van Arkel and Snoek showed, that eqn. (5.101) is in agreement with the experimental data.

In 1938, Böttcher[24] showed that the semi-empirical formula (5.101) follows immediately from Onsager's hypothesis, that only a part of the internal field directs the dipole.

§32. A correction to the Clausius-Mossotti equation

In section 27 it was shown, that it is possible to derive the Clausius-Mossotti equation by considering the internal field as composed of the cavity field and the reaction field of the induced dipole moment, and taking for the radius of the cavity Onsager's approximation (eqn. (4.5)):

$$\frac{4\pi}{3}Na^3 = 1.$$

It was noticed by Böttcher[25] in 1942, however, that for gases eqn. (4.5) gives a value for the radius of the cavity that is much greater than the distance at which two molecules can approach each other. Therefore, Böttcher proposed that a fixed radius, independent of the density, be used for each kind of molecule.

To calculate for this model the relation between ε and α, we start with eqn. (5.36), which holds independent of the assumption made for the radius a of the cavity:

$$(\boldsymbol{E}_{\mathrm{i}})_k = \frac{3\varepsilon}{2\varepsilon + 1 - 2\frac{\alpha_k}{a_k^3}(\varepsilon - 1)}\boldsymbol{E}. \tag{5.102}$$

Substituting this into eqn. (5.17):

$$\frac{\varepsilon - 1}{4\pi}\boldsymbol{E} = \sum N_k \alpha_k (\boldsymbol{E}_{\mathrm{i}})_k,$$

we find:

$$\frac{\varepsilon - 1}{12\pi\varepsilon} = \sum_k \frac{N_k \alpha_k}{2\varepsilon + 1 - 2\frac{\alpha_k}{a_k^3}(\varepsilon - 1)}. \tag{5.103}$$

Defining $u = \alpha/a^3$, we can write (5.103) for a pure compound as:

$$\frac{\varepsilon - 1}{12\pi\varepsilon} = \frac{N\alpha}{2\varepsilon + 1 - 2u(\varepsilon - 1)}. \tag{5.104}$$

From this equation the molar polarization $[P]$, defined by (5.40):

$$[P] = \frac{\varepsilon - 1}{\varepsilon + 2}\frac{M}{d},$$

can be calculated as:

$$[P] = \frac{4\pi}{3}N_{\mathrm{A}}\alpha\frac{9\varepsilon}{(\varepsilon + 2)\{2\varepsilon - 1 - 2u(\varepsilon - 1)\}}. \tag{5.105}$$

According to the Clausius-Mossotti theory, the molar polarization would be given by eqn. (5.41):

$$[P] = \frac{4\pi}{3} N_A \alpha.$$

Denoting this ideal value of the molar polarization as $[P]_0$, we can write (5.105) as:

$$[P] = h[P]_0, \tag{5.106}$$

where the correction factor h is given by:

$$h = \frac{9\varepsilon}{(\varepsilon + 2)\{2\varepsilon + 1 - 2u(\varepsilon - 1)\}}. \tag{5.107}$$

For $\varepsilon = 1$ we have $h = 1$ and for values of ε slightly different from 1 the factor h is virtually equal to 1. For larger values of ε, however, the correction is noticeable. In Fig. 28 the factor h is given as a function of ε for two values of u, *viz.* $u = 0.25$ and $u = 0.50$. As already mentioned in section 27, the ratio $u = \alpha/a^3$ is always smaller than 1.

It appears from Fig. 28 that the factor h initially increases with increasing ε, and reaches a maximum value h_m. The corresponding value ε_m of the dielectric constant is calculated from $\partial h/\partial\varepsilon = 0$ to be:

$$\varepsilon_m = \sqrt{\frac{1 + 2u}{1 - u}}. \tag{5.108}$$

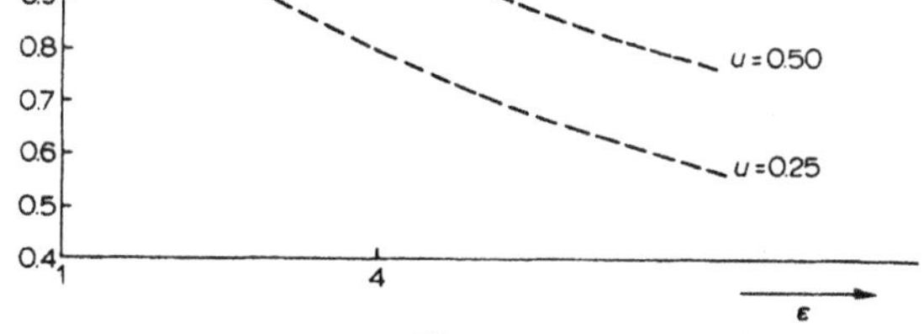

Fig. 28
The factor h for $u = 0.25$ and $u = 0.50$.

From (5.107) it follows, that the value $h = 1$ is obtained again, if:

$$\frac{\varepsilon - 1}{\varepsilon + 2} = u, \tag{5.109}$$

or:

$$\varepsilon = \frac{1 + 2u}{1 - u}. \tag{5.110}$$

This value of ε is equal to the internal dielectric constant of a homogeneous particle with the same polarizability and volume as the molecule. This appears from eqn. (2.101):

$$\alpha = \frac{\varepsilon - 1}{\varepsilon + 2} a^3.$$

The dielectric constant of the sample reaches the value of the internal dielectric constant if the density is so high that the radius of the molecular sphere is the same as the Onsager radius. Because a higher density and thus a higher dielectric constant could only be reached if the volume of a molecule were larger than the volume available for each molecule, in Fig. 28 the parts of the curves corresponding to $h < 1$ are dashed. In Table 20 the values of ε_m and h_m are given for some values of u.

From Fig. 28 and Table 20 we see that the variability of [P] increases with the value of u: for $u = 0.25$ the maximum increase of [P] is only 3 %, whereas for $u = 0.75$ it is as high as 35 %.

As an application of eqn. (5.105), we shall examine the dependence of the molar polarization of gaseous carbon dioxide on the pressure, as determined experimentally by Michels and Kleerekoper.[26] At $t = 50\,°C$ and $t = 100\,°C$ they found that [P] increases with increasing density, reaches a maximum for $\varepsilon = 1.35$, and then decreases (Fig. 29). This is expected from

TABLE 20

THE MAXIMUM OF h

$u = \alpha/a^3$	ε_m	h_m
0.25	1.41	1.031
0.50	2.00	1.125
0.75	3.16	1.351

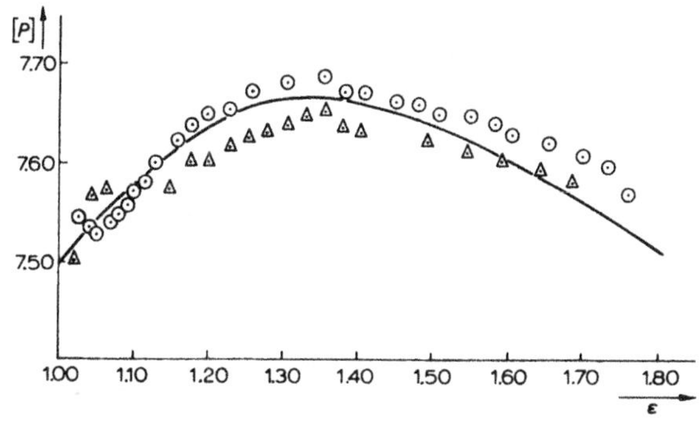

Fig. 29

The molar polarization of carbon dioxide as a function of ε. Measurements of Michels and Kleerekoper[26] ($\odot: t = 50°C$; $\triangle: t = 100°C$). The curve gives calculated values for $u = 0.215$.

eqn.(5.105) if we suppose α and a to be independent of pressure. If we apply eqn. (5.105) to the measurements of Michels and Kleerekoper, using constant values of α and a, we calculate from $\varepsilon_m = 1.35$ with the aid of (5.108), that $u = \alpha/a^3 = 0.215$. In Fig. 29 the curve gives the values of $[P]$ calculated with this value of u, and these prove to be in fair agreement with the measurements.

The fact that the points for different temperatures do not wholly coincide, is explicable because the assumption that α and a are independent of the pressure, is only an approximation. It is known[26] that both α and a decrease with increasing pressure. Although in the above application of (5.105) both effects partly annihilate each other, one may prefer to apply (5.105) in the way indicated by De Groot,[27] who uses values of a obtained from experimental data and calculates with the aid of (5.105) the polarizability α as a function of the pressure. For carbon dioxide the polarizability proves to be decreased by 18% at a pressure of 2000 atm, a value which is in agreement with the theoretical results of Michels, De Boer and Bijl.[28]

From the value $u = 0.215$ and the molar polarization at normal pressures, which gives $\alpha = 3.00$ Å^3, we calculate for the radius of the molecule: $a = 2.4$ Å. This value is in good agreement with the value 2.3 Å calculated from the molar volume of the liquid, and the value 2.0 Å calculated from the mutual separation of the molecules in the crystal.[29]

As mentioned in section 27, no deviations from the Clausius-Mossotti equation were found for hydrogen gas, even at a pressure of 1400 atm. This is in accordance with eqn. (5.105). In the case of hydrogen u is very small, and from Table 20 we see that h_m rapidly decreases with decreasing u. The change of $[P]$ predicted by (5.105) is in this case within the experimental error. For gases with large values of α/a^3 (*e.g.* oxygen, bromine, and particularly carbon disulfide), we can expect much larger deviations from the Clausius-Mossotti equation than in the case of carbon dioxide.

A method[25,30] for the calculation of both α and a for the molecule and which is also applicable for liquids, starts from eqn. (5.104):

$$\frac{\varepsilon - 1}{12\pi\varepsilon} = \frac{N\alpha}{2\varepsilon + 1 - 2u(\varepsilon - 1)}.$$

Inverting both members we have:

$$\frac{12\pi\varepsilon}{\varepsilon - 1} = \frac{2\varepsilon + 1 - 2u(\varepsilon - 1)}{N\alpha}, \tag{5.111}$$

or with $N = N_A d/M$:

$$\frac{12\pi\varepsilon d}{(\varepsilon - 1)(2\varepsilon + 1)} = \frac{M}{N_A\alpha} - \frac{M}{N_A a^3}\,\frac{2(\varepsilon - 1)}{2\varepsilon + 1}. \tag{5.112}$$

Thus by plotting $12\pi\varepsilon d/(\varepsilon - 1)(2\varepsilon + 1)$ against $2(\varepsilon - 1)/(2\varepsilon + 1)$ we find a straight line, cutting a length $M/N_A\alpha$ from the axis, and with slope $M/N_A a^3$.

As an example, we give in Fig. 30 the graph for carbon tetrachloride, based on values of ε given by Hartmann, Neumann and Rinck,[31] and on values of d derived from compressibility data given by Gibson and Loeffler.[32] We find $\alpha = 10.8$ Å^3 and $a = 3.05$ Å. The value $\alpha = 10.8$ Å^3 is in good agreement with the value $\alpha = 10.5$ Å^3 given by Stuart.[29] In the range of the measurements, the Onsager radius varies from $a = 3.27$ Å up to $a = 3.42$ Å. If we assume that at the highest density at which measurements were done the volume occupied by the molecules amounts to 64% of the total volume, which is the maximum value found for a chaotic distribution of hard spheres,[33] we calculate the molecular radius to be $a = 2.82$ Å. Hence, the effective radius in eqn. (5.112) lies in between the Onsager radius and the radius calculated for a chaotic distribution of hard spheres.

This method presupposes that for a certain value of the density, the dielectric constant is independent of the temperature and hence also of the pressure. For many non-polar liquids this condition is fulfilled. Theoretically, it is then possible to use only data obtained at 1 atm and at different tem-

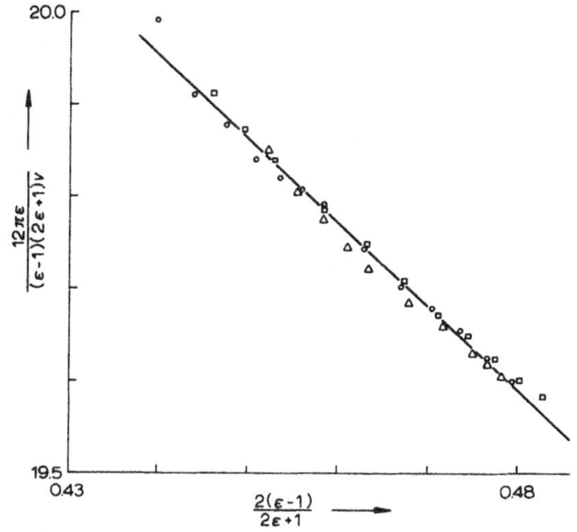

Fig. 30

The relation between $12\pi\varepsilon d/(\varepsilon - 1)(2\varepsilon + 1)$ and $2(\varepsilon - 1)/(2\varepsilon + 1)$ for liquid CCl_4. Based on dielectric measurements of Hartmann et al.[31] and compressibility data of Gibson and Loeffler.[32]

peratures, as was done by Pijpers.[34] One objection must be raised to this use of the method, however, because the supposition that the dielectric constant depends only on temperature by the temperature dependence of the density, is not checked in this way. For instance, Pijpers also applied this method for hexane, for which compound Mopsik[35] has demonstrated that at constant density the dielectric constant decreases with temperature, probably due to the fact that the hexane molecule has a very small permanent dipole moment.

§33. The correction to the Clausius-Mossotti equation for ellipsoidal molecules

For the derivation of the correction to the Clausius-Mossotti equation in the preceding section, it was assumed that the molecules have a spherical shape. It was pointed out by Sicotte,[36] however, that the influence of the shape of the molecules on this correction can be considerable. To investigate this influence, we shall assume that the shape of the molecules is ellipsoidal.

In section 30 we found that the average value of the moment induced by the directing field in an ellipsoidal molecule, is given by (5.81):

$$\overline{p_d} = \varepsilon E \ \overline{\alpha_\lambda/[(1-f_\lambda\alpha_\lambda)\{\varepsilon+(1-\varepsilon)A_\lambda\}]}. \tag{5.113}$$

For non-polar systems, this expression gives the average value of the total moment of a molecule, because the permanent moment is zero and there is no moment induced by the reaction field of this permanent moment. Hence:

$$\frac{\varepsilon-1}{4\pi}E = P = N\overline{p}_d = N\varepsilon E \ \overline{\alpha_\lambda/[(1-f_\lambda\alpha_\lambda)\{\varepsilon+(1-\varepsilon)A_\lambda\}]}. \tag{5.114}$$

As already remarked in section 30, the influence of the deviations of the spherical shape on the value of this expression is small, *i.e.*:

$$\overline{\alpha_\lambda/[(1-f_\lambda\alpha_\lambda)\{\varepsilon+(1-\varepsilon)A_\lambda\}]} \cong \frac{\alpha}{(1-f\alpha)\{\varepsilon+(1-\varepsilon)\frac{1}{3}\}}.$$

The influence of the deviations from the spherical shape on the derivative of $12\pi\varepsilon d/(\varepsilon-1)(2\varepsilon+1)$ with respect to $2(\varepsilon-1)/(2\varepsilon+1)$ is much larger, however, as appears from the following calculation.

From eqn. (5.114) we find:

$$\frac{12\pi\varepsilon d}{(\varepsilon-1)(2\varepsilon+1)} = \frac{M}{N_A}\frac{3}{2\varepsilon+1}\frac{1}{\overline{\alpha_\lambda/[(1-f_\lambda\alpha_\lambda)\{\varepsilon+(1-\varepsilon)A_\lambda\}]}}. \tag{5.115}$$

Using the variable $x = 2(\varepsilon-1)/(2\varepsilon+1)$, this can be written as:

$$\frac{12\pi\varepsilon d}{(\varepsilon-1)(2\varepsilon+1)}$$

$$= \frac{M}{N_A}\frac{1-x}{\overline{\alpha_\lambda/\left[1-\dfrac{3\alpha_\lambda A_\lambda(1-A_\lambda)3x/2(1-x)}{abc\left\{\dfrac{x+2}{2(1-x)}-\dfrac{3x}{2(1-x)}A_\lambda\right\}}\right]\left\{\dfrac{x+2}{2(1-x)}-\dfrac{3x}{2(1-x)}A_\lambda\right\}}}$$

$$= \frac{M}{N_A}\frac{1}{\overline{2\alpha_\lambda/\{x+2-3xA_\lambda-9\alpha_\lambda A_\lambda(1-A_\lambda)x/abc\}}}. \tag{5.116}$$

It is easily seen that for $\varepsilon \to 1$ or $x \to 0$ this expression approaches the value $M/N_A\bar{\alpha}$. It further appears that for $A_a = A_b = A_c = \frac{1}{3}$ and $\alpha_a = \alpha_b = \alpha_c = \alpha$, eqn. (5.116) changes into eqn. (5.112). To investigate the behaviour of (5.116) for other cases, values for the righthand member multiplied by $N_A abc/M$ are given in Table 21.

TABLE 21

VALUES OF $abc\left[2\alpha_\lambda\left\{x+2-3xA_\lambda-9\frac{\alpha_\lambda}{abc}A_\lambda(1-A_\lambda)x\right\}\right]^{-1}$ AS A FUNCTION OF x

x	homogeneous spheroids		$\alpha_a/\alpha_b = a/b$		CS_2
	$\varepsilon_i = 2$	$\varepsilon_i = 5$	$\bar{\alpha}/ab^2 = 0.25$	$\bar{\alpha}/ab^2 = 0.571$	
0	3.97	1.67	4.00	1.75	1.32
0.1	3.88	1.59	3.93	1.67	1.24
0.2	3.79	1.51	3.86	1.59	1.15
0.3	3.70	1.42	3.79	1.51	1.06
0.4	3.60	1.33	3.71	1.43	0.97
0.5		1.24		1.34	0.88
0.6		1.15		1.26	0.79
0.7		1.05		1.17	0.70

The following cases were taken:

1. Homogeneously polarizable spheroids with $a/b = a/c = 2$ and $\varepsilon_i = 2$ and $\varepsilon_i = 5$, respectively; the use of homogeneously polarizable spheroids makes this case comparable with the generalization of Onsager's theory for ellipsoidal molecules in section 30.
2. Spheroidal particles with $a/b = a/c = 2$ and with the same ratios of the principal polarizabilities $\alpha_a/\alpha_b = \alpha_a/\alpha_c = 2$, for the cases $\bar{\alpha}/ab^2 = 0.25$ and $\bar{\alpha}/ab^2 = 0.571$; the assumption of equal ratios between the axes of the particle and the principal polarizabilities was made by Sicotte and must be distinguished from the first case.
3. The CS_2-molecule; principal polarizabilities $\alpha_a = 15.14$ Å^3, $\alpha_b = \alpha_c = 5.54$ Å^3, according to Stuart,[29] and axes corresponding to the dimensions of the molecule as found from the Van der Waals radius of the sulfur atom and the interatomic distance C–S, both given by Pauling,[37] $a = 3.38$ Å, $b = 1.85$ Å.

The righthand member of eqn. (5.116) multiplied by $N_A abc/M$ reduces in the case of spherical particles to $a^3/\alpha - x$. In this case the derivative with respect to x is equal to -1. It appears from Table 21 that for ellipsoidal particles the derivative deviates considerably from this value. Therefore, it is necessary to correct for the deviations of the spherical shape if one wants to calculate the effective volume in the way indicated in section 32. The correction depends on the assumption made for the relation between the shape of the molecule and the ratio between the principal polarizabilities.*

§34. Mixtures of non-polar compounds

Both the Clausius-Mossotti equation and the corrected Clausius-Mossotti equation can be used for the calculation of the dielectric constant of mixtures of non-polar compounds.

Combining the Clausius-Mossotti equation for mixtures (5.29):

$$\frac{\varepsilon - 1}{\varepsilon + 2} = \frac{4\pi}{3} \sum_k N_k \alpha_k,$$

and the Clausius-Mossotti equation for pure compounds (5.30):

$$\frac{\varepsilon - 1}{\varepsilon + 2} = \frac{4\pi}{3} N\alpha,$$

one finds for the mixture:

$$\frac{\varepsilon_m - 1}{\varepsilon_m + 2} = \frac{4\pi}{3} \sum_k N_k \left[\frac{3}{4\pi N}\frac{\varepsilon - 1}{\varepsilon + 2}\right]_k = \sum_k \varphi_k \frac{\varepsilon_k - 1}{\varepsilon_k + 2}, \tag{5.117}$$

in which ε_m is the dielectric constant of the mixture and ε_k the dielectric constant of the k-th component in the pure state. φ_k represents the volume fraction of the k-th component: $\varphi_k = (N_k)_m/(N_k)_p$ where $(N_k)_m$ is the number of molecules of the k-th component in 1 cm^3 of the mixture and $(N_k)_p$ the number of molecules in 1 cm^3 of the pure compound. The quantities φ_k give the volumes which must be mixed to obtain 1 cm^3.

In the case of the corrected Clausius-Mossotti equation the procedure is more intricate. One then starts from eqn. (5.103):

$$\frac{\varepsilon_m - 1}{12\pi\varepsilon_m} = \sum_k \frac{N_k \alpha_k}{2\varepsilon_m + 1 - 2\frac{\alpha_k}{a_k^3}(\varepsilon_m - 1)}. \tag{5.118}$$

* If the correction for the deviations of the spherical shape is applied for carbon disulfide, however, the large deviations found by Mopsik[38] for this compound are only slightly diminished. The reason for this behaviour is not clear; it must be noted that for carbon disulfide the molar polarization in the liquid state is lower than the gas value.

In this equation, α_k and a_k must be determined from series of measurements on the pure components, as indicated in the previous sections. Subsequently, the implicit equation in ε_m must be solved, by calculating for various estimated values of ε_m the difference between both members of (5.118). It appears that the resulting values of ε_m differ very little from the values given by eqn. (5.117).[34]

Besides eqns. (5.117) and (5.118), which are based on molecular models, a number of empirical formulae[39–43] for the dielectric constant of non-polar mixtures have been suggested. Because the differences between the dielectric constants of various non-polar liquids are rather small, the results of the different formulae show only slight differences. This led Böttcher[44] to suggest instead of the more intricate formulas a simple additive relation:

$$\varepsilon_m = \sum_k \varphi_k \varepsilon_k . \tag{5.119}$$

The validity of the different expressions was investigated extensively by Pijpers.[34] In Tables 22, 23, and 24 his results are given for mixtures of carbon tetrachloride and benzene, carbon tetrachloride and cyclohexane, and benzene and cyclohexane, at 25 °C.

In the case of the carbon tetrachloride and benzene mixtures the difference between the various theoretical values is extremely small, due to the

TABLE 22

EXPERIMENTAL AND THEORETICAL VALUES OF THE DIELECTRIC CONSTANT OF MIXTURES OF CCl_4 AND BENZENE AT 25°C, ACCORDING TO PIJPERS[34]

Mole % CCl_4	Experimental	Theoretical	
		(5.117)	(5.119)
100.00	2.2285		
96.17	2.2317	2.2304	2.2304
84.74	2.2390	2.2344	2.2344
81.02	2.2409	2.2362	2.2362
48.56	2.2578	2.2507	2.2508
41.76	2.2613	2.2537	2.2537
27.85	2.2664	2.2602	2.2598
14.54	2.2718	2.2667	2.2666
5.01	2.2748	2.2719	2.2720
4.23	2.2750	2.2712	2.2721
0.00	2.2744		

TABLE 23

EXPERIMENTAL AND THEORETICAL VALUES OF THE DIELECTRIC CONSTANT OF MIXTURES OF CCl_4 AND CYCLOHEXANE AT 25°C, ACCORDING TO PIJPERS[34]

Mole % cyclohexane	Experimental	Theoretical	
		(5.117)	(5.119)
100.00	2.0174		
94.32	2.0277	2.0268	2.0269
88.33	2.0390	2.0374	2.0379
87.48	2.0395	2.0389	2.0394
77.34	2.0588	2.0574	2.0582
65.85	2.0814	2.0792	2.0804
63.07	2.0862	2.0837	2.0862
47.34	2.1190	2.1166	2.1180
30.39	2.1558	2.1542	2.1557
14.37	2.1938	2.1922	2.1932
5.73	2.2144	2.2134	2.2138
0.00	2.2285		

TABLE 24

EXPERIMENTAL AND THEORETICAL VALUES OF THE DIELECTRIC CONSTANT OF MIXTURES OF BENZENE AND CYCLOHEXANE AT 25°C, ACCORDING TO PIJPERS[34]

Mole % cyclohexane	Experimental	Theoretical	
		(5.117)	(5.119)
0.00	2.2744		
7.63	2.2482	2.2462	2.2466
12.44	2.2315	2.2292	2.2298
42.35	2.1409	2.1390	2.1394
43.78	2.1367	2.1352	2.1355
70.28	2.0729	2.0729	2.0726
85.51	2.0428	2.0425	2.0428
86.64	2.0421	2.0407	2.0403
98.95	2.0216	2.0193	2.0194
100.00	2.0174		

small difference between the dielectric constants of the components. The experimental values lie far higher than the range of the theoretical values. This is explained by the formation of molecular complexes of one CCl_4 and one benzene molecule, with a dipole moment of 0.2 *D*.

For mixtures of carbon tetrachloride and cyclohexane the differences between the theoretical values are markedly higher. Both for these mixtures and for the mixtures of benzene and cyclohexane, the experimental values are higher than one would expect from the Clausius-Mossotti equation. This effect is smaller than in the case of the carbon tetrachloride and benzene mixtures. It is interpreted as an indication that also in these mixtures polar molecular complexes may be formed.

It is clear that the occurrence of these polar molecular complexes is the limiting factor in the simple applicability of the continuum method to mixtures of non-polar compounds.

References

1. P. Langevin, *J. Phys.* (4) **4** (1905) 678.
2. P. Langevin, *Ann. Chim. Phys.* (8) **5** (1905) 70.
3. H. A. Lorentz, *The Theory of Electrons*, Dover, New York 1952, pp. 138, 306.
4. O. F. Mossotti, *Bibl. Univ. Modena* **6** (1847) 193, *Mem. di mathem. e fisica Modena*, **24** II (1850) 49.
5. R. Clausius, *Die mechanische Wärmetheorie* Vol. II, Braunschweig 1879, p. 62.
6. A. Michels, P. Sanders and A. Schipper, *Physica* **2** (1935) 753.
7. L. V. Lorenz, *Ann. Phys.* **11** (1880) 70.
8. H. A. Lorentz, *Ann. Phys.* **9** (1880) 641.
9. L. Onsager, *J. Am. Chem. Soc.* **58** (1936) 1486.
10. J. Timmermans, *Physico-chemical Constants of Pure Organic Compounds*, Elsevier, Amsterdam, Vol. I 1950, Vol. II 1965.
11. A. A. Maryott and E. R. Smith, *Table of Dielectric Constants of Pure Liquids*, N.B.S. Circular 514, Washington 1951.
12. R. D. Nelson Jr., D. R. Lide Jr., and A. A. Maryott, *Selected Values of Dipole Moments for Molecules in the Gas Phase*, N.S.R.D.S.-N.B.S. 10, Washington 1967.
13. S. O. Morgan and H. H. Lowry, *J. Phys. Chem.* **34** (1930) 2385.
14. A. V. Grosse, R. C. Wackher, and C. B. Linn, *J. Phys. Chem.* **44** (1940) 275.
15. J. Middelhoek, *Thesis* Leiden 1967.
16. K. Owen, O. R. Quayle, and E. M. Beavers, *J. Am. Chem. Soc.* **61** (1939) 900.
17. W. E. Doering, G. Cortes, and L. H. Knox, *J. Am. Chem. Soc.* **69** (1939) 1700.
18. Th. G. Scholte, *Thesis* Leiden 1950.
19. J. R. Weaver and R. W. Parry, *Inorg. Chem.* **5** (1966) 703.
20. F. Buckley and A. A. Maryott, *J. Res. Natl. Bur. St.* **53** (1954) 229.
21. J. A. Abbott and H. C. Bolton, *Trans. Faraday Soc.* **48** (1952) 422.
22. P. Debye, *Phys. Z.* **13** (1912) 97, *Coll. Papers*, Interscience, New York 1954, p. 173.
23. A. E. van Arkel and J. L. Snoek, *Phys. Z.* **33** (1932) 662, *ibid.* **35** (1934) 187, *Trans. Faraday Soc.* **30** (1934) 707.
24. C. J. F. Böttcher, *Physica* **5** (1938) 635.
25. C. J. F. Böttcher, *Physica* **9** (1942) 945.
26. A. Michels and L. Kleerekoper, *Physica* **6** (1939) 586.
27. S. R. de Groot, *Revue d'Optique* **28** (1949) 629.
28. A. Michels, J. de Boer and A. Bijl, *Physica* **4** (1937) 981.

29. H. A. Stuart, *Molekülstruktur*, 3rd ed., Springer, Berlin 1967.
30. C. J. F. Böttcher, *Rec. Trav. Chim.* **62** (1943) 325.
31. H. Hartmann, A. Neumann and G. Rinck, *Z. Phys. Chem. N.F.* **44** (1965) 204.
32. R. E. Gibson and O. H. Loeffler, *J. Am. Chem. Soc.* **63** (1941) 898.
33. G. D. Scott, *Nature* **188** (1960) 908.
34. F. W. Pijpers, *Thesis* Leiden 1958.
35. F. I. Mopsik, *J. Res. Natl. Bur. St.* **71A** (1967) 287.
36. Y. Sicotte, *J. Chim. Phys.* **63** (1966) 403.
37. L. Pauling, *The Nature of the Chemical Bond*, 3rd ed., Cornell University Press, Ithaca 1962.
38. F. I. Mopsik, *J. Chem. Phys.* **50** (1969) 2559.
39. J. H. Gladstone and J. Dale, *Phil Trans.* **153** (1863) 317.
40. Johst, *Ann. Phys.* **20** (1883) 47.
41. J. F. Eykman, *Recherches réfractométriques*, Haarlem 1919.
42. Dieterici, *Ann. Phys.* **67** (1922) 337.
43. K. Lichtenecker, *Phys. Z.* **25** (1924) 169, 193, 225.
44. C. J. F. Böttcher, *Thesis* Leiden 1940.

CHAPTER VI

STATISTICAL-MECHANICAL THEORIES OF THE DIELECTRIC CONSTANT

§35. Introduction

In contradistinction to the theories discussed in Chapter V, which use the continuum approach for the environment of the molecule, in this chapter the molecular structure of matter wi'l be taken into account explicitly. To this end we shall use the methods of statistical mechanics which provide a way of obtaining macroscopic quantities when the properties of the molecules and the molecular interactions are known.

In the statistical-mechanical theories of the dielectric constant simplified models are often used for the molecules and the intermolecular forces to make the calculations tractable. For example, a molecule is represented by an ideal dipole and a scalar polarizability or by an ideal dipole in the centre of a dielectric sphere, and the molecular interaction is taken to follow a hard-sphere or a Lennard-Jones potential. Even for these simplified models the calculations are often too difficult to perform, since the system comprises a very large number of molecules (of the order of 10^{23}) exerting long-range (dipole-dipole) forces on each other. Therefore, in many cases one has to be content with approximate expressions or series expansions with respect to the density.

All statistical-mechanical theories of the dielectric constant start from the consideration that the polarization $\mathscr{P}$, given in eqn. (2.43):

$$4\pi\mathscr{P} = \boldsymbol{D} - \boldsymbol{E},$$

is equal to the dipole density $\boldsymbol{P}$, when the influence of higher multipole densities may be neglected (compare sections 6 and 7). By definition, we may write for the dipole density $\boldsymbol{P}$ of a homogeneous system:

$$\boldsymbol{P}V = \langle \boldsymbol{M} \rangle, \tag{6.1}$$

where V is the volume of the dielectric under consideration and $\langle \boldsymbol{M} \rangle$ is its average total (dipole) moment (the brackets $\langle\,\rangle$ denote a statistical mechanical average). If we assume the system to be isotropic, we also have, according to eqn. (2.46):

$$\boldsymbol{D} = \varepsilon \boldsymbol{E}.$$

Thus we find:

$$(\varepsilon - 1)\boldsymbol{E} = \frac{4\pi}{V}\langle \boldsymbol{M} \rangle. \tag{6.2}$$

Since the dielectric is isotropic, $\langle \boldsymbol{M} \rangle$ will have the same direction as $\boldsymbol{E}$ and it will be sufficient to calculate the average component of $\boldsymbol{M}$ in the direction of $\boldsymbol{E}$. Using $\boldsymbol{e}$ to denote a unit vector in the direction of the field, we may therefore rewrite eqn. (6.2) in scalar form:

$$(\varepsilon - 1)E = \frac{4\pi}{V}\langle \boldsymbol{M} \cdot \boldsymbol{e} \rangle = \frac{4\pi}{V}\langle \boldsymbol{M} \rangle \cdot \boldsymbol{e}. \tag{6.3}$$

Another way of writing eqn. (6.3), which is often useful, results from the fact that $\boldsymbol{P}$ and $\langle \boldsymbol{M} \rangle$ contain in general also terms in higher powers of E than the first (see eqn. (2.47)). Thus, $(\varepsilon - 1)E/4\pi$ is the first term in a series development of P in powers of E, and must be set equal to the term linear in E of the series development of $\langle \boldsymbol{M} \cdot \boldsymbol{e} \rangle / V$ in powers of E. Since variations in V due to electrostriction do not appear in the linear term (see section 45), we may develop $\langle \boldsymbol{M} \cdot \boldsymbol{e} \rangle$ in a Taylor series, finding:

$$\varepsilon - 1 = \frac{4\pi}{V}\left(\frac{\partial}{\partial E}\langle \boldsymbol{M} \cdot \boldsymbol{e} \rangle\right)_{E=0}.$$

Rewriting with the external field E_0 instead of the Maxwell field E as the independent variable we obtain:

$$\varepsilon - 1 = \frac{4\pi}{V}\left(\frac{\partial E_0}{\partial E}\right)_{E=0}\left(\frac{\partial}{\partial E_0}\langle \boldsymbol{M} \cdot \boldsymbol{e} \rangle\right)_{E_0=0}. \tag{6.4}$$

It is clear that before an expression for the dielectric constant in terms of molecular quantities can be obtained two problems have to be solved.

1. E, the Maxwell field, has to be expressed as a function of the external field E_0. In some special cases electrostatic theory leads to a simple relation between E_0 and E, which can be used immediately. For instance, when the dielectric is contained between the plates of an ideal plane capacitor we have for E_0, the field due to the external charges:

$$E_0 = 4\pi\sigma = D = \varepsilon E \tag{6.5}$$

(compare eqns. (2.35) and (2.46)). When the shape of the dielectric is spherical, eqn. (2.71) gives:

$$E_0 = \frac{\varepsilon + 2}{3} E. \tag{6.6}$$

In the general case, however, $\boldsymbol{E}$ is given by the sum of the external field and the average field due to all molecules of the dielectric. It must be remarked that even for homogeneous $\boldsymbol{E}_0$ the Maxwell field $\boldsymbol{E}$ will not be homogeneous (except in the case of ellipsoids, *cf.* section 9d), thus making relations (6.1) to (6.4) unapplicable in the form given here.

2. $\langle \boldsymbol{M} \cdot \boldsymbol{e} \rangle$, expressed as a function of E_0, has to be calculated as the average of the sum of the dipole moments of all molecules for a given value for the external field. When the total number of particles in the dielectric is N, and the instantaneous dipole moment of the i-th particle is $\boldsymbol{m}_i$, we may write for the instantaneous total moment $\boldsymbol{M}$:

$$\boldsymbol{M} = \sum_{i=1}^{\mathrm{N}} \boldsymbol{m}_i. \tag{6.7}$$

Since N is very large, generally the summation in eqn. (6.7) cannot be performed directly, and appropriate approximations have to be found.

When the permanent dipole moment is zero, $\boldsymbol{m}_i$ can be expressed as a power series in the polarizability; then the summation can be performed provided that the series can be broken off in a suitable way (see section 37). In the case of gases the interaction between the molecules is small, and $\boldsymbol{m}_i$ can be approximated for instance by the moment of a single molecule plus the moment due to pair interactions; the summation is then reduced to multiplication with N (see section 38). When there is considerable interaction between the molecules (*e.g.* in liquids) the latter method is not applicable. In the case of polar systems, the N molecules are then separated in $\mathcal{N}$ molecules (constituting a small macroscopic sphere), which are treated exactly, and $\mathrm{N} - \mathcal{N}$ molecules which are treated as a continuum. In this way macroscopic methods can be used to calculate the contribution of the $\mathrm{N} - \mathcal{N}$ molecules to the total moment. By taking $\mathcal{N}$ sufficiently large, the calculation can be made as accurate as desired (see section 39). The evaluation of the moment of the $\mathcal{N}$ molecules for a number of special systems is treated in section 40.

§36. The 3N-dimensional vectors and tensors

Before we apply the gener. l considerations of the last section to some particular systems, we shall give a formalism in which a number of derivations can be made conveniently. To this end we shall follow the exposition of Mandel and Mazur.[1]

We consider a system of N identical molecules with permanent dipole strengths μ and scalar polarizabilities α. The i-th molecule is located at a point with radius vector $\boldsymbol{r}_i$ and has an instantaneous dipole moment $\boldsymbol{m}_i$. This dipole moment will be given by:

$$\boldsymbol{m}_i = \boldsymbol{\mu}_i + \alpha(\boldsymbol{E}_1)_i, \tag{6.8}$$

where $(\boldsymbol{E}_1)_i$, the local field at the position of the i-th molecule for a specified configuration of the other molecules, is given by:

$$(\boldsymbol{E}_1)_i = \boldsymbol{E}_0 - \sum_{j \neq i}^{N} \boldsymbol{T}_{ij} \cdot \boldsymbol{m}_j. \tag{6.9}$$

In this equation $\boldsymbol{E}_0$ is the external field and $-\boldsymbol{T}_{ij} \cdot \boldsymbol{m}_j$ is the field at $\boldsymbol{r}_i$ due to a dipole moment $\boldsymbol{m}_j$ at $\boldsymbol{r}_j$. According to eqns. (1.32) and (1.34) the dipole field tensor is given in this case by:

$$\boldsymbol{T}_{ij} = -\nabla\nabla \frac{1}{r_{ij}} = \frac{1}{r_{ij}^3}\left(\boldsymbol{I} - 3\frac{\boldsymbol{r}_{ij}\boldsymbol{r}_{ij}}{r_{ij}^2}\right), \tag{6.10}$$

where:

$$\boldsymbol{r}_{ij} = \boldsymbol{r}_i - \boldsymbol{r}_j. \tag{6.11}$$

Combining eqns. (6.8) and (6.9), we see that the dipole moment $\boldsymbol{m}_i$ depends on the positions, orientations, and magnitudes of the dipole moments of all other molecules.

An expression for the dipole moment of the i-th molecule can in principle be obtained by substitution of (6.9) into (6.8), writing out the $\boldsymbol{m}_j$ according to eqn. (6.8) with j replacing i, and repeating this procedure indefinitely. This leads to the following expression for $\boldsymbol{m}_i$:

$$\boldsymbol{m}_i = \boldsymbol{\mu}_i - \alpha \sum_{j \neq i}^{N} \boldsymbol{T}_{ij} \cdot \boldsymbol{\mu}_j + \alpha^2 \sum_{j \neq i}^{N} \sum_{k \neq j}^{N} \boldsymbol{T}_{ij} \cdot \boldsymbol{T}_{jk} \cdot \boldsymbol{\mu}_k - \cdots + \\ + \left[\alpha \boldsymbol{I} - \alpha^2 \sum_{j \neq i}^{N} \boldsymbol{T}_{ij} + \alpha^3 \sum_{j \neq i}^{N} \sum_{k \neq j}^{N} \boldsymbol{T}_{ij} \cdot \boldsymbol{T}_{jk} - \cdots\right] \cdot \boldsymbol{E}_0. \tag{6.12}$$

Thus, the dipole moment consists of two parts, one independent of the external field and the other a linear function of $\boldsymbol{E}_0$.

When written in this way, eqn. (6.12) is inconvenient in further derivations. To obtain a shorter notation, we now introduce a number of vectors in a 3N-dimensional space, which is a combination of N sub-spaces pertaining to each molecule. The 3N-dimensional vector $\mathbf{m}$ has N times 3 components; the first three of these components are equal to the components of $\boldsymbol{m}_1$, the second three components are equal to the components of $\boldsymbol{m}_2$, and so on. Hence the k-th component of $\boldsymbol{m}_i$ will be equal to the $[3(i-1)+k]$-th component of $\mathbf{m}$. In the same way we define 3N-dimensional vectors $\boldsymbol{\mu}$, $\mathbf{E}_1$ and $\mathbf{E}_0$, of which the components are equal to the components of the N vectors $\boldsymbol{\mu}_i$, and of the N vectors $(\boldsymbol{E}_1)_i$, and to the components of $\boldsymbol{E}_0$ repeated N times, respectively. A 3N $\times$ 3N-dimensional tensor $\mathbf{T}$ can be defined analogously, built up from the 3×3-dimensional dipole-dipole interaction tensors $\boldsymbol{T}_{ij}$, connected with the molecules i and j. When $i = j$ we define $\boldsymbol{T}_{ij}$ to be zero.

The combinations of eqns. (6.8) and (6.9) for all values of i may now be written as:

$$\mathbf{m} = \boldsymbol{\mu} + \alpha(\mathbf{E}_0 - \mathbf{T} \cdot \mathbf{m}). \tag{6.13}$$

This equation for the 3N-dimensional vector $\mathbf{m}$ can be solved formally, leading to:

$$\mathbf{m} = (\mathbf{I} + \alpha\mathbf{T})^{-1} \cdot (\boldsymbol{\mu} + \alpha\mathbf{E}_0). \tag{6.14}$$

In this equation $\mathbf{I}$ is the 3N $\times$ 3N-dimensional unit tensor. The tensor $(\mathbf{I} + \alpha\mathbf{T})^{-1}$ is the inverse of $\mathbf{I} + \alpha\mathbf{T}$, *i.e.*:

$$(\mathbf{I} + \alpha\mathbf{T}) \cdot (\mathbf{I} + \alpha\mathbf{T})^{-1} = \mathbf{I}. \tag{6.15}$$

It can be rewritten with the help of a formal series development:

$$(\mathbf{I} + \alpha\mathbf{T})^{-1} = \mathbf{I} - \alpha\mathbf{T} + \alpha^2\mathbf{T}^2 - \alpha^3\mathbf{T}^3 + \cdots, \tag{6.16}$$

where $\mathbf{T}^n$ is the n-times repeated inner product. This expression for $(\mathbf{I} + \alpha\mathbf{T})^{-1}$ is the 3N-dimensional analogon of the coefficients of the $\boldsymbol{\mu}_i, \boldsymbol{\mu}_j, \boldsymbol{\mu}_k, \ldots$ and of $\alpha\boldsymbol{E}_0$ in eqn. (6.12).

From eqn. (6.14) we see again that the dipole moment consists of two parts, one independent of the external field and the other linear in $\boldsymbol{E}_0$:

$$\mathbf{m} = \mathbf{m}^0 + \mathbf{A} \cdot \mathbf{E}_0, \tag{6.17}$$

with:

$$\mathbf{m}^0 = (\mathbf{I} + \alpha\mathbf{T})^{-1} \cdot \boldsymbol{\mu}, \tag{6.18}$$

and:

$$\mathbf{A} = \alpha(\mathbf{I} + \alpha\mathbf{T})^{-1}. \tag{6.19}$$

From eqn. (6.17) it appears that in the 3N-dimensional formalism the tensor **A** plays the rôle of a polarizability. Actually, each set of three rows of **A** represents the polarizability of a molecule corrected by terms due to the interaction with all other molecules (*cf.* eqns. (6.19) and (6.16)).

From eqn. (6.18) we see that the dipole moments of the molecules even in the absence of an external field are not equal to the permanent dipole moments. The difference is caused by the fact that the local field (*cf.* eqn. (6.9)) in the absence of an external field is not zero but equal to the field due to all other molecules.

For a fixed orientation of the *i*-th molecule, the average value of the local field in the absence of an external field is the reaction field of the dipole moment of the molecule (*cf.* eqn. (4.3)). The contribution to the dipole moment due to the local field in the absence of an external field can be written explicitly by noting that according to eqn. (6.15) we have:

$$\mathbf{I} = (\mathbf{I} + \alpha\mathbf{T}) \cdot (\mathbf{I} + \alpha\mathbf{T})^{-1} = (\mathbf{I} + \alpha\mathbf{T})^{-1} + \alpha\mathbf{T} \cdot (\mathbf{I} + \alpha\mathbf{T})^{-1},$$

or:

$$(\mathbf{I} + \alpha\mathbf{T})^{-1} = \mathbf{I} - \alpha\mathbf{T} \cdot (\mathbf{I} + \alpha\mathbf{T})^{-1},$$

and using (6.19):

$$\begin{aligned}(\mathbf{I} + \alpha\mathbf{T})^{-1} &= \mathbf{I} - \mathbf{T} \cdot \mathbf{A} \\ &= \mathbf{I} - \mathbf{A} \cdot \mathbf{T},\end{aligned} \tag{6.20}$$

since **T** and all powers of **T**, and thus **A** too, are symmetric. Substitution of (6.20) into (6.18) now leads to:

$$\mathbf{m}^0 = \boldsymbol{\mu} - \mathbf{A} \cdot \mathbf{T} \cdot \boldsymbol{\mu}. \tag{6.21}$$

In the general case, the local field, as given in eqn. (6.9), can be written in 3N-dimensional notation as:

$$\mathbf{E}_l = \mathbf{E}_0 - \mathbf{T} \cdot \mathbf{m}. \tag{6.22}$$

Using eqns. (6.17), (6.19), and (6.20), we may also write:

$$\begin{aligned}\mathbf{E}_l &= \mathbf{E}_0 - \mathsf{T}\cdot\mathbf{m}^0 - \mathsf{T}\cdot\mathsf{A}\cdot\mathbf{E}_0 \\ &= (\mathsf{I} + \alpha\mathsf{T})^{-1}\cdot\mathbf{E}_0 - \mathsf{T}\cdot\mathbf{m}^0 \\ &= \frac{1}{\alpha}\mathsf{A}\cdot\mathbf{E}_0 - \mathsf{T}\cdot\mathbf{m}^0.\end{aligned} \tag{6.23}$$

Thus the local field is built up of two parts, one proportional to the external field, the other a linear function of $\mathbf{m}^0$ and therefore of $\boldsymbol{\mu}$. If one molecule is selected and a statistical mechanical average is performed over the positions and orientations of all other molecules, eqn. (6.23) leads to an expression for the internal field.

If the local field at an arbitrary point $\boldsymbol{r}$ is considered, expressions (6.22) and (6.23) can still be used provided that as elements of the 3N × 3N-dimensional tensor $\boldsymbol{T}$ we take the dipole field tensor $\boldsymbol{T}$ connected with the field at point $\boldsymbol{r}$ due to molecule j at point $\boldsymbol{r}_j$ (*cf.* section 2c):

$$\boldsymbol{T} = \frac{1}{|\boldsymbol{r}-\boldsymbol{r}_j|^3}\left(\boldsymbol{I} - 3\frac{(\boldsymbol{r}-\boldsymbol{r}_j)(\boldsymbol{r}-\boldsymbol{r}_j)}{|\boldsymbol{r}-\boldsymbol{r}_j|^2}\right).$$

A statistical-mechanical average of these expressions over the positions and orientations of all molecules will then give the Maxwell field $\boldsymbol{E}$ (*cf.* section 6).

To perform the statistical mechanical calculations, an expression is required for the energy of the system as a function of the positions and orientations of the molecules. The electrostatic energy of a system of N molecules in an external field consists of three parts (*cf.* section 14): V_1, the potential energy of the dipoles in the external field; V_2, the potential energy of the dipoles in each others fields; and V_3, the work of polarization.

In the 3N-dimensional notation V_1 and V_2 are given by (*cf.* eqns. (3.9) and (3.118)):

$$V_1 = -\mathbf{m}\cdot\mathbf{E}_0, \tag{6.24}$$

$$V_2 = \tfrac{1}{2}\mathbf{m}\cdot\mathsf{T}\cdot\mathbf{m}. \tag{6.25}$$

The factor $\frac{1}{2}$ appears in eqn. (6.25) to avoid counting the interaction between each pair of dipoles twice.

The work of polarization is the work required to change the dipole moment from the value $\boldsymbol{\mu}_i$ to the value $\boldsymbol{m}_i$, summed over all molecules. According to eqns. (3.93) and (6.8), this is given by:

$$V_3 = \sum_{i=1}^{N}\int_{\boldsymbol{\mu}_i}^{\boldsymbol{m}_i}(\boldsymbol{E}_l)_i d\boldsymbol{m}_i = \sum_{i=1}^{N}\tfrac{1}{2}\alpha(\boldsymbol{E}_l)_i^2. \tag{6.26}$$

Using the 3N-dimensional notation this can be written as:

$$V_3 = \tfrac{1}{2}\alpha \mathbf{E}_1 \cdot \mathbf{E}_1. \tag{6.27}$$

If we now substitute (6.23) into (6.27) we find:

$$V_3 = \tfrac{1}{2}\alpha\left(\frac{1}{\alpha}\mathbf{A}\cdot\mathbf{E}_0 - \mathbf{T}\cdot\mathbf{m}^0\right)\cdot\left(\frac{1}{\alpha}\mathbf{A}\cdot\mathbf{E}_0 - \mathbf{T}\cdot\mathbf{m}^0\right)$$

$$= \tfrac{1}{2}\alpha\mathbf{m}^0\cdot\mathbf{T}\cdot\mathbf{T}\cdot\mathbf{m}^0 - \mathbf{m}^0\cdot\mathbf{T}\cdot\mathbf{A}\cdot\mathbf{E}_0 + \frac{1}{2\alpha}\mathbf{E}_0\cdot\mathbf{A}\cdot\mathbf{A}\cdot\mathbf{E}_0. \tag{6.28}$$

Substitution of (6.17) into (6.25) leads to:

$$V_2 = \tfrac{1}{2}\mathbf{m}^0\cdot\mathbf{T}\cdot\mathbf{m}^0 + \mathbf{m}^0\cdot\mathbf{T}\cdot\mathbf{A}\cdot\mathbf{E}_0 + \tfrac{1}{2}\mathbf{E}_0\cdot\mathbf{A}\cdot\mathbf{T}\cdot\mathbf{A}\cdot\mathbf{E}_0. \tag{6.29}$$

Combining (6.28) and (6.29) we see that the terms linear in $\boldsymbol{E}_0$ cancel each other, which results in:

$$V_2 + V_3 = \tfrac{1}{2}\boldsymbol{\mu}\cdot\mathbf{T}\cdot\mathbf{m}^0 + \tfrac{1}{2}\mathbf{E}_0\cdot\mathbf{A}\cdot\mathbf{E}_0. \tag{6.30}$$

To obtain this result, eqn. (6.18) and eqn. (6.19) rewritten with the help of eqn. (6.20), have been used.

The total electrostatic energy $V_{el} = V_1 + V_2 + V_3$ is now given by the sum of (6.24) and (6.30). After using (6.17) we find:

$$V_{el} = \tfrac{1}{2}\boldsymbol{\mu}\cdot\mathbf{T}\cdot\mathbf{m}^0 - \mathbf{m}^0\cdot\mathbf{E}_0 - \tfrac{1}{2}\mathbf{E}_0\cdot\mathbf{A}\cdot\mathbf{E}_0. \tag{6.31}$$

The sum of the electrostatic energy and the non-electrical energy V_0 (*e.g.* London-Van der Waals interaction energy, *cf.* section 16) is the total energy U of the dielectric, which must be used in statistical-mechanical averaging:

$$U = V_0 + V_{el}. \tag{6.32}$$

For the average of a quantity like $\boldsymbol{M}\cdot\boldsymbol{e}$, which is a function of the positions and orientations of all molecules, we may now write:

$$\langle \boldsymbol{M}\cdot\boldsymbol{e}\rangle = \frac{\int dX\, \boldsymbol{M}\cdot\boldsymbol{e}\exp(-U/kT)}{\int dX \exp(-U/kT)}, \tag{6.33}$$

where X denotes the set of position and orientation variables of all molecules. This expression for the average can be obtained from the general ensemble average by integrating over the momenta.

The integration over the momenta results in a weight factor, which is contained in our notation dX. For example, if we take spherical coordinates r, θ, φ for the position of a molecule and integrate over the conjugated momenta, we obtain a weight factor $r^2 \sin\theta$, in the integration over the coordinates of that molecule. Thus, in this case $dX = r^2 \sin\theta \, dr \, d\theta \, d\varphi$, which is the expression for a volume element in spherical coordinates. Another example is furnished by eqn. (5.10), where the factor $\sin\theta$ could have been obtained by integrating over the momentum in the general expression for the average of $\cos\theta$.

With the help of the expression for $\langle \boldsymbol{M} \cdot \boldsymbol{e} \rangle$ in eqn. (6.33), eqn. (6.4) can be evaluated. Denoting the denominator in eqn. (6.33) by Z:

$$Z = \int dX \exp(-U/kT), \tag{6.34}$$

and differentiating eqn. (6.33) with respect to E_0, we find for eqn. (6.4):

$$\begin{aligned}
\varepsilon - 1 = \frac{4\pi}{V}\left(\frac{\partial E_0}{\partial E}\right)_{E=0} \Bigg[& \frac{1}{Z}\int dX \frac{\partial \boldsymbol{M}\cdot\boldsymbol{e}}{\partial E_0} \exp(-U/kT) - \\
& - \frac{1}{Z}\frac{1}{kT}\int dX\, \boldsymbol{M}\cdot\boldsymbol{e}\frac{\partial U}{\partial E_0}\exp(-U/kT) + \\
& + \frac{1}{Z^2}\frac{1}{kT}\int dX\, \boldsymbol{M}\cdot\boldsymbol{e}\exp(-U/kT)\int dX \frac{\partial U}{\partial_{E_0}}\exp(-U/kT)\Bigg]_{E_0=0} \\
= \frac{4\pi}{V}\left(\frac{\partial E_0}{\partial E}\right)_{E=0} \Bigg[& \left\langle \frac{\partial \boldsymbol{M}\cdot\boldsymbol{e}}{\partial E_0}\right\rangle_0 - \frac{1}{kT}\left\langle \boldsymbol{M}\cdot\boldsymbol{e}\frac{\partial U}{\partial E_0}\right\rangle_0 + \\
& + \frac{1}{kT}\left\langle \boldsymbol{M}\cdot\boldsymbol{e}\right\rangle_0 \left\langle \frac{\partial U}{\partial E_0}\right\rangle_0\Bigg].
\end{aligned} \tag{6.35}$$

In this expression we use the subscript 0 to indicate that the statistical-mechanical average is taken in the absence of an external field. The average moment of a dielectric is zero in the absence of an external field, so that we have:

$$\varepsilon - 1 = \frac{4\pi}{V}\left(\frac{\partial E_0}{\partial E}\right)_{E=0}\left[\left\langle \frac{\partial \boldsymbol{M}\cdot\boldsymbol{e}}{\partial E_0}\right\rangle_0 - \frac{1}{kT}\left\langle \boldsymbol{M}\cdot\boldsymbol{e}\frac{\partial U}{\partial E_0}\right\rangle_0\right]. \tag{6.36}$$

With the help of eqns. (6.31) and (6.32) the differentiation $\partial U/\partial E_0$ can be performed explicitly. To obtain a shorter notation we introduce a 3N-dimensional vector $\mathbf{e}$ with components equal to the N times repeated set of components of the unit vector $\boldsymbol{e}$ in the direction of $\boldsymbol{E}_0$. We may then write:

$$\mathbf{E}_0 = E_0\mathbf{e}. \tag{6.37}$$

Substituting expression (6.37) into eqn. (6.31), we find for the derivative of U with respect to E_0:

$$\frac{\partial U}{\partial E_0} = -\frac{\partial}{\partial E_0}(E_0\mathbf{m}^0 \cdot \mathbf{e} + \tfrac{1}{2}E_0^2\mathbf{e} \cdot \mathbf{A} \cdot \mathbf{e})$$
$$= -\mathbf{m}^0 \cdot \mathbf{e} - E_0\mathbf{e} \cdot \mathbf{A} \cdot \mathbf{e} = -\mathbf{m} \cdot \mathbf{e} = -\boldsymbol{M} \cdot \boldsymbol{e}. \tag{6.38}$$

In this derivation eqn. (6.17) for $\mathbf{m}$ has been used; the equality of $\mathbf{m} \cdot \mathbf{e}$ and $\boldsymbol{M} \cdot \boldsymbol{e}$ can be seen immediately by writing out the components.

The result of this derivation is analogous to eqn. (3.84) for the free energy of a dielectric. In eqn. (6.38), however, the relation between the energy and the moment of a specified configuration of molecules is given, whereas eqn. (3.84) is a relation between macroscopic quantities, *i.e.* quantities which are averages over all configurations.

Using eqns. (6.17) and (6.37) again, we can write for the derivative $\partial \boldsymbol{M} \cdot \boldsymbol{e}/\partial E_0$:

$$\frac{\partial \boldsymbol{M} \cdot \boldsymbol{e}}{\partial E_0} = \frac{\partial \mathbf{m} \cdot \mathbf{e}}{\partial E_0} = \mathbf{e} \cdot \mathbf{A} \cdot \mathbf{e}. \tag{6.39}$$

Substituting (6.38) and (6.39) into (6.36), we obtain:

$$\varepsilon - 1 = \frac{4\pi}{V}\left(\frac{\partial E_0}{\partial E}\right)_{E=0}\left[\langle \mathbf{e} \cdot \mathbf{A} \cdot \mathbf{e}\rangle_0 + \frac{1}{kT}\langle(\boldsymbol{M} \cdot \boldsymbol{e})^2\rangle_0\right]. \tag{6.40}$$

The last term in eqn. (6.40) can be further simplified by noting that the angle θ between the moment $\boldsymbol{M}$ and the external field $\mathbf{E}_0$ has an arbitrary value when the external field is zero. Thus we may write:

$$\langle(\boldsymbol{M} \cdot \boldsymbol{e})^2\rangle_0 = \langle M^2 \cos^2\theta\rangle_0$$
$$= \langle M^2\rangle_0\langle\cos^2\theta\rangle_0$$
$$= \tfrac{1}{3}\langle M^2\rangle_0, \tag{6.41}$$

since according to eqn. (5.7):

$$\langle\cos^2\theta\rangle_0 = \int_{\theta=0}^{\theta=\pi} \cos^2\theta \, \tfrac{1}{2}\sin\theta \, d\theta$$
$$= -\tfrac{1}{2}[\tfrac{1}{3}\cos^3\theta]_{\theta=0}^{\theta=\pi} = \tfrac{1}{3}. \tag{6.42}$$

Substituting (6.41) into (6.40), we have:

$$\varepsilon - 1 = 4\pi N \left(\frac{\partial E_0}{\partial E}\right)_{E=0} \left[\frac{1}{\mathrm{N}}\langle \mathbf{e}\cdot\mathbf{A}\cdot\mathbf{e}\rangle_0 + \frac{1}{3kT\mathrm{N}}\langle M^2\rangle_0\right], \tag{6.43}$$

where $N = \mathrm{N}/V$ is the number density.

Eqn. (6.43) is a general expression for the dielectric constant of a homogeneous, isotropic dielectric. For the case of a very dilute gas, we may neglect all interaction between the molecules and put $\langle \mathbf{e}\cdot\mathbf{A}\cdot\mathbf{e}\rangle_0 = \mathrm{N}\alpha$ (*cf.* eqn. (6.19)) and $\langle M^2\rangle_0 = \sum_{i,j=1}^{\mathrm{N}} \langle \boldsymbol{\mu}_i\cdot\boldsymbol{\mu}_j\rangle_0 = \sum_{i,j=1}^{\mathrm{N}} \mu^2\delta_{ij} = \mathrm{N}\mu^2$. We then also have $E = E_0$ (*cf.* section 31), so that we obtain:

$$\varepsilon - 1 = 4\pi N\left(\alpha + \frac{\mu^2}{3kT}\right), \tag{6.44}$$

which is the Debye equation for gases as given in eqn. (5.100).

When the interactions between the molecules cannot be neglected, the statistical-mechanical averages in eqn. (6.43) have to be performed explicitly. In the following sections a number of important cases will be treated.

§37. The Clausius-Mossotti equation

When the permanent dipole moment is zero, only the induced moments have to be taken into account and the calculation of the statistical-mechanical averages becomes relatively simple. In this case only the term with $\langle \mathbf{e}\cdot\mathbf{A}\cdot\mathbf{e}\rangle$ remains in expression (6.43) for the dielectric constant. We shall not calculate this term directly, but give a different version of the original derivation which is more convenient for the present purpose.

First, we shall treat the case when the correlation between the molecules can be neglected. According to the considerations developed in section 27, the Clausius-Mossotti equation (5.30) should be valid in these circumstances.

Taking N identical molecules with scalar polarizability α and no permanent dipole moment in an external field $\boldsymbol{E}_0$, and using the 3N-dimensional formalism, we find for the dipole moment of the system according to eqn. (6.13):

$$\mathbf{m} = \alpha\mathbf{E}_0 - \alpha\,\mathbf{T}\cdot\mathbf{m}. \tag{6.45}$$

The average of the component of the total moment in the direction of $\mathbf{E}_0$ is now given, with the help of the 3N-dimensional vector $\mathbf{e}$, by:

$$\langle \boldsymbol{M} \cdot \boldsymbol{e} \rangle = \langle \mathbf{m} \cdot \mathbf{e} \rangle = \alpha \mathbf{E}_0 \cdot \mathbf{e} - \alpha \langle \mathbf{e} \cdot \mathbf{T} \cdot \mathbf{m} \rangle. \tag{6.46}$$

At very low densities the interaction-term $\langle \mathbf{e} \cdot \mathbf{T} \cdot \mathbf{m} \rangle$ can be neglected since the elements of $\mathbf{T}$ are inversely proportional to the third power of the intermolecular distances (*cf.* eqn. (6.10)). We then obtain eqn. (6.44) for the case $\mu = 0$.

To evaluate $\langle \mathbf{e} \cdot \mathbf{T} \cdot \mathbf{m} \rangle$ we shall as a first approximation replace the moments of the molecules by their averages, thus neglecting the variations of the induced moment due to the motions of the molecules. Then we may write:

$$\langle \mathbf{e} \cdot \mathbf{T} \cdot \mathbf{m} \rangle = \left\langle \sum_{i,j=1}^{N} \boldsymbol{e} \cdot \boldsymbol{T}_{ij} \cdot \boldsymbol{m}_j \right\rangle = \left\langle \sum_{i,j=1}^{N} \boldsymbol{e} \cdot \boldsymbol{T}_{ij} \cdot \boldsymbol{e} \langle \boldsymbol{m}_j \cdot \boldsymbol{e} \rangle \right\rangle, \tag{6.47}$$

since the only non-zero component of $\langle \boldsymbol{m}_j \rangle$ is in the direction of $\boldsymbol{e}$. The average moment $\langle \boldsymbol{m}_j \cdot \boldsymbol{e} \rangle$ is independent of the index j, since all possible positions of the molecule are taken into account in the average. So $\langle \boldsymbol{m}_j \cdot \boldsymbol{e} \rangle$ may be taken outside the summation, as well as outside the brackets:

$$\langle \mathbf{e} \cdot \mathbf{T} \cdot \mathbf{m} \rangle = \langle \sum_{i,j=1}^{N} \boldsymbol{e} \cdot \boldsymbol{T}_{ij} \cdot \boldsymbol{e} \rangle \langle \boldsymbol{m}_j \cdot \boldsymbol{e} \rangle = \langle \mathbf{e} \cdot \mathbf{T} \cdot \mathbf{e} \rangle \langle \boldsymbol{m}_j \cdot \boldsymbol{e} \rangle. \tag{6.48}$$

Substituting eqn. (6.48) into eqn. (6.46) and summing over j, we obtain:

$$\sum_{j=1}^{N} \langle \mathbf{m} \cdot \mathbf{e} \rangle = \sum_{j=1}^{N} \alpha \mathrm{N} E_0 - \alpha \sum_{j=1}^{N} \langle \mathbf{e} \cdot \mathbf{T} \cdot \mathbf{e} \rangle \langle \boldsymbol{m}_j \cdot \boldsymbol{e} \rangle,$$

or:

$$\mathrm{N} \langle \mathbf{m} \cdot \mathbf{e} \rangle = \alpha \mathrm{N}^2 E_0 - \alpha \langle \mathbf{e} \cdot \mathbf{T} \cdot \mathbf{e} \rangle \langle \mathbf{m} \cdot \mathbf{e} \rangle,$$

so that we have:

$$\langle \mathbf{m} \cdot \mathbf{e} \rangle = \alpha \mathrm{N} E_0 - \frac{\alpha}{\mathrm{N}} \langle \mathbf{e} \cdot \mathbf{T} \cdot \mathbf{e} \rangle \langle \mathbf{m} \cdot \mathbf{e} \rangle. \tag{6.49}$$

Solving eqn. (6.49) for $\langle \mathbf{m} \cdot \mathbf{e} \rangle$, we find:

$$\langle \mathbf{m} \cdot \mathbf{e} \rangle = \frac{\alpha \mathrm{N} E_0}{1 + \frac{\alpha}{\mathrm{N}} \langle \mathbf{e} \cdot \mathbf{T} \cdot \mathbf{e} \rangle}. \tag{6.50}$$

With the help of eqn. (6.3) we may also write, since $\langle \mathbf{m} \cdot \mathbf{e} \rangle = \langle \boldsymbol{M} \cdot \boldsymbol{e} \rangle$:

$$\frac{\varepsilon - 1}{4\pi} VE = \frac{\alpha \mathrm{N} E_0}{1 + \frac{\alpha}{\mathrm{N}} \langle \mathbf{e} \cdot \mathbf{T} \cdot \mathbf{e} \rangle}. \tag{6.51}$$

Eqn. (6.51) is an expression for the dielectric constant of a non-polar dielectric when the variations of the induced moments are neglected. To obtain an explicit expression, the average of the dipole-dipole interaction tensor must be calculated and a relation must be given between E and E_0. Both the average and the relation between E and E_0 depend on the shape of the dielectric; the resulting expression for the dielectric constant must of course be independent of the shape. We shall perform the calculations for the case of a spherical dielectric *in vacuo* and for a plane slab of dielectric between the plates of an ideal capacitor. Both cases will lead to the same expression for the dielectric constant.

The calculation of the average dipole-dipole interaction tensor proceeds as follows. The average $\langle \mathbf{e} \cdot \mathbf{T} \cdot \mathbf{e} \rangle$ is written out as a double summation of the averages of 3-dimensional tensors $\boldsymbol{T}_{ij}$. These averages will not depend on the indices i and j (except that they are zero when $i = j$, according to the definition), because all possible pairs of positions are taken into account in the average. Thus we may write:

$$\begin{aligned} \langle \mathbf{e} \cdot \mathbf{T} \cdot \mathbf{e} \rangle &= \sum_{\substack{i,j=1 \\ i \neq j}}^{\mathrm{N}} \langle \boldsymbol{e} \cdot \boldsymbol{T}_{ij} \cdot \boldsymbol{e} \rangle \\ &= \mathrm{N}(\mathrm{N} - 1) \langle \boldsymbol{e} \cdot \boldsymbol{T}_{ij} \cdot \boldsymbol{e} \rangle \\ &= \mathrm{N}^2 \langle \boldsymbol{e} \cdot \boldsymbol{T}_{ij} \cdot \boldsymbol{e} \rangle, \end{aligned} \tag{6.52}$$

since N is very large.

The function $\boldsymbol{e} \cdot \boldsymbol{T}_{ij} \cdot \boldsymbol{e}$ depends only on the positions of molecules i and j. Therefore, we may integrate over all orientation and position coordinates except the position vectors $\boldsymbol{r}_i$ and $\boldsymbol{r}_j$, which leads to:

$$\langle \mathbf{e} \cdot \mathbf{T} \cdot \mathbf{e} \rangle = \frac{\mathrm{N}^2}{V^2} \iiint_V \iiint_V \boldsymbol{e} \cdot \boldsymbol{T}_{ij} \cdot \boldsymbol{e} \exp\left(-\bar{U}_{ij}(\boldsymbol{r}_i, \boldsymbol{r}_j)/kT\right) d\boldsymbol{r}_i d\boldsymbol{r}_j. \tag{6.53}$$

Here, $\bar{U}_{ij}(\boldsymbol{r}_i, \boldsymbol{r}_j)$ is the average energy of the i-th molecule when the positions of the i-th and j-th molecules are held constant, or the average interaction energy of the i-th and the j-th molecule. In a homogeneous system it is a function of the distance $r = |\boldsymbol{r}_j - \boldsymbol{r}_i|$ only; in such cases the exponential function is often called the radial distribution function $g(r)$:

$$g(r) = \exp(-\bar{U}_{ij}(r)/kT). \tag{6.54}$$

The derivation of eqn. (6.53) has been given by Kirkwood[2] in a general way. When the average of a function $f(\boldsymbol{r}_i, \boldsymbol{r}_j)$ of the positions of molecules i and j has to be calculated, the integration can be carried out in two steps:

$$\langle f(\boldsymbol{r}_i, \boldsymbol{r}_j)\rangle = \frac{\int \mathrm{d}X\, f(\boldsymbol{r}_i, \boldsymbol{r}_j)\exp(-U(X)/kT)}{\int \mathrm{d}X \exp(-U(X)/kT)}$$

$$= \frac{1}{V^2}\iiint_V \iiint_V f(\boldsymbol{r}_i, \boldsymbol{r}_j)\frac{V^2 \int \mathrm{d}[X/\boldsymbol{r}_i, \boldsymbol{r}_j]\exp(-U(X)/kT)}{\int \mathrm{d}X \exp(-U(X)/kT)}\boldsymbol{dr}_i\boldsymbol{dr}_j$$

$$= \frac{1}{V^2}\iiint_V \iiint_V f(\boldsymbol{r}_i, \boldsymbol{r}_j)\exp(-\bar{U}_{ij}(\boldsymbol{r}_i, \boldsymbol{r}_j)/kT)\boldsymbol{dr}_i\boldsymbol{dr}_j. \tag{6.55}$$

Here, $\mathrm{d}[X/\boldsymbol{r}_i, \boldsymbol{r}_j]$ denotes an integration over all coordinates except the position vectors $\boldsymbol{r}_i$ and $\boldsymbol{r}_j$, and $\bar{U}_{ij}(\boldsymbol{r}_i, \boldsymbol{r}_j)$ is defined by:

$$\exp(-\bar{U}_{ij}(\boldsymbol{r}_i, \boldsymbol{r}_j)/kT) = \frac{V^2 \int \mathrm{d}[X/\boldsymbol{r}_i, \boldsymbol{r}_j]\exp(-U(X)/kT)}{\int \mathrm{d}X \exp(-U(X)/kT)}. \tag{6.56}$$

To obtain some insight in the nature of $\bar{U}_{ij}$, we take of both sides of eqn. (6.56) the gradient at the position of the i-th molecule:

$$-\frac{1}{kT}\exp(-\bar{U}_{ij}/kT)\nabla_i\bar{U}_{ij} = -\frac{V^2}{kT}\frac{\int \mathrm{d}[X/\boldsymbol{r}_i, \boldsymbol{r}_j]\nabla_i U \exp(-U/kT)}{\int \mathrm{d}X \exp(-U/kT)}.$$

Substitution of eqn. (6.56) and simplification leads to:

$$-\nabla_i\bar{U}_{ij} = \frac{\int \mathrm{d}[X/\boldsymbol{r}_i, \boldsymbol{r}_j](-\nabla_i U)\exp(-U/kT)}{\int \mathrm{d}[X/\boldsymbol{r}_i, \boldsymbol{r}_j]\exp(-U/kT)}.$$

Thus we see that $-\nabla_i\bar{U}_{ij}$ is the average force on molecule i when positions $\boldsymbol{r}_i$ and $\boldsymbol{r}_j$ are held constant, *i.e.* $\bar{U}_{ij}$ is the average energy of molecule i when the values of $\boldsymbol{r}_i$ and $\boldsymbol{r}_j$ are specified. When $\boldsymbol{r}_i$ is not very near the surface of the system, $\bar{U}_{ij}$ will depend only on the distance $r = |\boldsymbol{r}_j - \boldsymbol{r}_i|$ between the molecules at $\boldsymbol{r}_i$ and $\boldsymbol{r}_j$.

Taking the origin at the position of the i-th molecule, we may simplify eqn. (6.53) since $\boldsymbol{T}_{ij}$ and $\overline{U}_{ij}$ are functions of the distance only. Carrying out the integration over $\boldsymbol{r}_i$, we find:

$$\langle \mathbf{e} \cdot \mathbf{T} \cdot \mathbf{e} \rangle = \frac{\mathrm{N}^2}{V} \iiint_V \boldsymbol{e} \cdot \boldsymbol{T}_{ij} \cdot \boldsymbol{e} \exp(-\overline{U}_{ij}(r)/kT) \mathrm{d}v. \tag{6.57}$$

For a number of special cases the form of $\overline{U}_{ij}(r)$ has been determined theoretically or experimentally.[3] For the evaluation of eqn. (6.57), however, the exact form of $\overline{U}_{ij}(r)$ is not required. It is sufficient to use the following properties, which can be derived from general considerations:

1. $\overline{U}_{ij}$ becomes infinitely large when r goes to zero,
2. $\overline{U}_{ij}$ becomes zero when r becomes infinitely large.

Since $\exp(-\overline{U}_{ij}(r)/kT)$ becomes zero when r goes to zero, we may exclude from the integration in eqn. (6.57) a small sphere around the origin with radius r_0 and volume v:

$$\langle \mathbf{e} \cdot \mathbf{T} \cdot \mathbf{e} \rangle = \frac{\mathrm{N}^2}{V} \iiint_{V-v} \boldsymbol{e} \cdot \boldsymbol{T}_{ij} \cdot \boldsymbol{e} \exp(-\overline{U}_{ij}(r)/kT) \mathrm{d}v. \tag{6.58}$$

We now add and subtract the integral of the dipole-dipole interaction tensor:

$$\langle \mathbf{e} \cdot \mathbf{T} \cdot \mathbf{e} \rangle = \frac{\mathrm{N}^2}{V} \iiint_{V-v} \boldsymbol{e} \cdot \boldsymbol{T}_{ij} \cdot \boldsymbol{e}\, \mathrm{d}v + \frac{\mathrm{N}^2}{V} \iiint_{V-v} \boldsymbol{e} \cdot \boldsymbol{T}_{ij} \cdot \boldsymbol{e}[\exp(-\overline{U}_{ij}/kT) - 1] \mathrm{d}v. \tag{6.59}$$

In the second term of the righthand member the integration may be extended to infinity in view of the fact that the factor $[\exp(-\overline{U}_{ij}/kT) - 1]$ will be zero even before the boundaries of the dielectric have been reached. We may now write:

$$\iiint_{V-v} \boldsymbol{e} \cdot \boldsymbol{T}_{ij} \cdot \boldsymbol{e}[\exp(-\overline{U}_{ij}/kT) - 1] \,\mathrm{d}v$$

$$= \iiint_{\infty-v} \boldsymbol{e} \cdot \boldsymbol{T}_{ij} \cdot \boldsymbol{e}[\exp(-\overline{U}_{ij}/kT) - 1] \,\mathrm{d}v. \tag{6.60}$$

Using eqn. (6.10) for $\boldsymbol{T}_{ij}$ and taking spherical coordinates with the polar axis in the direction of $\boldsymbol{e}$, *i.e.* $\boldsymbol{e} \cdot \boldsymbol{r} = r \cos\theta$, we find:

$$\iiint_{V-v} \boldsymbol{e} \cdot \boldsymbol{T}_{ij} \cdot \boldsymbol{e}[\exp(-\overline{U}_{ij}/kT) - 1] \,\mathrm{d}v =$$

$$= \int_{r_0}^{\infty}\int_0^{2\pi}\int_0^{\pi} \frac{1 - 3\cos^2\theta}{r}[\exp(-\bar{U}_{ij}/kT) - 1]\sin\theta\, d\theta\, d\varphi\, dr$$

$$= \int_{r_0}^{\infty}\int_0^{2\pi}\left\{\int_0^{\pi} P_0(\cos\theta)P_2(\cos\theta)\sin\theta\, d\theta\right\}\frac{2}{r}[\exp(-\bar{U}_{ij}/kT) - 1]\, d\varphi\, dr$$

$$= 0. \qquad (6.61)$$

Here we have used the table of Legendre polynomials (Appendix III, section 4) and the fact that the Legendre polynomials are orthogonal (eqn. (A3.27)).

Substitution of eqn. (6.61) into eqn. (6.59) now gives:

$$\langle \mathbf{e}\cdot\mathbf{T}\cdot\mathbf{e}\rangle = \frac{N^2}{V}\iiint_{V-v} \boldsymbol{e}\cdot\boldsymbol{T}_{ij}\cdot\boldsymbol{e}\, dv. \qquad (6.62)$$

The integration in the righthand member can be simplified by applying the tensor analogon for Gauss' theorem (*cf.* eqn. (A1.31)):

$$\iiint \nabla \boldsymbol{A}\, dv = \oiint \boldsymbol{A}\boldsymbol{dS}. \qquad (6.63)$$

Since $\boldsymbol{T}_{ij}$ is given by $-\nabla\nabla(1/r)$ according to eqn. (6.10), we find in this case:

$$\boldsymbol{e}\cdot\left[\iiint_{V-v} \boldsymbol{T}_{ij}\, dv\right]\cdot\boldsymbol{e} = \boldsymbol{e}\cdot\left[-\iint_S \nabla\frac{1}{r}\boldsymbol{dS} - \iint_s \nabla\frac{1}{r}\boldsymbol{dS}\right]\cdot\boldsymbol{e}, \qquad (6.64)$$

where S denotes the surface of the dielectric and s the surface of the small sphere around molecule i. Taking spherical coordinates with the polar axis in the direction of $\boldsymbol{e}$, the second integral in the righthand member of (6.64) leads to:

$$-\boldsymbol{e}\cdot\left[\iint_s \nabla\frac{1}{r}\boldsymbol{dS}\right]\cdot\boldsymbol{e} = \int_0^{2\pi}\int_0^{\pi} \frac{\boldsymbol{e}\cdot\boldsymbol{r}_0}{r_0^3}r_0^2\sin\theta\left(-\frac{\boldsymbol{r}_0}{r_0}\cdot\boldsymbol{e}\right) d\theta\, d\varphi$$

$$= -2\pi\int_0^{\pi}\cos^2\theta\sin\theta\, d\theta = -\frac{4\pi}{3}. \qquad (6.65)$$

Here we have used the fact that the surface element $\boldsymbol{dS}$ is in the direction of the outward normal of the surface, *i.e.* in this case in the direction of $-\boldsymbol{r}_0/r_0$.

When the dielectric has a spherical shape, the same calculation can be made for the first integral in the righthand member of eqn. (6.64). Since the outward normal is now in the same direction as the radius vector, we obtain:

$$-\boldsymbol{e}\cdot\left[\iint\limits_{S\,(\text{sphere})} \nabla\frac{1}{r}\; d\boldsymbol{S}\right]\cdot\boldsymbol{e} = \frac{4\pi}{3}. \tag{6.66}$$

Substitution of eqns. (6.65) and (6.66) into the righthand member of (6.64) gives zero, so that we find:

$$\langle \mathbf{e}\cdot\mathbf{T}\cdot\mathbf{e}\rangle_{\text{sphere}} = 0. \tag{6.67}$$

Substituting this into eqn. (6.51) and using eqn. (6.6), we now have:

$$\frac{\varepsilon - 1}{\varepsilon + 2} = \frac{4\pi}{3}\frac{\mathrm{N}}{V}\alpha, \tag{6.68}$$

which is the Clausius-Mossotti equation (*cf.* eqn. (5.30), where $N = \mathrm{N}/V$).

When the dielectric is contained between the plates of a plane capacitor, we find, using the coordinate system of Fig. 31:

$$-\boldsymbol{e}\cdot\left[\iint\limits_{S(\text{capac.})} \nabla\frac{1}{r}d\boldsymbol{S}\right]\cdot\boldsymbol{e} = \iint\limits_{S_1}\frac{\boldsymbol{e}\cdot\boldsymbol{r}_1}{r_1^3}\boldsymbol{e}\cdot d\boldsymbol{S}_1 + \iint\limits_{S_2}\frac{\boldsymbol{e}\cdot\boldsymbol{r}_2}{r_2^3}\boldsymbol{e}\cdot d\boldsymbol{S}_2. \tag{6.69}$$

Taking polar coordinates in the planes of the capacitor plates, we have:

$$d\boldsymbol{S}_1 = -\boldsymbol{e}\rho_1\,\mathrm{d}\varphi_1\,\mathrm{d}\rho_1, \tag{6.70}$$

$$d\boldsymbol{S}_2 = \boldsymbol{e}\rho_2\,\mathrm{d}\varphi_2\,\mathrm{d}\rho_2, \tag{6.71}$$

$$\boldsymbol{e}\cdot\boldsymbol{r}_1 = -a_1, \tag{6.72}$$

$$\boldsymbol{e}\cdot\boldsymbol{r}_2 = a_2. \tag{6.73}$$

Substitution of eqns. (6.70), (6.71), (6.72), and (6.73) into eqn. (6.69) and use of the relations:

$$r_1^2 = a_1^2 + \rho_1^2, \tag{6.74}$$

$$r_2^2 = a_2^2 + \rho_2^2, \tag{6.75}$$

gives:

$$-\boldsymbol{e}\cdot\left[\iint\limits_{S(\text{capac.})} \nabla\frac{1}{r}d\boldsymbol{S}\right]\cdot\boldsymbol{e} = 2\pi\int_0^\infty \frac{a_1\rho_1\,\mathrm{d}\rho_1}{(a_1^2+\rho_1^2)^{3/2}} + 2\pi\int_0^\infty \frac{a_2\,\rho_2\,\mathrm{d}\rho_2}{(a_2^2+\rho_2^2)^{3/2}}. \tag{6.76}$$

Since $\frac{\mathrm{d}}{\mathrm{d}\rho}1/(a^2+\rho^2)^{1/2} = -\rho/(a^2+\rho^2)^{3/2}$, this leads to:

$$-\boldsymbol{e}\cdot\left[\iint\limits_{S(\text{capac.})} \nabla\frac{1}{r}d\boldsymbol{S}\right]\cdot\boldsymbol{e} = 2\pi + 2\pi = 4\pi. \tag{6.77}$$

Substituting eqns. (6.65) and (6.77) into (6.64) and the resulting expression into (6.62), we find:

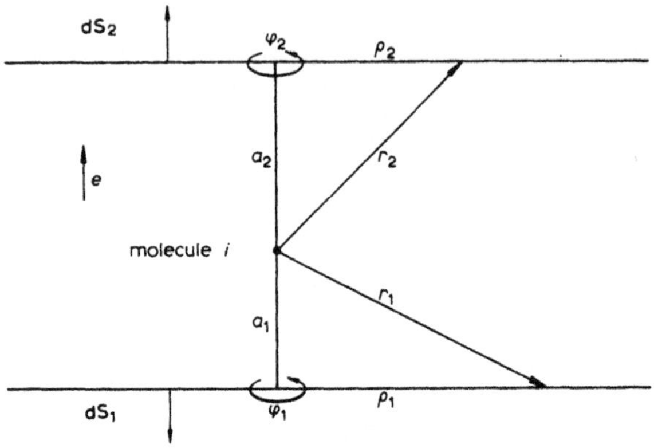

Fig. 31

$$\langle \mathbf{e} \cdot \mathbf{T} \cdot \mathbf{e} \rangle_{\text{capac.}} = \frac{8\pi}{3} \frac{N^2}{V}. \tag{6.78}$$

Expression (6.78) for the case of a plane capacitor can be substituted into the general relation (6.51). Using also eqn. (6.5) for the external field, we find:

$$\frac{\varepsilon - 1}{4\pi} V = \frac{N\alpha\varepsilon}{1 + \frac{8\pi}{3} \frac{N}{V} \alpha}. \tag{6.79}$$

Solving this expression for $\frac{4\pi}{3} \frac{N}{V} \alpha$, we obtain:

$$\frac{\varepsilon - 1}{\varepsilon + 2} = \frac{4\pi}{3} \frac{N}{V} \alpha,$$

which is the Clausius-Mossotti equation (compare eqn. (6.68)).

When the dielectric has a shape differing from a sphere or a plane slab, the calculation of the average dipole-dipole interaction tensor becomes very difficult. The same holds for the determination of the relation between E_0 and E. The resulting expressions, however, will always lead to the Clausius-Mossotti equation as long as the approximation (6.47) is used.

According to eqn. (6.23), the local field in a non-polar dielectric can be written as:

$$\mathbf{E}_l = (\mathbf{I} + \alpha \mathbf{T})^{-1} \cdot \mathbf{E}_0, \tag{6.80}$$

or:

$$(\mathbf{I} + \alpha \mathbf{T}) \cdot \mathbf{E}_l = \mathbf{E}_0. \tag{6.81}$$

Since the internal field is the average local field, it can be calculated from eqn. (6.80). When we use the approximation that the induced moment is constant (*cf.* eqn. (6.47)), the local field inducing this moment must also be taken constant. In this case we may write, starting from eqn. (6.81):

$$(\mathrm{N} + \alpha\langle \mathbf{e}\cdot\mathbf{T}\cdot\mathbf{e}\rangle)\langle E_l\rangle = \mathrm{N}E_0. \tag{6.82}$$

With the help of eqns. (6.67) and (6.6) we now have for the case of a spherical dielectric:

$$E_i = \langle E_l\rangle = \frac{\varepsilon + 2}{3}E. \tag{6.83}$$

This is the Lorentz formula for the internal field (*cf.* eqn. (5.22)). When the dielectric is a plane slab between capacitor plates, eqns. (6.78), (6.68), and (6.5) lead to the same expression for E_i.

From this derivation we see that the remark made in section 27, *i.e.* that the internal field is given exactly by the Lorentz field when the distribution of the surrounding molecules is completely random, refers to the case that there is no correlation between the positions and the induced moments of the molecules.

To obtain some insight into the approximations made in the derivation of the Clausius-Mossotti equation, we shall derive a more general expression for the dielectric constant. If we compare the general equation (6.46) with the expression (6.49) which is valid for the special case that the induced moments of the molecules are constant (the Clausius-Mossotti approximation), we see that the average moment $\langle \mathbf{m}\cdot\mathbf{e}\rangle$ can be written as the sum of the Clausius-Mossotti average moment and a fluctuation term:

$$\begin{aligned}\langle \mathbf{m}\cdot\mathbf{e}\rangle &= \alpha\mathbf{E}_0\cdot\mathbf{e} - \alpha\langle \mathbf{e}\cdot\mathbf{T}\cdot\mathbf{m}\rangle - \frac{\alpha}{\mathrm{N}}\langle \mathbf{e}\cdot\mathbf{T}\cdot\mathbf{e}\rangle\langle \mathbf{m}\cdot\mathbf{e}\rangle + \\ &\quad + \frac{\alpha}{\mathrm{N}}\langle \mathbf{e}\cdot\mathbf{T}\cdot\mathbf{e}\rangle\langle \mathbf{m}\cdot\mathbf{e}\rangle \\ &= \langle \mathbf{m}\cdot\mathbf{e}\rangle_{\mathrm{CM}} - \alpha\left[\langle \mathbf{e}\cdot\mathbf{T}\cdot\mathbf{m}\rangle - \frac{1}{\mathrm{N}}\langle \mathbf{e}\cdot\mathbf{T}\cdot\mathbf{e}\rangle\langle \mathbf{m}\cdot\mathbf{e}\rangle\right].\end{aligned} \tag{6.84}$$

In this equation the subscript CM has been added to the brackets to distinguish the uncorrelated average from the general average. The fluctuations represented in the second term of the last member of eqn. (6.84) are sometimes called translational fluctuations (see *e.g.* Kirkwood[4]), because when a molecule traverses its path the induced moment fluctuates around its average value.

The fluctuation term in eqn. (6.84) can be represented as a power series in α by successive substitutions of the expression:

$$\mathbf{m} = \frac{1}{N}\langle \mathbf{m}\cdot\mathbf{e}\rangle_{CM}\mathbf{e} - \alpha\left[\mathbf{T}\cdot\mathbf{m} - \frac{1}{N}\langle \mathbf{T}\cdot\mathbf{e}\rangle\langle \mathbf{m}\cdot\mathbf{e}\rangle\right]. \tag{6.85}$$

This expression is analogous to eqn. (6.84) and can be derived by combining eqns. (6.45) and (6.49). Substituting eqn. (6.85) into the fluctuation term in eqn. (6.84) and collecting the terms in α and α^2, we obtain:

$$\begin{aligned}\langle \mathbf{m}\cdot\mathbf{e}\rangle = \langle \mathbf{m}\cdot\mathbf{e}\rangle_{CM} - &\\ - \alpha\Bigg[&\frac{1}{N}\langle \mathbf{e}\cdot\mathbf{T}\cdot\mathbf{e}\rangle\langle \mathbf{m}\cdot\mathbf{e}\rangle_{CM} - \frac{1}{N}\langle \mathbf{e}\cdot\mathbf{T}\cdot\mathbf{e}\rangle\langle \mathbf{m}\cdot\mathbf{e}\rangle_{CM}\Bigg] + \\ + \alpha^2\Bigg[&\langle \mathbf{e}\cdot\mathbf{T}\cdot\mathbf{T}\cdot\mathbf{m}\rangle - \frac{1}{N}\langle \mathbf{e}\cdot\mathbf{T}\rangle\cdot\langle \mathbf{T}\cdot\mathbf{e}\rangle\langle \mathbf{m}\cdot\mathbf{e}\rangle - \\ &- \frac{1}{N}\langle \mathbf{e}\cdot\mathbf{T}\cdot\mathbf{e}\rangle\langle \mathbf{e}\cdot\mathbf{T}\cdot\mathbf{m}\rangle + \frac{1}{N^2}\langle \mathbf{e}\cdot\mathbf{T}\cdot\mathbf{e}\rangle\langle \mathbf{e}\cdot\mathbf{T}\cdot\mathbf{e}\rangle\langle \mathbf{m}\cdot\mathbf{e}\rangle\Bigg],\end{aligned}$$

or:

$$\langle \mathbf{m}\cdot\mathbf{e}\rangle = \langle \mathbf{m}\cdot\mathbf{e}\rangle_{CM} + \alpha^2\left[\langle \mathbf{e}\cdot\mathbf{T}\cdot\mathbf{T}\cdot\mathbf{m}\rangle - \frac{1}{N}\langle \mathbf{e}\cdot\mathbf{T}\cdot\mathbf{e}\rangle\langle \mathbf{e}\cdot\mathbf{T}\cdot\mathbf{m}\rangle\right], \tag{6.86}$$

since $\langle \mathbf{e}\cdot\mathbf{T}\rangle\cdot\langle \mathbf{T}\cdot\mathbf{e}\rangle = \frac{1}{N}\langle \mathbf{e}\cdot\mathbf{T}\cdot\mathbf{e}\rangle\langle \mathbf{e}\cdot\mathbf{T}\cdot\mathbf{e}\rangle$.

Substituting eqn. (6.85) into the fluctuation term of eqn. (6.86), we then have:

$$\begin{aligned}\langle \mathbf{m}\cdot\mathbf{e}\rangle = \langle \mathbf{m}\cdot\mathbf{e}\rangle_{CM} + \alpha^2\Bigg[&\frac{1}{N}\langle \mathbf{e}\cdot\mathbf{T}\cdot\mathbf{T}\cdot\mathbf{e}\rangle\langle \mathbf{m}\cdot\mathbf{e}\rangle_{CM} - \frac{1}{N^2}\langle \mathbf{e}\cdot\mathbf{T}\cdot\mathbf{e}\rangle^2\langle \mathbf{m}\cdot\mathbf{e}\rangle_{CM}\Bigg] - \\ - \alpha^3\Bigg[&\langle \mathbf{e}\cdot\mathbf{T}\cdot\mathbf{T}\cdot\mathbf{T}\cdot\mathbf{m}\rangle - \frac{1}{N}\langle \mathbf{e}\cdot\mathbf{T}\cdot\mathbf{T}\rangle\cdot\langle \mathbf{m}\cdot\mathbf{e}\rangle - \\ &- \frac{1}{N}\langle \mathbf{e}\cdot\mathbf{T}\cdot\mathbf{e}\rangle\langle \mathbf{e}\cdot\mathbf{T}\cdot\mathbf{T}\cdot\mathbf{m}\rangle + \\ &+ \frac{1}{N^2}\langle \mathbf{e}\cdot\mathbf{T}\cdot\mathbf{e}\rangle\langle \mathbf{e}\cdot\mathbf{T}\rangle\ \langle \mathbf{T}\cdot\mathbf{e}\rangle\langle \mathbf{m}\cdot\mathbf{e}\rangle\Bigg].\end{aligned} \tag{6.87}$$

This procedure can be repeated indefinitely, so that we obtain a series of fluctuation terms in powers of α, each multiplied by $\langle \mathbf{m}\cdot\mathbf{e}\rangle_{CM}$. Thus, we may write:

$$\langle \mathbf{m}\cdot\mathbf{e}\rangle = \langle \mathbf{m}\cdot\mathbf{e}\rangle_{CM} + S\langle \mathbf{m}\cdot\mathbf{e}\rangle_{CM}, \tag{6.88}$$

with:

$$S = \alpha^2\left[\frac{1}{N}\langle \mathbf{e}\cdot\mathbf{T}\cdot\mathbf{T}\cdot\mathbf{e}\rangle - \frac{1}{N^2}\langle \mathbf{e}\cdot\mathbf{T}\cdot\mathbf{e}\rangle^2\right] + O(\alpha^3). \tag{6.89}$$

A closed expression for S can be obtained in the following way. From eqn. (6.85) we derive:

$$(\mathbf{I} + \alpha\mathbf{T})\cdot\mathbf{m} = \frac{1}{N}\langle \mathbf{m}\cdot\mathbf{e}\rangle_{CM}\mathbf{e} + \frac{\alpha}{N}\langle \mathbf{T}\cdot\mathbf{e}\rangle\langle \mathbf{m}\cdot\mathbf{e}\rangle,$$

or:

$$\mathbf{m} = (\mathbf{I} + \alpha\mathbf{T})^{-1}\cdot\left[\frac{1}{N}\langle \mathbf{m}\cdot\mathbf{e}\rangle_{CM}\mathbf{e} + \frac{\alpha}{N}\langle \mathbf{T}\cdot\mathbf{e}\rangle\langle \mathbf{m}\cdot\mathbf{e}\rangle\right]. \tag{6.90}$$

The average component $\langle \mathbf{m}\cdot\mathbf{e}\rangle$ is now given by:

$$\langle \mathbf{m}\cdot\mathbf{e}\rangle = \langle \mathbf{e}\cdot(\mathbf{I} + \alpha\mathbf{T})^{-1}\rangle\cdot\left[\frac{1}{N}\langle \mathbf{m}\cdot\mathbf{e}\rangle_{CM}\mathbf{e} + \frac{\alpha}{N}\langle \mathbf{T}\cdot\mathbf{e}\rangle\langle \mathbf{m}\cdot\mathbf{e}\rangle\right]. \tag{6.91}$$

Solving for $\langle \mathbf{m}\cdot\mathbf{e}\rangle$ we find:

$$\langle \mathbf{m}\cdot\mathbf{e}\rangle = \frac{1}{N}\,\frac{\langle \mathbf{e}\cdot(\mathbf{I} + \alpha\mathbf{T})^{-1}\cdot\mathbf{e}\rangle}{1 - \frac{\alpha}{N}\langle \mathbf{e}\cdot(\mathbf{I} + \alpha\mathbf{T})^{-1}\rangle\cdot\langle \mathbf{T}\cdot\mathbf{e}\rangle}\langle \mathbf{m}\cdot\mathbf{e}\rangle_{CM}. \tag{6.92}$$

Comparing this with eqn. (6.88) we see that:

$$S = \frac{\langle \mathbf{e}\cdot(\mathbf{I} + \alpha\mathbf{T})^{-1}\cdot\mathbf{e}\rangle}{N - \alpha\langle \mathbf{e}\cdot(\mathbf{I} + \alpha\mathbf{T})^{-1}\rangle\cdot\langle \mathbf{T}\cdot\mathbf{e}\rangle} - 1. \tag{6.93}$$

By developing $(\mathbf{I} + \alpha\mathbf{T})^{-1}$ in a power series (eqn. (6.16)) and rewriting as a power series in α, this expression for S can be brought into the form of eqn. (6.89).

With the help of eqn. (6.88) a correction to the Clausius-Mossotti equation can be found. According to eqn. (6.3), we may always write, since $\langle \boldsymbol{M}\cdot\boldsymbol{e}\rangle = \langle \mathbf{m}\cdot\mathbf{e}\rangle$:

$$\langle \mathbf{m}\cdot\mathbf{e}\rangle = \frac{\varepsilon - 1}{4\pi}VE. \tag{6.94}$$

In the Clausius-Mossotti approximation we replace $\langle \mathbf{m}\cdot\mathbf{e}\rangle$ by $\langle \mathbf{m}\cdot\mathbf{e}\rangle_{CM}$ and use eqn. (6.68) in the righthand member of eqn. (6.94) to obtain:

$$\langle \mathbf{m}\cdot\mathbf{e}\rangle_{CM} = \frac{\varepsilon + 2}{3}N\alpha E. \tag{6.95}$$

Substitution of eqns. (6.94) and (6.95) into eqn. (6.88) now leads to the expression:

$$\frac{\varepsilon - 1}{\varepsilon + 2} = \frac{4\pi}{3} \frac{\mathrm{N}}{V} \alpha(1 + S), \tag{6.96}$$

where S is the correction term, given by eqn. (6.89) or (6.93).

Eqn. (6.96) is a generalization of the Clausius-Mossotti equation for isotropic (spherical) molecules with a scalar polarizability. It was first given by De Boer, Van der Maesen and Ten Seldam;[5] the advantage of this expression over expressions given by other authors[4,6,7] lies in the fact that the Clausius-Mossotti equation gives a good description of the values of ε obtained experimentally for non-polar dielectrics, so that the deviations from the Clausius-Mossotti equation will in general be small.

We shall limit ourselves to a discussion of the evaluation of the first term of S, which is proportional to α^2. In this term three types of averages of the dipole-dipole interaction tensor occur. The average $\langle \mathbf{e} \cdot \mathbf{T} \cdot \mathbf{e} \rangle$ has already been given in eqn. (6.57):

$$\langle \mathbf{e} \cdot \mathbf{T} \cdot \mathbf{e} \rangle = \frac{\mathrm{N}^2}{V} \iiint\limits_V \boldsymbol{e} \cdot \boldsymbol{T}_{ij} \cdot \boldsymbol{e} \exp(-\bar{U}_{ij}(r)/kT) \mathrm{d}v.$$

The average $\langle \mathbf{e} \cdot \mathbf{T} \cdot \mathbf{T} \cdot \mathbf{e} \rangle$ is split up as follows:

$$\begin{aligned} \langle \mathbf{e} \cdot \mathbf{T} \cdot \mathbf{T} \cdot \mathbf{e} \rangle &= \sum_{i=1}^{\mathrm{N}} \sum_{\substack{j=1 \\ j \neq i}}^{\mathrm{N}} \sum_{\substack{k=1 \\ k \neq j}}^{\mathrm{N}} \langle \boldsymbol{e} \cdot \boldsymbol{T}_{ij} \cdot \boldsymbol{T}_{jk} \cdot \boldsymbol{e} \rangle \\ &= \sum_{i=1}^{\mathrm{N}} \sum_{\substack{j=1 \\ j \neq i}}^{\mathrm{N}} \langle \boldsymbol{e} \cdot \boldsymbol{T}_{ij} \cdot \boldsymbol{T}_{ji} \cdot \boldsymbol{e} \rangle + \sum_{i=1}^{\mathrm{N}} \sum_{\substack{j=1 \\ j \neq i}}^{\mathrm{N}} \sum_{\substack{k=1 \\ k \neq i,j}}^{\mathrm{N}} \langle \boldsymbol{e} \cdot \boldsymbol{T}_{ij} \cdot \boldsymbol{T}_{jk} \cdot \boldsymbol{e} \rangle \\ &= \mathrm{N}(\mathrm{N} - 1) \langle \boldsymbol{e} \cdot \boldsymbol{T}_{ij} \cdot \boldsymbol{T}_{ji} \cdot \boldsymbol{e} \rangle + \\ &\quad + \mathrm{N}(\mathrm{N} - 1)(\mathrm{N} - 2) \langle \boldsymbol{e} \cdot \boldsymbol{T}_{ij} \cdot \boldsymbol{T}_{jk} \cdot \boldsymbol{e} \rangle. \end{aligned} \tag{6.97}$$

Introducing the average triplet interaction energy $\bar{U}_{ijk}(\boldsymbol{r}_i, \boldsymbol{r}_j, \boldsymbol{r}_k)$ and denoting the vectors $\boldsymbol{r}_j - \boldsymbol{r}_i$ by $\boldsymbol{r}$ and $\boldsymbol{r}_k - \boldsymbol{r}_i$ by $\boldsymbol{r}'$ we find expressions analogous to eqn. (6.57):

$$\langle \boldsymbol{e} \cdot \boldsymbol{T}_{ij} \cdot \boldsymbol{T}_{ji} \cdot \boldsymbol{e} \rangle = \frac{1}{V} \iiint\limits_V \boldsymbol{e} \cdot \boldsymbol{T}_{ij} \cdot \boldsymbol{T}_{ji} \cdot \boldsymbol{e} \exp(-\bar{U}_{ij}(r)/kT) \mathrm{d}v, \tag{6.98}$$

and:

$$\langle \boldsymbol{e} \cdot \boldsymbol{T}_{ij} \cdot \boldsymbol{T}_{jk} \cdot \boldsymbol{e} \rangle = \frac{1}{V^2} \iiint\limits_V \iiint\limits_V \boldsymbol{e} \cdot \boldsymbol{T}_{ij} \cdot \boldsymbol{T}_{jk} \cdot \boldsymbol{e} \exp(-\bar{U}_{ijk}(\boldsymbol{r}, \boldsymbol{r}')/kT) \mathrm{d}v\, \mathrm{d}v'. \tag{6.99}$$

To evaluate the averages in eqns. (6.98) and (6.99) it is necessary to know the average pair and triplet interaction energies. In most cases it is sufficient to use the so-called superposition approximation introduced by Kirkwood[2] for the triplet interaction energy:

$$\bar{U}_{ijk}(\boldsymbol{r}, \boldsymbol{r}') = \bar{U}_{ij}(r) + \bar{U}_{ik}(r') + \bar{U}_{jk}(|\boldsymbol{r}' - \boldsymbol{r}|). \tag{6.100}$$

The average pair interaction energy can be taken equal to a suitable intermolecular potential at low densities. In general, it can be expressed as a power series in the density which can be broken off after a few terms in the case of gases. In the case of liquids it may be possible to use some other approximation (see *e.g.* Yvon's approximation for the radial distribution function[6]); this approximation can also be expressed as a power series in the density.

In this way the first term of S can be calculated as a power series in the number density $N = \mathrm{N}/V$:

$$\frac{\varepsilon - 1}{\varepsilon + 2} = \frac{4\pi}{3} N\alpha[1 + \alpha^2(b_1 N + b_2 N^2 + \cdots) + O(\alpha^3)\cdots], \tag{6.101}$$

where b_1 and b_2 are coefficients that can be calculated when a suitable intermolecular potential has been chosen. De Boer, Van der Maesen and Ten Seldam[5] give numerical results, taking for the intermolecular potential a Lennard-Jones or a square-well potential. Both Kirkwood[4] and Brown[7] use a rigid-sphere intermolecular potential with and without Van der Waals attractive forces. They obtain equivalent expressions for the coefficients of the first and the second power of the number density; when a rigid-sphere intermolecular potential is chosen, this leads to:

$$b_1 = \frac{8\pi}{3\sigma^3}, \qquad b_2 = -\tfrac{5}{3}\pi^2, \tag{6.102}$$

where σ is the diameter of the rigid spheres.

The introduction of Van der Waals attractive forces leads to small correction terms:

$$b_1 = \frac{8\pi}{3\sigma^3}\left(1 + \frac{C}{3kT\sigma^6}\right), \quad b_2 = -\tfrac{5}{3}\pi^2\left(1 + 0.007\frac{C}{kT\sigma^6}\right), \tag{6.103}$$

where C is the proportionality constant of the Van der Waals attractive potential, which can be determined with the help of the Van der Waals' equation of state (see Kirkwood[4]) or from the second virial coefficient.

It is of interest to compare these expressions with the correction to the Clausius-Mossotti equation which has been derived in section 32 with the help of the continuum approach. To be able to do this the different expressions will have to be brought into the same form. Eqn. (5.105) must therefore be rewritten in the form of a power series in the polarizability and the number density.

Abbreviating $(4\pi/3)N\alpha = z$ and using $u = \alpha/a^3$ (*cf.* section 32), we may write eqn. (5.104), which is equivalent to eqn. (5.105), as:

$$\varepsilon - 1 = \frac{9\varepsilon z}{2\varepsilon + 1 - 2u(\varepsilon - 1)}. \tag{6.104}$$

Solving this equation for ε, we find:

$$\varepsilon = \frac{1 - 4u + 9z + 3\sqrt{1 + 2(1 - 4u)z + 9z^2}}{4(1 - u)}. \tag{6.105}$$

Developing the square root up to powers of z^3, eqn. (6.105) leads to:

$$\varepsilon = 1 + 3z + 3(1 + 2u)z^2 + 3(-1 + 2u + 8u^2)z^3 + \cdots. \tag{6.106}$$

With the help of this relation we obtain for $(\varepsilon - 1)/(\varepsilon + 2)$ after developing $(\varepsilon + 2)^{-1}$ in powers of z, and neglecting terms in powers of z higher than the third:

$$\frac{\varepsilon - 1}{\varepsilon + 2} = z[1 + 2uz + (8u^2 - 2u - 2)z^2 + \cdots]. \tag{6.107}$$

Since we are interested in the coefficient of α^2 in the correction factor, we must neglect the terms in uz^2 and u^2z^2. We then have:

$$\frac{\varepsilon - 1}{\varepsilon + 2} = z[1 + 2uz - 2z^2 + \cdots]. \tag{6.108}$$

Using the same abbreviation $z = (4\pi/3)N\alpha$, eqn. (6.101) leads in the case of a rigid-sphere intermolecular potential (eqn. (6.102)) to:

$$\frac{\varepsilon - 1}{\varepsilon + 2} = z\left[1 + 2\frac{\alpha}{\sigma^3}z - \frac{15}{16}z^2 + \cdots\right]. \tag{6.109}$$

Comparing eqns. (6.108) and (6.109), we see that the coefficients of z will be equal when a, the radius of the cavity, is equal to σ, the diameter of the rigid sphere, or equivalently the radius of the sphere of exclusion around the rigid sphere. This is plausible, since in the continuum model the polarizable surroundings of the molecule begin at a distance a from the centre, while for rigid spheres the centres of these spheres, where the polarizability is located, cannot approach closer than σ.

When Van der Waals attractive forces between the rigid spheres are introduced the coefficients b_1 and b_2 change slightly (*cf.* eqn. (6.103)). Comparing the expressions (6.102) and (6.103), we see that the introduction of attractive forces at constant value of σ leads to a decrease in the value of a which must be used to obtain the same value for the coefficient of z in the series development for $(\varepsilon - 1)/(\varepsilon + 2)$. Thus we expect the values of a determined from the experimental values of the dielectric constant and the density to be somewhat lower than σ. On the other hand, they will be always larger than $\frac{1}{2}\sigma$, the radius of the molecule (*cf.* section 32).

The expressions of Kirkwood and Brown (*cf.* eqns. (6.101), (6.102), and (6.103)) and the numerical results of De Boer, Van der Maesen and Ten Seldam for the corrected Clausius-Mossotti equation, as well as the correction derived in section 32 with the help of the continuum approach, all predict an initial increase of the Clausius-Mossotti expression $(\varepsilon - 1)/(\varepsilon + 2)N$ with increasing density, followed by a decrease. This is in qualitative agreement with the experimental results for non-polar gases (with the exception of helium, where quantum-mechanical effects dominate[8,9]); for liquids the density is so high that only the decrease can be observed (see *e.g.* Mopsik[10]). In the region of liquid densities the development in powers of the density cannot be broken off after a few terms, so that a quantitative comparison is not possible. For gases such a comparison is warranted. It appears from Fig. 32 that even for the noble gas argon, which is the simplest possible case

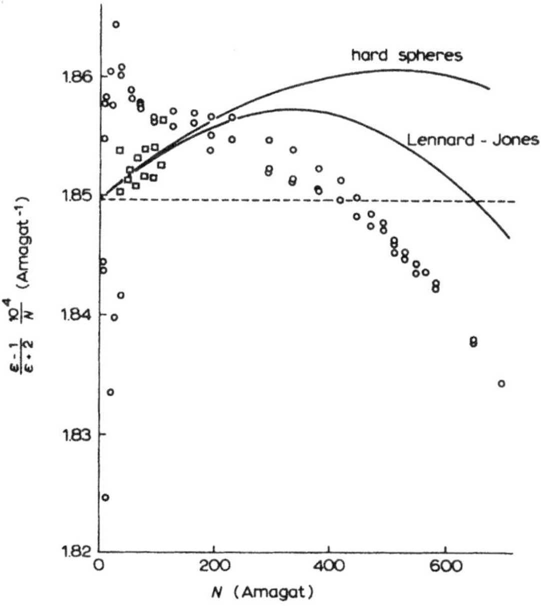

Fig. 32
Values of the Clausius-Mossotti expression as a function of density for argon. Measurements of Michels, Ten Seldam, and Overdijk[11] (○) and Oudemans[12] (□). The curves show the calculated values for a hard-sphere potential and a Lennard-Jones 6-12 potential. The Amagat unit is the unit used by Michels, Ten Seldam, and Overdijk: 1 Amagat = 4.4660 10^{-5} mole/cm^3.

for which experimental results are available,[5,11] there occur at high densities significant deviations from the theoretical results, as calculated with the help of a number of realistic intermolecular potentials.

In 1936, Michels, Michels and Bijl[13] suggested that the polarizability of a molecule is a function of the density. This suggestion is plausible because in principle, the movability of the elementary charges in a molecule will depend on the presence of neighbouring molecules. De Groot and Ten Seldam[14] developed a quantum-mechanical theory that predicts a decrease of the polarizability at high densities. Jansen and Mazur[15] put forward a quantum-mechanical perturbation theory predicting an additional increase of the Clausius-Mossotti expression at low densities. Although the experimental values of the dielectric constant and the density are not sufficiently accurate to allow a definite conclusion (*cf.* Fig. 32), such an increase at low densities, above the effect due to translational fluctuations, is not found.

The theory of Jansen and Mazur was refined by Jansen and Solem[16] and Jansen[17]; according to their results for the noble gases, the effects of the change in polarizability will be of the same order as the effects of the translational fluctuations and in the same direction. Careful measurements done by Orcutt and Cole[18] show that for the noble gases helium, argon, and krypton at low densities the Clausius-Mossotti expression has a lower value than that predicted on the grounds of translational fluctuations alone. Consequently the effect of the polarizability should be negative. This result was derived by Heinrichs[19] with the help of a quantum-mechanical perturbation theory. It is not clear on what point the theory of Jansen and his co-workers and that of Heinrichs diverge, although Lindner and Kromhout[20] suggest that the fundamental theory is sound and that additional approximations are responsible for the differing results. Until these difficulties have been resolved, we shall limit ourselves to the conclusion that according to the experimental results the change in polarizability due to increasing density decreases the dielectric constant. In view of the magnitude of this effect for the noble gases, it can be neglected in many other cases, where other effects play a more important rôle.

When the molecules are not spherical, other effects neglected up till now must be taken into account. A non-spherical molecule generally has a non-zero quadrupole moment that induces dipole moments in the surrounding molecules. In this way it contributes to the polarization of the dielectric, and thus to the dielectric constant. This effect will be important when the quadrupole moment of the molecule is large, *e.g.* in the case of CO_2 (*cf.* section 5, Table 3). Further, when the molecule is not spherical, the anisotropy of the polarizability will also play a rôle. This leads to the so-called rotational fluctuations (Kirkwood[4]), since now the average $\langle \boldsymbol{\alpha} \cdot \boldsymbol{T} \rangle$ is not equal to the product of the averages $\langle \boldsymbol{\alpha} \rangle \cdot \langle \boldsymbol{T} \rangle$ (*cf.* eqn. (6.46)). Even when the anisotropy of the polarizability is large, this effect is rather small. Of course, the quadrupole moment and the anisotropy of the polarizability also influence the intermolecular potential and thus indirectly the dielectric constant. This effect is small and can almost always be neglected.

An evaluation of the effects of the quadrupole moment and the rotational fluctuations has been given by Jansen[17] in a treatment which is a generalization of the work of Buckingham and Pople[21] and Zwanzig.[22] An extension of the expression for the quadrupole effect for axially symmetric molecules to the general case has been given by Johnston and Cole.[23] Their results are expressed as coefficients of a development of the Clausius-Mossotti expression $(\varepsilon - 1)/(\varepsilon + 2)N$ in powers of the number density. This procedure for polar as well as for non-polar gases will be treated in the next section.

§38. The dielectric virial coefficients of gases

In the last section we treated the Clausius-Mossotti expression as a power series in the polarizability α. It is also possible to develop this expression as a power series in the density, which is especially useful in the case of gases, where the density is small and the power series will converge rapidly. This method will prove to be applicable for polar as well as non-polar gases.

The development in powers of the density is analogous to the so-called virial series for the equation of state for non-ideal gases:

$$pV = A(T) + B(T)N + C(T)N^2 + \cdots, \tag{6.110}$$

where $N = \mathrm{N}/V$ is the number density. The temperature-dependent coefficients A, B, C, ... are called virial coefficients. When the density goes to zero, the gas becomes ideal and the ideal-gas law $pV = \mathrm{N}kT$ is valid. Applying this to eqn. (6.110) we find for the first virial coefficient:

$$A(T) = \mathrm{N}kT. \tag{6.111}$$

The higher virial coefficients can be determined directly by experiment, or from the expansion of a semi-empirical equation of state (*e.g.* Van der Waals' equation) in powers of the number density. It is also possible to derive statistical-mechanical expressions for the virial coefficients.[24,25] An advantage in the theoretical calculations is the fact that the second virial coefficient depends only on two-particle interactions, the third virial coefficient only on three-particle interactions, and so on. For instance the general expression for the second virial coefficient is:

$$B(T) = \frac{NkT}{2} \iiint\limits_V \iiint\limits_V [1 - \exp(-U_{ij}(\boldsymbol{r}_i, \boldsymbol{r}_j)/kT)] d\boldsymbol{r}_i d\boldsymbol{r}_j, \qquad (6.112)$$

where $\boldsymbol{r}_i$ and $\boldsymbol{r}_j$ are the radius vectors of the representative molecules i and j, and U_{ij} is the intermolecular potential.

When the molecules possess a permanent dipole moment, the intermolecular potential is given by a superposition of the non-electrostatic energy (*e.g.* a Lennard-Jones potential) and the electrostatic energy (*cf.* eqn. (3.120)). This type of intermolecular potential is called a Stockmayer potential, since Stockmayer[26] was the first to calculate second virial coefficients with it. He supposed the molecules to be non-polarizable; a more refined calculation, including effects due to the polarizability of the molecule, was given by Buckingham and Pople.[27]

For the development of the Clausius-Mossotti expression in powers of the number density we may write:

$$\frac{\varepsilon - 1}{\varepsilon + 2} \frac{1}{N} = \mathscr{A}(T) + \mathscr{B}(T)N + \mathscr{C}(T)N^2 + \cdots, \qquad (6.113)$$

where $\mathscr{A}(T)$, $\mathscr{B}(T)$, $\mathscr{C}(T)$, ... are the so-called dielectric virial coefficients. From the power series we can derive the following definitions for these coefficients:

$$\mathscr{A}(T) = \lim_{N \to 0} \frac{\varepsilon - 1}{\varepsilon + 2} \frac{1}{N}, \qquad (6.114)$$

$$\mathscr{B}(T) = \lim_{N \to 0} \frac{1}{N} \left(\frac{\varepsilon - 1}{\varepsilon + 2} \frac{1}{N} - \mathscr{A} \right), \qquad (6.115)$$

$$\mathscr{C}(T) = \lim_{N \to 0} \left(\frac{1}{N} \right)^2 \left(\frac{\varepsilon - 1}{\varepsilon + 2} \frac{1}{N} - \mathscr{A} - \mathscr{B}N \right), \qquad (6.116)$$

and so on.

To obtain expressions for the dielectric virial coefficients we shall use eqn. (6.43). For convenience we choose a dielectric with a spherical shape. Then we may write, according to eqns. (6.6) and (6.43):

$$\frac{\varepsilon - 1}{\varepsilon + 2}\frac{1}{N} = \frac{4\pi}{3}\left(\frac{1}{\mathrm{N}}\langle \mathbf{e}\cdot\mathbf{A}\cdot\mathbf{e}\rangle_0 + \frac{1}{3kT\mathrm{N}}\langle M^2\rangle_0\right). \tag{6.117}$$

According to the definition of the first dielectric virial coefficient $\mathscr{A}$ given in eqn. (6.114) $\mathscr{A}$ can be found by calculating the righthand member of eqn. (6.117) for the case of a very dilute gas. Then we have $\varepsilon + 2 = 3$, so that with the help of eqn. (6.44) we obtain:

$$\mathscr{A}(T) = \frac{4\pi}{3}\left(\alpha + \frac{\mu^2}{3kT}\right), \tag{6.118}$$

where α is the polarizability and μ the dipole strength of an isolated molecule.

Using (6.117) and (6.118) in the defining expression (6.115) we find for the second dielectric virial coefficient:

$$\mathscr{B}(T) = \lim_{N\to 0}\frac{4\pi}{3N}\left\{\frac{1}{\mathrm{N}}\langle \mathbf{e}\cdot\mathbf{A}\cdot\mathbf{e}\rangle_0 - \alpha + \frac{1}{3kT}\left(\frac{1}{\mathrm{N}}\langle M^2\rangle_0 - \mu^2\right)\right\}. \tag{6.119}$$

Since the limit $N \to 0$ is taken, only two-particle interactions have to be taken into account. To evaluate this expression we start from eqn. (6.13):

$$\mathbf{m} = \boldsymbol{\mu} + \alpha(\mathbf{E}_0 - \mathbf{T}\cdot\mathbf{m}).$$

In the case of two particles this equation can be solved explicitly. Considering only the molecules 1 and 2, we have:

$$\boldsymbol{m}_1 = \boldsymbol{\mu}_1 + \alpha\boldsymbol{E}_0 - \alpha\boldsymbol{T}_{12}\cdot\boldsymbol{m}_2, \tag{6.120}$$

$$\boldsymbol{m}_2 = \boldsymbol{\mu}_2 + \alpha\boldsymbol{E}_0 - \alpha\boldsymbol{T}_{21}\cdot\boldsymbol{m}_1. \tag{6.121}$$

The dipole-dipole interaction tensor is, according to eqn. (1.32), given by:

$$\boldsymbol{T}_{12} = \boldsymbol{T}_{21} = \boldsymbol{T} = \frac{1}{r^3}\left(\boldsymbol{I} - \frac{3}{r^2}\boldsymbol{rr}\right), \tag{6.122}$$

where $\boldsymbol{r} = \boldsymbol{r}_2 - \boldsymbol{r}_1$ is the distance between the two representative molecules. The simultaneous equations (6.120) and (6.121) can be solved following the procedure outlined in section 16. Substituting (6.121) into (6.120) we find:

$$\boldsymbol{m}_1 - \alpha^2\boldsymbol{T}\cdot\boldsymbol{T}\cdot\boldsymbol{m}_1 = \boldsymbol{\mu}_1 + \alpha\boldsymbol{E}_0 - \alpha\boldsymbol{T}\cdot\boldsymbol{\mu}_2 - \alpha^2\boldsymbol{T}\cdot\boldsymbol{E}_0. \tag{6.123}$$

Writing out this equation with the help of (6.122), we obtain an equation in $\boldsymbol{m}_1$ and $\boldsymbol{m}_1\cdot\boldsymbol{r}$. Inner multiplication of both sides of this equation with $\boldsymbol{r}$

leads to an expression for $\boldsymbol{m}_1 \cdot \boldsymbol{r}$:

$$\boldsymbol{m}_1 \cdot \boldsymbol{r} = (1 - 4\alpha^2/r^6)^{-1}\{\boldsymbol{\mu}_1 \cdot \boldsymbol{r} + \alpha(1 + 2\alpha/r^3)\boldsymbol{E}_0 \cdot \boldsymbol{r} + 2\alpha\boldsymbol{\mu}_2 \cdot \boldsymbol{r}/r^3\}. \quad (6.124)$$

This expression for $\boldsymbol{m}_1 \cdot \boldsymbol{r}$ can be substituted back into eqn. (6.123). In this way we find (*cf.* eqns. (3.134) and (3.135)):

$$\boldsymbol{m}_1 = \frac{\alpha}{1 + \alpha/r^3}\left\{\boldsymbol{E}_0 + \frac{3\alpha/r^5}{1 - 2\alpha/r^3}(\boldsymbol{E}_0 \cdot \boldsymbol{r})\boldsymbol{r}\right\} + $$
$$+ \frac{1}{1 - \alpha^2/r^6}\left\{\boldsymbol{\mu}_1 + \frac{3\alpha^2/r^8}{1 - 4\alpha^2/r^6}(\boldsymbol{\mu}_1 \cdot \boldsymbol{r})\boldsymbol{r}\right\} +$$
$$+ \frac{\alpha/r^3}{1 - \alpha^2/r^6}\left\{-\boldsymbol{\mu}_2 + \frac{1 - 2\alpha^2/r^6}{1 - 4\alpha^2/r^6}\frac{3}{r^2}(\boldsymbol{\mu}_2 \cdot \boldsymbol{r})\boldsymbol{r}\right\}, \quad (6.125)$$

and a corresponding expression for $\boldsymbol{m}_2$. Eqn. (6.125) can be written as the sum of the permanent and the induced moment of particle 1 when alone, and the contributions to the moment due to the interaction with particle 2:

$$\boldsymbol{m}_1 = \boldsymbol{\mu}_1 + \alpha\boldsymbol{E}_0 + \frac{\alpha^2/r^3}{1 + \alpha/r^3}\left\{-\boldsymbol{E}_0 + \frac{1}{1 - 2\alpha/r^3}\frac{3}{r^2}(\boldsymbol{E}_0 \cdot \boldsymbol{r})\boldsymbol{r}\right\} +$$
$$+ \frac{\alpha/r^3}{1 - \alpha^2/r^6}\left\{-\boldsymbol{\mu}_2 + \frac{1 - 2\alpha^2/r^6}{1 - 4\alpha^2/r^6}\frac{3}{r^2}(\boldsymbol{\mu}_2 \cdot \boldsymbol{r})\boldsymbol{r}\right\} +$$
$$+ \frac{\alpha^2/r^6}{1 - \alpha^2/r^6}\left\{\boldsymbol{\mu}_1 + \frac{1}{1 - 4\alpha^2/r^6}\frac{3}{r^2}(\boldsymbol{\mu}_1 \cdot \boldsymbol{r})\boldsymbol{r}\right\}. \quad (6.126)$$

From this expression for $\boldsymbol{m}_1$ we infer that in the case where only two--particle interactions have to be taken into account in the average, the moment $\boldsymbol{m}_i$ of particle i is given by $\boldsymbol{\mu}_i + \alpha\boldsymbol{E}_0$ plus the moments due to the interaction with all other particles separately, as given in (6.126):

$$\boldsymbol{m}_i = \boldsymbol{\mu}_i + \alpha\boldsymbol{E}_0 +$$
$$+ \sum_{k \neq i}^{N}\left[\frac{\alpha^2/r_{ik}^3}{1 + \alpha/r_{ik}^3}\left\{-\boldsymbol{E}_0 + \frac{1}{1 - 2\alpha/r_{ik}^3}\frac{3}{r_{ik}^2}(\boldsymbol{E}_0 \cdot \boldsymbol{r}_{ik})\boldsymbol{r}_{ik}\right\} +\right.$$
$$+ \frac{\alpha/r_{ik}^3}{1 - \alpha^2/r_{ik}^6}\left\{-\boldsymbol{\mu}_k + \frac{1 - 2\alpha^2/r_{ik}^6}{1 - 4\alpha^2/r_{ik}^6}\frac{3}{r_{ik}^2}(\boldsymbol{\mu}_k \cdot \boldsymbol{r}_{ik})\boldsymbol{r}_{ik}\right\} +$$
$$\left.+ \frac{\alpha^2/r_{ik}^6}{1 - \alpha^2/r_{ik}^6}\left\{\boldsymbol{\mu}_i + \frac{1}{1 - 4\alpha^2/r_{ik}^6}\frac{3}{r_{ik}^2}(\boldsymbol{\mu}_i \cdot \boldsymbol{r}_{ik})\boldsymbol{r}_{ik}\right\}\right]. \quad (6.127)$$

The coefficient of E_0 in expression (6.127) can be used to evaluate $\langle \mathbf{e} \cdot \mathbf{A} \cdot \mathbf{e} \rangle_0$ in eqn. (6.119) with the help of eqns. (6.7) and (6.39). Since after averaging the terms due to the interaction are the same for all values of k, we may write, replacing N − 1 by N since N is very large:

$$\frac{1}{\mathrm{N}}\langle \mathbf{e} \cdot \mathbf{A} \cdot \mathbf{e} \rangle_0 = \frac{1}{\mathrm{N}} \sum_{i=1}^{\mathrm{N}} \left[\alpha + \mathrm{N} \left\langle \frac{\alpha^2/r^3}{1 + \alpha/r^3} \left\{ -1 + \frac{3(\boldsymbol{e} \cdot \boldsymbol{r})^2/r^2}{1 - 2\alpha/r^3} \right\} \right\rangle_0 \right]. \tag{6.128}$$

Writing $\boldsymbol{e} \cdot \boldsymbol{r} = r \cos \theta$ and using $\langle \cos^2 \theta \rangle_0 = \frac{1}{3}$ (eqn. (6.42)), eqn. (6.128) becomes:

$$\frac{1}{\mathrm{N}}\langle \mathbf{e} \cdot \mathbf{A} \cdot \mathbf{e} \rangle_0 = \alpha + \mathrm{N} \left\langle \frac{2\alpha^3/r^6}{(1 + \alpha/r^3)(1 - 2\alpha/r^3)} \right\rangle_0. \tag{6.129}$$

The evaluation of $(1/\mathrm{N})\langle M^2 \rangle_0$ in eqn. (6.119) is more complicated, because of the occurrence of products $\boldsymbol{m}_i \cdot \boldsymbol{m}_j$. Since the average must be calculated for $E_0 = 0$, we may take $\boldsymbol{m}_i$ equal to $\boldsymbol{m}_i^0$; writing out the summations, we obtain:

$$\begin{aligned} \frac{1}{\mathrm{N}}\langle M^2 \rangle_0 &= \frac{1}{\mathrm{N}} \sum_{i=1}^{\mathrm{N}} \sum_{j=1}^{\mathrm{N}} \langle \boldsymbol{m}_i^0 \cdot \boldsymbol{m}_j^0 \rangle_0 \\ &= \langle \boldsymbol{m}_i^0 \cdot \boldsymbol{m}_i^0 \rangle_0 + \mathrm{N} \langle \boldsymbol{m}_i^0 \cdot \boldsymbol{m}_j^0 \rangle_0^{j \neq i}, \end{aligned} \tag{6.130}$$

since N − 1 ≃ N.

Using for the contribution to expression (6.127) due to the interaction in the absence of an external field the abbreviation $\boldsymbol{p}_{ik}^0$:

$$\begin{aligned} \boldsymbol{p}_{ik}^0 = {} & \frac{\alpha/r_{ik}^3}{1 - \alpha^2/r_{ik}^6} \left\{ -\boldsymbol{\mu}_k + \frac{1 - 2\alpha^2/r_{ik}^6}{1 - 4\alpha^2/r_{ik}^6} \frac{3}{r_{ik}^2} (\boldsymbol{\mu}_k \cdot \boldsymbol{r}_{ik}) \boldsymbol{r}_{ik} \right\} + \\ & + \frac{\alpha^2/r_{ik}^6}{1 - \alpha^2/r_{ik}^6} \left\{ \boldsymbol{\mu}_i + \frac{1}{1 - 4\alpha^2/r_{ik}^6} \frac{3}{r_{ik}^2} (\boldsymbol{\mu}_i \cdot \boldsymbol{r}_{ik}) \boldsymbol{r}_{ik} \right\}, \end{aligned} \tag{6.131}$$

we may write:

$$\boldsymbol{m}_i^0 = \boldsymbol{\mu}_i + \sum_{k \neq i} \boldsymbol{p}_{ik}^0. \tag{6.132}$$

The interaction between the molecules need only be taken into account for pairs of molecules, so that all triplet and higher interactions can be neglected. Therefore, we find:

$$\langle \boldsymbol{m}_i^0 \cdot \boldsymbol{m}_i^0 \rangle_0 = \langle (\boldsymbol{\mu}_i + \sum_{k \neq i} \boldsymbol{p}_{ik}^0) \cdot (\boldsymbol{\mu}_i + \sum_{j \neq i} \boldsymbol{p}_{ij}^0) \rangle_0$$

$$= \langle \boldsymbol{\mu}_i \cdot \boldsymbol{\mu}_i \rangle_0 + \sum_{j \neq i} \langle \boldsymbol{\mu}_i \cdot \boldsymbol{p}_{ij}^0 \rangle_0 + \sum_{k \neq i} \langle \boldsymbol{\mu}_i \cdot \boldsymbol{p}_{ik}^0 \rangle_0 +$$

$$+ \sum_{j \neq i} \sum_{k \neq i} \langle \boldsymbol{p}_{ij}^0 \cdot \boldsymbol{p}_{ik}^0 \rangle_0$$

$$= \mu^2 + 2\text{N} \langle \boldsymbol{\mu}_i \cdot \boldsymbol{p}_{ij}^0 \rangle_0^{j \neq i} + \text{N} \langle \boldsymbol{p}_{ij}^0 \cdot \boldsymbol{p}_{ij}^0 \rangle_0^{j \neq i}, \qquad (6.133)$$

where all terms with $k \neq j$ have been neglected. In the same way we obtain:

$$\langle \boldsymbol{m}_i^0 \cdot \boldsymbol{m}_j^0 \rangle_0^{j \neq i} = \langle (\boldsymbol{\mu}_i + \sum_{k \neq i} \boldsymbol{p}_{ik}^0) \cdot (\boldsymbol{\mu}_j + \sum_{l \neq j} \boldsymbol{p}_{jl}^0) \rangle_0^{j \neq i}$$

$$= \langle \boldsymbol{\mu}_i \cdot \boldsymbol{\mu}_j \rangle_0^{j \neq i} + \sum_{k \neq i} \langle \boldsymbol{\mu}_j \cdot \boldsymbol{p}_{ik}^0 \rangle_0^{j \neq i} +$$

$$+ \sum_{l \neq j} \langle \boldsymbol{\mu}_i \cdot \boldsymbol{p}_{jl}^0 \rangle_0^{j \neq i} + \sum_{k \neq i} \sum_{l \neq j} \langle \boldsymbol{p}_{ik}^0 \cdot \boldsymbol{p}_{jl}^0 \rangle_0^{j \neq i}$$

$$= \langle \boldsymbol{\mu}_i \cdot \boldsymbol{\mu}_j \rangle_0^{j \neq i} + \langle \boldsymbol{\mu}_j \cdot \boldsymbol{p}_{ij}^0 \rangle_0^{j \neq i} + \langle \boldsymbol{\mu}_i \cdot \boldsymbol{p}_{ji}^0 \rangle_0^{j \neq i} + \langle \boldsymbol{p}_{ij}^0 \cdot \boldsymbol{p}_{ji}^0 \rangle_0^{j \neq i}. \qquad (6.134)$$

Eqns. (6.133) and (6.134) can be substituted into eqn. (6.130), leading to:

$$\frac{1}{\text{N}} \langle M^2 \rangle_0 = \mu^2 + \text{N}[\langle \boldsymbol{\mu}_i \cdot \boldsymbol{\mu}_j \rangle_0 + 2 \langle \boldsymbol{\mu}_i \cdot \boldsymbol{p}_{ij}^0 \rangle_0 +$$

$$+ \langle \boldsymbol{\mu}_i \cdot \boldsymbol{p}_{ji}^0 \rangle_0 + \langle \boldsymbol{\mu}_j \cdot \boldsymbol{p}_{ij}^0 \rangle_0 + \langle \boldsymbol{p}_{ij}^0 \cdot \boldsymbol{p}_{ij}^0 \rangle_0 + \langle \boldsymbol{p}_{ij}^0 \cdot \boldsymbol{p}_{ji}^0 \rangle_0]. \qquad (6.135)$$

Here, the condition $j \neq i$ should be understood.

Using (6.131) and the coordinate system of Fig. 20 (p. 117) with $\boldsymbol{r}_{ij} = -\boldsymbol{r}_{ji} = \boldsymbol{r}$, we can work out eqn. (6.135). Using also the abbreviations A_n as introduced in eqn. (3.133):

$$A_n = 1 - n\alpha^2/r^6, \qquad (6.136)$$

we obtain after some rearrangements:

$$\frac{1}{\text{N}} \langle M^2 \rangle_0 = \mu^2 + \text{N}\mu^2 \Bigg[\langle \cos \gamma \rangle_0 +$$

$$+ \left\langle \frac{\alpha/r^3}{A_1} \left\{ -2(1 + \cos \gamma) + \frac{3A_2}{A_4} (\cos \theta_i + \cos \theta_j)^2 \right\} \right\rangle_0 +$$

$$+ \left\langle \frac{\alpha^2/r^6}{A_1^2} \left\{ (2A_1 + 1)(1 + \cos \gamma) + \frac{6A_1}{A_4} \cos^2 \theta_i + \right. \right.$$

$$\left. \left. + \frac{6A_1A_4 + 9A_2^2 - 6A_2A_4}{A_4^2} \cos \theta_i \cos \theta_j + \frac{9A_2^2 - 6A_2A_4}{A_4^2} \cos^2 \theta_j \right\} \right\rangle_0 +$$

$$+ \left\langle \frac{\alpha^3/r^9}{A_1^2} \left\{ -2(1 + \cos \gamma) + \frac{9A_2 + 3A_2A_4 - 3A_4}{A_4^2} (\cos \theta_i + \cos \theta_j)^2 \right\} \right\rangle_0 +$$

$$+\left\langle \frac{\alpha^4/r^{12}}{A_1^2}\left\{(1+\cos\gamma)+\frac{9+6A_4}{A_4^2}(\cos^2\theta_i+\cos\theta_i\cos\theta_j)\right\}\right\rangle_0\Bigg]. \quad (6.137)$$

An expression symmetric in i and j can be obtained by repeating the derivation with j and i instead of i and j as the representative molecules, adding the result to expression (6.137) and dividing by two. The more usual form for $\langle M^2\rangle_0$ is then obtained:

$$\frac{1}{\mathrm{N}}\langle M^2\rangle_0 = \mu^2 + \mathrm{N}\mu^2\Bigg[\langle\cos\gamma\rangle_0 +$$

$$+\left\langle \frac{\alpha/r^3}{A_1}\left\{-2(1+\cos\gamma)+\frac{3A_2}{A_4}(\cos\theta_i+\cos\theta_j)^2\right\}\right\rangle_0 +$$

$$+\left\langle \frac{\alpha^2/r^6}{A_1^2}\left\{(2A_1+1)(1+\cos\gamma)+\right.\right.$$

$$\left.\left.+\frac{9A_2^2+6A_1A_4-6A_2A_4}{2A_4^2}(\cos\theta_i+\cos\theta_j)^2\right\}\right\rangle_0 +$$

$$+\left\langle \frac{\alpha^3/r^9}{A_1^2}\left\{-2(1+\cos\gamma)+\frac{9A_2+3A_2A_4-3A_4}{A_4^2}(\cos\theta_i+\cos\theta_j)^2\right\}\right\rangle_0 +$$

$$+\left\langle \frac{\alpha^4/r^{12}}{A_1^2}\left\{(1+\cos\gamma)+\frac{9+6A_4}{2A_4^2}(\cos\theta_i+\cos\theta_j)^2\right\}\right\rangle_0\Bigg]. \quad (6.138)$$

The averages in expressions (6.129) and (6.138) are averages of functions of r, γ, θ_i, and θ_j or, according to eqn. (3.119), of r, θ_i, φ_i, θ_j and φ_j. Therefore, we may integrate over all orientation and position coordinates except $\boldsymbol{r}_i$ and $\boldsymbol{r}_j$ (*cf.* eqn. (6.53)). Subsequent integration over the position of the i-th molecule leads to a factor $V/4\pi$, so that we obtain:

$$\langle f(r,\theta_i,\varphi_i,\theta_j,\varphi_j\rangle_0$$

$$= \frac{1}{4\pi V}\int_0^\infty\int_0^\pi\int_0^{2\pi}\int_0^\pi\int_0^{2\pi} f\exp\left(-\frac{U_{ij}}{kT}\right) r^2\sin\theta_i\sin\theta_j\,\mathrm{d}r\,\mathrm{d}\theta_i\,\mathrm{d}\varphi_i\,\mathrm{d}\theta_j\,\mathrm{d}\varphi_j, \quad (6.139)$$

where U_{ij} is the interaction energy of the molecules i and j in the absence of other molecules and in the absence of an external field. U_{ij} can be given as the sum of the non-electrostatic interaction energy V_0 (*e.g.* a Lennard-Jones interaction potential) and the electrostatic interaction energy. Applying eqn. (3.136) with all charges taken equal to zero, and using the coordinate system of Fig. 20 we thus have:

$$U_{ij} = V_0 + \frac{\mu^2}{A_1 r^3}\left\{\cos\gamma - \frac{3A_2}{A_4}\cos\theta_i\cos\theta_j - \right.$$

$$\left. - \frac{\alpha}{2r^3}\left(2 + \frac{3}{A_4}(\cos^2\theta_i + \cos^2\theta_j)\right)\right\}. \tag{6.140}$$

When expressions (6.129), (6.138), and (6.140) are used in actual calculations of $\mathscr{B}$, they are first developed as a series in α/r^3. These series converge rapidly when r is not too small, and are usually broken off after the term in α^2/r^6 (*cf.* the magnitude of the corresponding terms in eqn. (3.137)). An advantage of this procedure is that it leads to the disappearance of the spurious singularities at $r^3 = \alpha$ and $r^3 = 2\alpha$, where A_1 and A_4, respectively, become zero. Since for all molecules the experimental values of $(4\pi/3)\alpha$ are smaller than the molecular volume (*cf.* section 10), these singularities will occur when r is smaller than the distance of closest approach. In that case the description of the molecular behaviour in terms of ideal dipoles and point polarizabilities breaks down (*cf.* section 16) and eqns. (6.129), (6.138), and (6.140) may not be applied. Actually, due to the repulsion at short distances, V_0 will become very large when r becomes small, so that this region of r-values will not be important for the integration.

Performing the series development in eqn. (6.129) up to terms in α^3 and in eqn. (6.138) up to terms in α^2, and substituting the resulting expressions into eqn. (6.119), we obtain:

$$\mathscr{B} = \lim_{N\to 0}\frac{4\pi}{3N}\mathrm{N}\left[\langle 2\alpha^3/r^6\rangle_0 + \frac{\mu^2}{3kT}\left\{\langle\cos\gamma\rangle_0 + \right.\right.$$

$$+ \left\langle \frac{\alpha}{r^3}\{-2(1+\cos\gamma) + 3(\cos\theta_i + \cos\theta_j)^2\}\right\rangle_0 +$$

$$\left.\left. + \left\langle \frac{\alpha^2}{r^6}\{3(1+\cos\gamma) + \tfrac{9}{2}(\cos\theta_i + \cos\theta_j)^2\}\right\rangle_0\right\}\right]. \tag{6.141}$$

Using (6.139), we obtain:

$$\mathscr{B} = -\tfrac{1}{3}\int_0^\infty\int_0^\pi\int_0^{2\pi}\int_0^\pi\int_0^{2\pi}\left[\frac{2\alpha^3}{r^6} + \right.$$

$$+ \frac{\mu^2}{3kT}\left\{\cos\gamma + \frac{\alpha}{r^3}\{-2(1+\cos\gamma) + 3(\cos\theta_i + \cos\theta_j)^2\} + \right.$$

$$+ \frac{\alpha^2}{r^6}\{3(1 + \cos\gamma) + \tfrac{9}{2}(\cos\theta_i + \cos\theta_j)^2\}\Bigg\}\Bigg]\exp(-U_{ij}/kT)\cdot$$

$$\cdot\, r^2 \sin\theta_i \sin\theta_j \, dr\, d\theta_i\, d\varphi_i\, d\theta_j\, d\varphi_j. \tag{6.142}$$

It is clear that in the integrand of eqn. (6.142) the first term (in α^3) will be much smaller than the second term (in μ^2) when μ is not very small. In the case of non-polar gases, however, it is the only remaining term; this is the reason why it has been included in eqn. (6.142).

In the case of non-polar gases the corrections to the Clausius-Mossotti equation due to translational fluctuations derived in section 37 should agree with the expression based on dielectric virial coefficients. When $\mu = 0$, eqn. (6.142) reduces to:

$$\mathscr{B} = \frac{32\pi^2\alpha^3}{3}\int_0^\infty \frac{1}{r^4}\exp(-U_{ij}/kT)\,dr, \tag{6.143}$$

since U_{ij} will only depend on the intermolecular distance and not on orientations. When a rigid-sphere intermolecular potential is chosen, we have:

$$\left.\begin{aligned} U_{ij} &= \infty \quad \text{for} \quad r < \sigma \\ U_{ij} &= 0 \quad \text{for} \quad r \geq \sigma \end{aligned}\right\}. \tag{6.144}$$

Substituting (6.144) into (6.143), we obtain upon integration:

$$\mathscr{B} = \frac{32\pi^2\alpha^3}{9\sigma^3}. \tag{6.145}$$

We can also calculate $\mathscr{B}$ by comparing eqn. (6.101) for the corrections to the Clausius-Mossotti equation with eqn. (6.113), where $\mathscr{B}$ is defined. The former expression is a development in powers of α, with coefficients which are calculated as series in the density; in the latter expression the coefficients of a power series in the density are given as a power series in α. Comparison of the corresponding terms now gives for $\mathscr{B}$:

$$\mathscr{B} = \frac{4\pi}{3}\alpha^3 b_1. \tag{6.146}$$

Substituting the rigid-sphere value of b_1 (eqn. (6.102)), we obtain:

$$\mathscr{B} = \frac{32\pi^2\alpha^3}{9\sigma^3},$$

which is the same result as given in eqn. (6.145).

It is clear that in principle the terms of $\mathscr{B}$ in α^4 and higher powers of α, as well as the higher dielectric virial coefficients, can also be correlated with the terms in the correction to the Clausius-Mossotti equation as given in section 37.

When an appropriate form for U_{ij} is chosen, the integrations in eqn. (6.142) can be performed (in closed form or numerically), and the values of $\mathscr{B}$ obtained in this way can be compared with the experimental values. Usually, the experimental values of $\mathscr{B}$ are given as molar quantities, *i.e.* as the coefficient in a series development in the molar density instead of the number density. In that case we write, denoting the molar density by $n = \text{N}/N_A V$:

$$\frac{\varepsilon - 1}{\varepsilon + 2}\frac{1}{n} = \mathscr{A}' + \mathscr{B}'n + \mathscr{C}'n^2 + \cdots, \tag{6.147}$$

and thus:

$$\mathscr{A}' = N_A\mathscr{A},$$
$$\mathscr{B}' = N_A^2\mathscr{B},$$
$$\mathscr{C}' = N_A^3\mathscr{C}, \text{ etc.}, \tag{6.148}$$

where N_A is Avogadro's number.

The results of the measurement of $\mathscr{B}'$ for a number of non-polar gases are given in Table 25. The theoretical values of $\mathscr{B}'$ given in this Table are calculated with the help of a Lennard-Jones intermolecular potential and Buckingham's method of integration.[27–29] For the choice of the Lennard-Jones parameters in these calculations we refer to the original publications (see Table 25).

TABLE 25

THE SECOND DIELECTRIC VIRIAL COEFFICIENT OF SOME NON-POLAR GASES

Gas	Temperature of measurement	$\mathscr{B}'$ (experimental) ($cm^6/mole^2$)	$\mathscr{B}'$ (theoretical) ($cm^6/mole^2$)	Ref.
Helium (He)	50°C	-0.05 ± 0.05	0.03	18
		-0.10 ± 0.05		
Neon (Ne)	50°C	-0.32 ± 0.10	0.05	18
Argon (Ar)	50°C	0.40 ± 0.20	2.0	18
Krypton (Kr)	50°C	5.7 ± 0.3	6.3	18
Hydrogen (H_2)	50°C	0.22 ± 0.20	0.04	18
		-0.15 ± 0.15		
		-0.02 ± 0.05		
Nitrogen (N_2)	50°C	0.65 ± 0.30	1.8	18
		0.83 ± 0.20		
Methane (CH_4)	42°C	7.3 ± 1.8	–	8
Carbon dioxide (CO_2)	50°C	42.1 ± 0.8	8.4	30, 9
		40.7 ± 0.7		
Ethene (C_2H_4)	50°C	46.4 ± 1.2	–	30
		43.2 ± 1.2		

It is clear that the theoretical values of $\mathscr{B}'$ all lie in the range 1–10 cm^6/mole^2. The experimental values fall into two sets: for the noble gases and the simple molecules H_2 and N_2 the experimental values are lower (sometimes only slightly) than the theoretical values, while for the less symmetrical molecules CO_2 and C_2H_4 the experimental values are much higher. Methane (CH_4) lies in between these two sets, and is a special case.

The lowering of the values of the Clausius-Mossotti expression corresponding with the lowering of the experimental values of $\mathscr{B}'$ has already been mentioned in section 37 (p. 230). This effect is explained as the result of a decrease of the polarizability as the density increases, although the theories put forward on this point are contradictory.[14–17,19,20]

The high values of $\mathscr{B}'$ in the case of carbon dioxide and ethene cannot be explained from the anisotropy of the polarizability. The influence of the anisotropy is too small,[31] while the effect of rotational fluctuations can always be neglected.[4]

To explain the high values of $\mathscr{B}'$ found for carbon dioxide and ethene the quadrupole moment of these molecules must be taken into account.* The contribution of the quadrupole moment to $\mathscr{B}$ can be calculated in the same way as the contributions due to the permanent dipole moments.

A quadrupole moment with quadrupole strength q induces dipole moments in the neighbouring molecules that are proportional to αq. These moments are directed by the external field, which leads to a term proportional to $(\alpha q)^2/3kT$ in $\mathscr{B}$. The calculation for the general case has been given by Johnston and Cole[23]; for axially symmetric molecules a simpler expression can be derived.[17] When the anisotropy of the polarizability is neglected, this expression reduces to:

$$\mathscr{B}_q = \frac{16\pi^2\alpha^2 q_{33}^2}{3kT} \int_0^\infty \frac{1}{r^6} \exp(-U_{ij}/kT)\, dr, \tag{6.149}$$

where $q_{33} = 2q$ is used to characterize the quadrupole moment (*cf.* section 5, eqn. (1.137)).

Taking for the absolute value of the quadrupole strength 2.3 10^{-26} e.s.u.

* The H_2 and N_2 molecules also possess a quadrupole moment, leading to contributions of 0.02 and 2.4 cm^6/mole^2 to $\mathscr{B}'$, respectively.[18] These contributions enlarge the theoretical value of $\mathscr{B}'$, however, and only worsen the agreement between theoretical and experimental values.

TABLE 26

EXPERIMENTAL AND CALCULATED VALUES OF THE SECOND DIELECTRIC VIRIAL COEFFICIENT OF CARBON DIOXIDE

Temperature	Experimental values of $\mathscr{B}'$ Ref. 9	Ref. 30	Theoretical values, with $q = 2.3\ 10^{-26}$ e.s.u.
50°C	40.0	41.4	40.8
100°C	36.5	34.7	34.9
150°C	—	30.0	31.0

for CO_2 and using a Lennard-Jones potential, Oudemans[32] calculated the value of the sum of the contributions to $\mathscr{B}'$ due to the translational fluctuations and the quadrupole moment (according to eqns. (6.143) and (6.149)) for a number of temperatures (see Table 26). The agreement between calculated and experimental temperature dependence is quite good in view of the experimental and theoretical uncertainties. The absolute value $2.3\ 10^{-26}$ e.s.u. for q compares favourably with values obtained with other methods.[33,34]

The experimental values of $\mathscr{B}'$ for ethene cannot be explained in the same way as was done for carbon dioxide with the help of the effect of a non-zero quadrupole moment, because these values appear to increase with increasing temperature. Johnston and Cole[23] suggest that this behaviour may be caused by the high quadrupole-quadrupole interaction energy, which favours configurations in which the quadrupole-induced dipole moments are antiparallel. When the temperature is increased, these configurations become less important, leading to higher values of $\mathscr{B}'$.

In the case of methane also a temperature-dependent second dielectric virial coefficient is found. In view of the symmetry, however, there cannot be a contribution due to the quadrupole moment in this case. But there is a non-zero octupole moment that can induce dipole moments in the neighbouring molecules.

Using a scalar polarizability α and characterizing the octupole moment by its octupole strength u (*cf.* section 5, p. 45), we find that the contribution $\mathscr{B}_u$ to the second dielectric virial coefficient is given by:[8]

$$\mathscr{B}_u = \frac{320\pi^2(\alpha u)^2}{9kT} \int_0^\infty \frac{1}{r^8} \exp(-U_{ij}/kT)\, dr. \qquad (6.150)$$

Assuming that the second dielectric virial coefficient is completely given by $\mathscr{B}_u$ and using a Lennard-Jones potential, Johnston, Oudemans and Cole[8] calculate from eqn. (6.150) a value $u = 3.4\ 10^{-34}$ e.s.u. for the octupole strength, which can be compared with the value $u = 1.8\ 10^{-34}$ e.s.u. given in section 5 (p. 57).

A further extension of the theory of the second dielectric virial coefficient for non-polar gases appears inappropriate at the moment in view of the scarcity of accurate experimental results.

For polar gases even fewer accurate measurements are available than for non-polar gases. A fortunate circumstance is that the effects are usually much larger, so that larger inaccuracies can be accepted. As an illustration, we give in Table 27 a number of accurate values for some halogenated methanes, as measured by Sutter and Cole.[35] Both positive and negative values for $\mathscr{B}'$ are found.

Buckingham and Pople[28] showed that integration of eqn. (6.142) for $\mathscr{B}$

TABLE 27

THE SECOND DIELECTRIC VIRIAL COEFFICIENT OF SOME HALOGENATED METHANES, AS MEASURED BY SUTTER AND COLE[35]

Substance	Molecular dimensions* a(Å)	b(Å)	b/a	Temperature (°C)	$\mathscr{B}'$ (cm^6/mole^2)
Fluoroform (CF_3H)	4.3	5.0	1.2	50.0	1125 ± 52
				96.3	903 ± 20
				143.3	704 ± 10
Trichlorofluoromethane (CCl_3F)	5.3	6.1	1.2	96.3	215 ± 30
Chlorotrifluoromethane ($CClF_3$)	5.5	5.0	0.9	50.0	60.0 ± 4.7
				96.3	53.5 ± 5.0
				143.3	46.8 ± 6.0
Fluoromethane (CFH_3)	4.5	4.0	0.9	50.0	−1307 ± 37
				96.3	−606 ± 30
				143.3	−331 ± 66
Chloromethane ($CClH_3$)	5.2	4.0	0.8	50.0	−4470 ± 200
				96.3	−2517 ± 50
				131.6	−1696 ± 60

* a and b denote dimensions of the molecule, in the direction of the symmetry axis and perpendicular to it, respectively. They have been estimated with the help of molecular models, using effective radii given by Briegleb[36] and interatomic distances and angles given by Stuart.[37] Briegleb's values for the atomic radii differ from the more generally used values given by Pauling[38] only for the fluorine atom where Briegleb's value seems more realistic.

using a Lennard-Jones or a comparable central-force intermolecular potential, will always lead to positive values for $\mathscr{B}$. They suggest that in fact the repulsive forces between molecules are shape-dependent; assuming a simple orientational dependence for the non-electrostatic intermolecular potential V_0, they obtain negative values for $\mathscr{B}$. In view of this suggestion estimates for the molecular dimensions are also given in Table 27. There appears to be a tendency for $\mathscr{B}'$ to decrease as the molecular shape changes from plate-like (oblate spheroids) to rod-like (prolate spheroids).

In section 16 it was shown that the anti-parallel configuration of two dipoles will be more favourable than the parallel configuration when the intermolecular distance can be sufficiently smaller in the first case (*cf.* Fig. 21). For plate-like molecules with the dipole moment in the direction of the symmetry-axis, such as CF_3H, the distance will be shortest in a parallel configuration. For molecules like CFH_3 and $CClH_3$, however, which have a rod-like shape, the dipole moment lies in the direction of the longest axis, and the anti-parallel configuration will be more favourable.

This leads to a preponderance of configurations with a low dipole moment, *i.e.* to negative values of $\mathscr{B}$. In this way the values of $\mathscr{B}'$ in Table 27 can be explained qualitatively.

It is rather difficult to express these ideas in a quantitative theory. Sutter and Cole[35] found that the suggestion of Buckingham and Pople,[28] *i.e.* that a simple dependence on orientation be introduced in V_0, leads to a rough agreement in the case of CFH_3, $CClH_3$ and $CClF_3$. The values for fluoroform could not, however, be explained in this way. More information about the orientational dependence of intermolecular potentials is required before further progress can be made. A first step in this direction are the calculations on a model for linear molecules given by Sweet and Steele,[39] which result in negative values for the second dielectric virial coefficient. A particularly promising development appears to be the refinement of this model by Chang,[40] who calculated values of $\mathscr{B}'$ for fluoromethane and fluoroform which are in good agreement with the experimental values.

The theory of the dielectric virial coefficients can be extended in several different directions. Expressions can be obtained for the second dielectric virial coefficient in the case of mixtures of gases.[41,42] A general formalism in which the effects due to multipole moments of any order can be taken into account has been given by Kielich.[43]

It is also possible to obtain explicit expressions for the third dielectric virial coefficient in the same way as has been done for the second dielectric virial coefficient in this section.[44] Since there are very few measurements which lead to significant values for the third dielectric virial coefficient, there is no reason to derive these expressions here. General expressions for the virial development of the Clausius-Mossotti expression have been given by Hill[45]; attempts to obtain specific results have been made by Kaufman and Watson[46] and for the case of non-polar systems by Isihara and Hanks.[47] Although the cluster-expansion technique used by these authors leads to interesting expressions, at this moment it is not possible to develop them to a form suitable for comparison with experiment.

§39. The dielectric constant of pure dipole liquids

A virial development of the Clausius-Mossotti expression $(\varepsilon - 1)/(\varepsilon + 2)N$, as applied for gases in the last section is no longer suitable at liquid densities. There is, however, another method for reducing the number of molecules that have to be taken into account in the statistical-mechanical averages, devised by Kirkwood[48] and further developed by Fröhlich.[49] This method consists of taking a region with $\mathcal{N}$ molecules which are treated explicitly; the remaining $\mathrm{N} - \mathcal{N}$ molecules are considered to form a continuum and are treated as such. The approximations in this method can be made as small as necessary by taking $\mathcal{N}$ sufficiently large. If this value of $\mathcal{N}$ is still manageable in the calculations, the method can be used to introduce the molecular interactions into the calculation of the dielectric constant of polar liquids.

First we shall consider the idealized case that the polarizability of the molecules can be neglected, so that only the permanent dipole moments have to be taken into account. Following Fröhlich, we take a sphere of volume V, containing $\mathcal{N}$ molecules. For convenience in the calculations we suppose that it is embedded in its own material, which extends to infinity. The material outside the sphere can be treated as a continuum with dielectric constant ε. In this case the external field working on the sphere is the cavity field (eqn. (2.70)):

$$\boldsymbol{E}_0 = \boldsymbol{E}_c = \frac{3\varepsilon}{2\varepsilon + 1}\boldsymbol{E}, \tag{6.151}$$

where $\boldsymbol{E}$ is the Maxwell field in the material outside the sphere, (*cf.* the definition of the fields in the dielectric in section 9a). Eqn. (6.151) can be substituted into eqn. (6.43); remembering that the molecules are non-polarizable, so that $\langle \mathbf{e} \cdot \mathbf{A} \cdot \mathbf{e} \rangle_0 = 0$, we obtain:

$$\varepsilon - 1 = \frac{4\pi}{V} \frac{3\varepsilon}{2\varepsilon + 1} \frac{\langle M^2 \rangle_0}{3kT}. \tag{6.152}$$

The average of the square of the total moment can be calculated as follows. First we note that in the non-polarizable case we have:

$$\boldsymbol{M} = \sum_{i=1}^{\mathcal{N}} \boldsymbol{\mu}_i, \tag{6.153}$$

so that we may write:

$$\langle M^2 \rangle_0 = \sum_{i=1}^{\mathcal{N}} \frac{\int \mathrm{d}X^{\mathcal{N}} \boldsymbol{\mu}_i \cdot \boldsymbol{M} \exp(-U/kT)}{\int \mathrm{d}X^{\mathcal{N}} \exp(-U/kT)}. \tag{6.154}$$

In this equation we have added the superscript $\mathcal{N}$ to $\mathrm{d}X$ to emphasize that the integration is performed over the positions and orientations of $\mathcal{N}$ molecules. The integration in the numerator of eqn. (6.154) can be carried out in two steps. Since $\boldsymbol{\mu}_i$ is a function of the orientation of the i-th molecule only, the integration over the positions and orientations of all other molecules, denoted as $\mathcal{N} - i$, can be carried out first. In this way we obtain (apart from a normalizing factor) the average moment of the sphere in the field of the i-th dipole with fixed orientation. The average moment, denoted by $\boldsymbol{M}_i^*$, can be written as:

$$\boldsymbol{M}_i^* = \frac{\int \mathrm{d}X^{\mathcal{N}-i} \boldsymbol{M} \exp(-U/kT)}{\int \mathrm{d}X^{\mathcal{N}-i} \exp(-U/kT)}. \tag{6.155}$$

The average moment $\boldsymbol{M}_i^*$ is a function of the position and orientation of the i-th molecule only. Expression (6.155) for $\boldsymbol{M}_i^*$ can be substituted into eqn. (6.154). Denoting the positional and orientational coordinates of the i-th molecule by X^i and using a weight factor $p(X^i)$ given by:

$$p(X^i) = \frac{\int \mathrm{d}X^{\mathcal{N}-i} \exp(-U/kT)}{\int \mathrm{d}X^{\mathcal{N}} \exp(-U/kT)}, \tag{6.156}$$

we obtain:

$$\langle M^2\rangle_0 = \sum_{i=1}^{\mathcal{N}} \int p(X^i)\boldsymbol{\mu}_i \cdot \boldsymbol{M}_i^* \, \mathrm{d}X^i,$$

$$= \mathcal{N} \int p(X^i)\boldsymbol{\mu}_i \cdot \boldsymbol{M}_i^* \, \mathrm{d}X^i, \qquad (6.157)$$

since after integration over the positions and orientations of molecule i, the resulting expression will not depend on the value of i.

Before we substitute the expression for $\langle M^2\rangle_0$ into eqn. (6.152), we note that it is possible to rewrite expression (6.157) in a suggestive form. According to eqns. (6.155) and (6.153) $\boldsymbol{M}_i^*$ can be written as the sum of moments $\boldsymbol{\mu}_j$, averaged with the orientation of the i-th dipole held fixed. Denoting the angle between the orientations of the i-th and the j-th dipole by θ_{ij}, this leads to:

$$\boldsymbol{\mu}_i \cdot \boldsymbol{M}_i^* = \mu^2 \sum_{j=1}^{\mathcal{N}} \frac{\int \mathrm{d}X^{\mathcal{N}-i} \cos\theta_{ij} \exp(-U/kT)}{\int \mathrm{d}X^{\mathcal{N}-i} \exp(-U/kT)},$$

and thus to:

$$\langle M^2\rangle_0 = \mathcal{N}\mu^2 \sum_{j=1}^{\mathcal{N}} \int p(X^i) \frac{\int \mathrm{d}X^{\mathcal{N}-i} \cos\theta_{ij} \exp(-U/kT)}{\int \mathrm{d}X^{\mathcal{N}-i} \exp(-U/kT)} \mathrm{d}X^i.$$

This expression for $\langle M^2\rangle_0$ can be abbreviated by introducing the average of $\cos\theta_{ij}$, defined as:

$$\langle \cos\theta_{ij}\rangle = \int p(X^i) \frac{\int \mathrm{d}X^{\mathcal{N}-i} \cos\theta_{ij} \exp(-U/kT)}{\int \mathrm{d}X^{\mathcal{N}-i} \exp(-U/kT)} \mathrm{d}X^i. \qquad (6.158)$$

We then write:

$$\langle M^2\rangle_0 = \mathcal{N}\mu^2 \sum_{j=1}^{\mathcal{N}} \langle \cos\theta_{ij}\rangle. \qquad (6.159)$$

If we now substitute eqn. (6.157) or its rewritten form, eqn. (6.159), into eqn. (6.152), we find after some rearrangement, and using $N = \mathcal{N}/V$ for the number density:

$$\frac{(\varepsilon - 1)(2\varepsilon + 1)}{12\pi\varepsilon} = \frac{N}{3kT}\int p(X^i)\boldsymbol{\mu}_i \cdot \boldsymbol{M}_i^* \, \mathrm{d}X^i$$

$$= \frac{N}{3kT}\mu^2 \sum_{j=1}^{\mathcal{N}} \langle\cos\theta_{ij}\rangle. \tag{6.160}$$

Eqn. (6.160) can be compared with expression (5.60) for the case of a pure dipole liquid with non-polarizable molecules ($\alpha = 0$):

$$\frac{(\varepsilon - 1)(2\varepsilon + 1)}{12\pi\varepsilon} = \frac{N}{3kT}\mu^2. \tag{6.161}$$

This expression was derived in Chapter V with the help of the continuum approach, in which a sphere containing only one molecule is used. If in the same way the sphere containing $\mathcal{N}$ molecules in eqn. (6.160) is restricted to one molecule, one has $\boldsymbol{M}_i^* = \boldsymbol{\mu}_i$ and the only term in the summation is $\cos\theta_{ii}$, which is equal to 1. Therefore, eqn. (6.161) is a special case of the more general eqn. (6.160).

When the sphere contains more than one molecule, the value of $\boldsymbol{M}_i^*$ can be different from $\boldsymbol{\mu}_i$. When the number of molecules included in the sphere increases, $\boldsymbol{M}_i^*$ reaches a limiting value, so that $\boldsymbol{M}_i^*$ will be independent of $\mathcal{N}$ as long as $\mathcal{N}$ exceeds a certain minimum value. According to eqn. (2.94) the dipole moment $\boldsymbol{M}$ of a sphere in the field of an arbitrary charge distribution within it, is given by:

$$\boldsymbol{M} = -\frac{\varepsilon - 1}{\varepsilon + 2}\boldsymbol{m}, \tag{6.162}$$

where $\boldsymbol{m}$ is the dipole moment of the charge distribution. Since this expression for $\boldsymbol{M}$ does not depend on the radius of the sphere, the dipole moment of a spherical shell in the field of a point dipole within the inner sphere must be zero. This conclusion will also hold if the sphere is not *in vacuo*, as was supposed in the derivation of eqn. (2.94), but embedded in a dielectric, even with the same dielectric constant as the sphere itself. Thus the addition of a number of molecules contained in a spherical shell to the original number $\mathcal{N}$ will not change the moment of the sphere as long as the spherical shell can be treated macroscopically. From this argument we conclude that the deviations of $\boldsymbol{M}_i^*$ from the value $\boldsymbol{\mu}_i$ are the result of molecular interactions between the i-th molecule and its neighbours.

It is well known that liquids are characterized by short-range order and long-range disorder. The correlations between the orientations (and also between the positions) due to the short-range ordering will lead to values of $\boldsymbol{M}_i^*$ differing from $\boldsymbol{\mu}_i$. For this reason Kirkwood[48] introduced a correlation factor g which accounted for the deviations of

$$\int p(X^i)\boldsymbol{\mu}_i \cdot \boldsymbol{M}_i^* \, \mathrm{d}X^i = \mu^2 \sum_{j=1}^{\mathcal{N}} \langle \cos \theta_{ij} \rangle$$

from the value μ^2:

$$g = \frac{1}{\mu^2} \int p(X^i)\boldsymbol{\mu}_i \cdot \boldsymbol{M}_i^* \, \mathrm{d}X^i = \sum_{j=1}^{\mathcal{N}} \langle \cos \theta_{ij} \rangle. \tag{6.163}$$

With the help of this definition, eqn. (6.160) may be written as:

$$\frac{(\varepsilon - 1)(2\varepsilon + 1)}{12\pi\varepsilon} = \frac{N}{3kT} g\mu^2. \tag{6.164}$$

When there is no more correlation between the molecular orientations than can be accounted for with the help of the continuum method, one has $g = 1$, and eqn. (6.164) reduces to Onsager's equation for the non-polarizable case, *i.e.* eqn. (5.67) with $\varepsilon_\infty = 1$ (see eqn. (6.161)).

An approximate expression for the Kirkwood correlation factor can be derived by taking only nearest-neighbours interactions into account. In that case the sphere is shrunk to contain only the i-th molecule and its z nearest neighbours. We then have:

$$\boldsymbol{M}_i^* = \frac{\int \mathrm{d}X^{\mathcal{N}-i} \left(\boldsymbol{\mu}_i + \sum_{j=1}^{z} \boldsymbol{\mu}_j \right) \exp(-U/kT)}{\int \mathrm{d}X^{\mathcal{N}-i} \exp(-U/kT)}, \tag{6.165}$$

with $\mathcal{N} = z + 1$. Substituting this into eqn. (6.163) and using the fact that the material is isotropic, we obtain:

$$g = 1 + \sum_{j=1}^{z} \int p(X^i) \, \mathrm{d}X^i \frac{\int \mathrm{d}X^{\mathcal{N}-i} \cos \theta_{ij} \exp(-U/kT)}{\int \mathrm{d}X^{\mathcal{N}-i} \exp(-U/kT)}. \tag{6.166}$$

Since after averaging the result of the integration will not depend on the value of j, all terms in the summation are equal and we may write:

$$g = 1 + z\langle \cos \theta_{ij} \rangle. \tag{6.167}$$

Since $\cos\theta_{ij}$ depends only on the orientations of the two molecules, all other coordinates can be integrated out and we may write:

$$\langle\cos\theta_{ij}\rangle = \frac{\int d\omega_i\, d\omega_j \cos\theta_{ij} \exp(-\bar{U}_{\omega_i\omega_j}/kT)}{\int d\omega_i\, d\omega_j \exp(-\bar{U}_{\omega_i\omega_j}/kT)}, \tag{6.168}$$

where ω_i and ω_j denote the orientational coordinates of the i-th and j-th molecules, and $\bar{U}_{\omega_i\omega_j}$ is a rotational intermolecular interaction energy, averaged over all positions and the orientations of all other molecules.

It is clear from eqn. (6.167) that g will be different from 1 when $\langle\cos\theta_{ij}\rangle \neq 0$, *i.e.* when there is correlation between the orientations of neighbouring molecules. When the molecules tend to direct themselves with parallel dipole moments, $\langle\cos\theta_{ij}\rangle$ will be positive and g will be larger than 1. When the molecules prefer an ordering with anti-parallel dipoles, g will be smaller than 1.

It is possible to obtain a better approximation for g by extending the number of neighbours taken into account. In section 40 this will be done for some special cases.

When the molecules are polarizable, the method of derivation as given above for non-polarizable molecules, is no longer applicable. If we would follow this method of derivation the total moment $\boldsymbol{M}$ of the sphere will be given by:

$$\boldsymbol{M} = \sum_{i=1}^{\mathcal{N}} (\boldsymbol{\mu}_i + \boldsymbol{p}_i), \tag{6.169}$$

where $\boldsymbol{p}_i$ is the induced moment of the i-th molecule. The induced moment $\boldsymbol{p}_i$ is a function of the positions and orientations of all other molecules, because according to eqn. (3.89) and (6.8) it is given by:

$$\boldsymbol{p}_i = \boldsymbol{m}_i - \boldsymbol{\mu}_i = \alpha(\boldsymbol{E}_1)_i, \tag{6.170}$$

where the local field $(\boldsymbol{E}_1)_i$ depends on the positions and orientations of all other molecules. For this reason, the total moment $\boldsymbol{m}_i = \boldsymbol{\mu}_i + \boldsymbol{p}_i$ cannot be held fixed without imposing a constraint on the positions and orientations of all other molecules. Therefore it is not possible to perform the integrations in $\langle M^2\rangle_0$ in two steps, as was done in the non-polarizable case.

Although it is possible to keep the position and the orientation of the permanent moment $\boldsymbol{\mu}_i$ fixed, the resulting expressions are rather complicated and can only be evaluated by introducing further approximations. Before we give an outline of this procedure, we shall first show that the polarizability of the molecules can also approximately be taken into account by the introduction of a more simplified model.

For the representation of a dielectric with dielectric constant ε, consisting of polarizable molecules with a permanent dipole moment, Fröhlich[49] introduced a continuum with dielectric constant ε_∞ in which point dipoles with a moment $\boldsymbol{\mu}_d$ are embedded. In this model each molecule is replaced by a point dipole $\boldsymbol{\mu}_d$ having the same non-electrostatic interactions with the other point dipoles as the molecules had, while the polarizability of the molecules can be imagined to be smeared out to form a continuum with dielectric constant ε_∞.

Because of the introduction of ε_∞ to represent the induced polarization (*cf.* section 27, p. 172), it is not possible to use eqn. (6.43) as it stands. To derive an expression which can be used, we shall split up the polarization $\boldsymbol{P}$ in two parts, the induced polarization $\boldsymbol{P}_{in}$ and the orientation polarization $\boldsymbol{P}_{or}$.* The induced polarization is equal to the polarization of the continuum with dielectric constant ε_∞, so that we can write (*cf.* eqn. (2.46)):

$$\boldsymbol{P}_{in} = \frac{\varepsilon_\infty - 1}{4\pi}\boldsymbol{E}. \tag{6.171}$$

The orientation polarization is given by the dipole density due to the dipoles $\boldsymbol{\mu}_d$. If we consider a sphere with volume V containing $\mathcal{N}$ dipoles (as we did in the non-polarizable case), we can write:

$$\boldsymbol{P}_{or} = \frac{1}{V}\langle \boldsymbol{M}_d \cdot \boldsymbol{e}\rangle, \tag{6.172}$$

where:

$$\boldsymbol{M}_d = \sum_{i=1}^{\mathcal{N}} (\boldsymbol{\mu}_d)_i. \tag{6.173}$$

$\langle \boldsymbol{M}_d \cdot \boldsymbol{e}\rangle$, the average component in the direction of the field, of the moment

* The division of $\boldsymbol{P}$ into $\boldsymbol{P}_{in}$ and $\boldsymbol{P}_{or}$ must be distinguished from the division into $\boldsymbol{P}_\alpha$ and $\boldsymbol{P}_\mu$ in eqn. (5.3). Apart from the contribution of the permanent dipoles ($\boldsymbol{P}_\mu$), $\boldsymbol{P}_{or}$ includes also that part of $\boldsymbol{P}_\alpha$ which is due to the presence of the permanent dipoles.

due to the dipoles in the sphere, is given by an expression analogous to eqn. (6.33):

$$\langle \boldsymbol{M}_{\mathrm{d}} \cdot \boldsymbol{e} \rangle = \frac{\int \mathrm{d}X^{\mathcal{N}} \boldsymbol{M}_{\mathrm{d}} \cdot \boldsymbol{e} \exp(-U/kT)}{\int \mathrm{d}X^{\mathcal{N}} \exp(-U/kT)}. \tag{6.174}$$

In this expression U is the energy of the dipoles in the sphere. This energy consists of three parts:

1. the energy of the dipoles in the external field;
2. the electrostatic interaction energy of the dipoles;
3. the non-electrostatic interaction energy between the molecules which is responsible for the short-range correlation between orientations and positions of the molecules (*cf.* p. 122).

The external field in this model is equal to the field within a spherical cavity filled with a continuum with dielectric constant ε_∞, while the cavity is situated in a dielectric with dielectric constant ε. This field will be called the Fröhlich field $\boldsymbol{E}_{\mathrm{F}}$. According to eqn. (2.69) it is given by:

$$\boldsymbol{E}_{\mathrm{F}} = \frac{3\varepsilon}{2\varepsilon + \varepsilon_\infty} \boldsymbol{E}. \tag{6.175}$$

Thus the energy of the dipoles in the external field can be written as $-\boldsymbol{M}_{\mathrm{d}} \cdot \boldsymbol{E}_{\mathrm{F}}$.

The derivation of eqn. (6.43) started from eqn. (6.4). To obtain eqn. (6.4), the first derivative of P with respect to E was identified with $(\varepsilon - 1)/4\pi$. Proceeding in the same way, we obtain in this case:

$$\frac{\varepsilon - 1}{4\pi} = \left(\frac{\partial}{\partial E} (P_{\mathrm{in}} + P_{\mathrm{or}}) \right)_{E=0},$$

or after substitution of (6.171) and (6.172) and rearrangement:

$$\varepsilon - \varepsilon_\infty = \frac{4\pi}{V} \left(\frac{\partial}{\partial E} \langle \boldsymbol{M}_{\mathrm{d}} \cdot \boldsymbol{e} \rangle \right)_{E=0}.$$

We now rewrite with E_{F} instead of E as the independent variable:

$$\varepsilon - \varepsilon_\infty = \frac{4\pi}{V} \left(\frac{\partial E_{\mathrm{F}}}{\partial E} \right)_{E=0} \left(\frac{\partial}{\partial E_{\mathrm{F}}} \langle \boldsymbol{M}_{\mathrm{d}} \cdot \boldsymbol{e} \rangle \right)_{E_{\mathrm{F}}=0}. \tag{6.176}$$

This equation has the same form as eqn. (6.4).

The differentiation with respect to E_F in expression (6.176) can be worked out in the same way as in section 36 (*cf.* eqns. (6.35)–(6.43)). Since in this case we have $\partial U/\partial E_F = -\boldsymbol{M}_d \cdot \boldsymbol{e}$, we obtain:

$$\varepsilon - \varepsilon_\infty = \frac{4\pi}{V}\left(\frac{\partial E_F}{\partial E}\right)_{E=0} \frac{\langle M_d^2 \rangle_0}{3kT}. \tag{6.177}$$

Eqn. (6.177) is the expression for the dielectric constant in Fröhlich's model, that corresponds with eqn. (6.43). The separation between the effects due to the dipoles $\boldsymbol{\mu}_d$ and the effects due to the continuum with dielectric constant ε_∞, however, differs from the separation in eqn. (6.43) between $\langle \mathbf{e} \cdot \mathbf{A} \cdot \mathbf{e} \rangle$ and $\langle M^2 \rangle_0$.

With the help of eqn. (6.175) for $\boldsymbol{E}_F$ we can write:

$$\varepsilon - \varepsilon_\infty = \frac{4\pi}{V} \frac{3\varepsilon}{2\varepsilon + \varepsilon_\infty} \frac{\langle M_d^2 \rangle_0}{3kT}, \tag{6.178}$$

or, after rearrangement:

$$\langle M_d^2 \rangle_0 = \frac{kTV}{4\pi} \frac{(\varepsilon - \varepsilon_\infty)(2\varepsilon + \varepsilon_\infty)}{\varepsilon}. \tag{6.179}$$

Eqn. (6.178) is analogous to eqn. (6.152) for the case that the dipoles are embedded in a continuum with dielectric constant ε_∞ instead of in a vacuum.

Eqn. (6.178) can be derived directly from eqn. (6.152) by replacing ε with $\varepsilon' = \varepsilon/\varepsilon_\infty$, the dielectric constant relative to ε_∞ instead of relative to vacuum ($\varepsilon = 1$). The average of the square of the total moment is replaced by $\langle M_d^2 \rangle_0/\varepsilon_\infty$, since the energy of a dipole in a continuum with dielectric constant ε_∞ is equal to $1/\varepsilon_\infty$ times the energy *in vacuo*, and $\langle M_d^2 \rangle_0$ is derived from the energy of the dipoles. After some rearrangements eqn. (6.178) is obtained.

The average $\langle M_d^2 \rangle_0$ in eqns. (6.178) or (6.179) can be evaluated in the same way as has been done for $\langle M^2 \rangle_0$ in eqn. (6.152). Instead of eqn. (6.164) we now obtain:

$$\frac{(\varepsilon - \varepsilon_\infty)(2\varepsilon + \varepsilon_\infty)}{12\pi\varepsilon} = \frac{N}{3kT} g\mu_d^2. \tag{6.180}$$

The moment $\boldsymbol{\mu}_d$ can be connected with the moment $\boldsymbol{\mu}$ of the molecule in the gas phase in the following way. In terms of the simplified model evaporation consists in the disengagement of small spheres with dielectric constant ε_∞ and a permanent dipole moment $\boldsymbol{\mu}_d$ in the centre. The moment $\boldsymbol{m}$ of such a sphere *in vacuo* consists of the permanent moment $\boldsymbol{\mu}_d$ and the moment induced by $\boldsymbol{\mu}_d$ in the surrounding dielectric as given by eqn. (2.94) with the substitution $\sum_i \boldsymbol{m}_i = \boldsymbol{\mu}_d$. We thus obtain for the moment $\boldsymbol{m}$:

$$\boldsymbol{m} = \boldsymbol{\mu}_d - \frac{\varepsilon_\infty - 1}{\varepsilon_\infty + 2}\boldsymbol{\mu}_d = \frac{3}{\varepsilon_\infty + 2}\boldsymbol{\mu}_d. \tag{6.181}$$

Obviously, $\boldsymbol{m}$ must be set equal to the moment $\boldsymbol{\mu}$ of the molecule in the gas phase. In this way we find:

$$\boldsymbol{\mu}_d = \frac{\varepsilon_\infty + 2}{3}\boldsymbol{\mu}. \tag{6.182}$$

Eqn. (6.182) can also be derived by following the reversed procedure, *i.e.* by considering the incorporation of a molecule in the medium. If a small sphere with permanent moment $\boldsymbol{\mu}$ and internal dielectric constant ε_∞ is embedded in a dielectric with dielectric constant ε_∞, the moment of the sphere will be enlarged by the reaction field to $\boldsymbol{\mu}^*$. According to eqn. (4.36) with the substitutions $\varepsilon = n_D^2 = \varepsilon_\infty$, we have for $\boldsymbol{\mu}^*$:

$$\boldsymbol{\mu}^* = \frac{(2\varepsilon_\infty + 1)(\varepsilon_\infty + 2)}{9\varepsilon_\infty}\boldsymbol{\mu}. \tag{6.183}$$

The external moment $\boldsymbol{\mu}_e$ of such a sphere is obtained with the help of eqn. (4.26):

$$\boldsymbol{\mu}_e = \frac{3\varepsilon_\infty}{2\varepsilon_\infty + 1}\boldsymbol{\mu}^* = \frac{\varepsilon_\infty + 2}{3}\boldsymbol{\mu}. \tag{6.184}$$

According to the definition of the external moment, $\boldsymbol{\mu}_e$ must be set equal to $\boldsymbol{\mu}_d$, so that we obtain eqn. (6.182).

Substituting eqn. (6.182) into eqn. (6.180), we obtain, after rearrangement:

$$g\mu^2 = \frac{9kT}{4\pi N}\,\frac{(\varepsilon - \varepsilon_\infty)(2\varepsilon + \varepsilon_\infty)}{\varepsilon(\varepsilon_\infty + 2)^2}. \tag{6.185}$$

Eqn. (6.185) is called the Kirkwood-Fröhlich equation. It can be seen as the generalization of the Onsager equation (eqn. (5.67)) for the case that specific molecular interactions occur.

In the derivation of the Kirkwood-Fröhlich equation a continuum with dielectric constant ε_∞ in which permanent dipoles μ_d are embedded, is used as a model for a pure dipole liquid, instead of a set of particles characterized by a polarizability, dipole moment, etc., in their centres. The use of a continuum with dielectric constant ε_∞ to represent the induced polarization implies the neglect of the correlations between positions and induced moments of the molecules, in the same way as in the case of the Clausius-Mossotti equation (see section 37). Since the deviations of the Clausius-

Mossotti equation are rather small, it may be supposed that the deviations of the Kirkwood-Fröhlich equation due to the neglect of the correlations mentioned above will also be small, in any case smaller than the effects due to the specific molecular interactions. This supposition can be checked by calculating g according to eqn. (6.163) or eqn. (6.167) and comparing the result with the value for g derived from the experimental values of ε, ε_∞, N, T, and μ, with the help of eqn. (6.185). If the molecular distribution in the liquid were sufficiently well known, the theoretical calculation of g could be carried through completely. Instead, it is only possible to derive qualitatively the behaviour of the Kirkwood correlation factor g starting from the specific molecular interactions, which is in accordance with the experimentally observed behaviour. In section 40 this will be shown for a number of cases. In this way it appears that the Kirkwood-Fröhlich equation can be used, even in the absence of a complete justification, as a convenient method of introducing the specific molecular interactions in the calculation of the dielectric constant of pure dipole liquids.

The approximations in the derivation of the Kirkwood-Fröhlich equation are analogous to the neglect of translational fluctuations in the derivation of the Clausius-Mossotti equation. This can be shown by deriving the Kirkwood-Fröhlich equation in another way described by Cole.[50] A slightly different version of this derivation has been given by Buckingham.[51]

We assume the dielectric to be a sphere with volume V, containing N molecules with polarizability α and permanent dipole strength μ. Therefore the general equation (6.43) with the field given by eqn. (6.6) may be applied. Neglecting translational fluctuations in the moments induced by the external field, the induced polarization given by $\langle \mathbf{e} \cdot \mathsf{A} \cdot \mathbf{e} \rangle_0$ in eqn. (6.43) can be set equal to the Clausius-Mossotti value for a dielectric with dielectric constant ε_∞, *i.e.*:

$$\langle \mathbf{e} \cdot \mathsf{A} \cdot \mathbf{e} \rangle_0 = \frac{3V}{4\pi} \frac{\varepsilon_\infty - 1}{\varepsilon_\infty + 2}. \tag{6.186}$$

The orientation polarization, given by $\langle M^2 \rangle_0 / 3kT\mathrm{N}$ in eqn. (6.43), is obtained from the linear term of a series development of $\langle \boldsymbol{M} \cdot \boldsymbol{e} \rangle$ in powers of E_0.

For the evaluation of $\langle M^2 \rangle_0$ the moments of the molecules in the absence of an external field must be used. From eqn. (6.12) we see that the only remaining terms, denoted by $\boldsymbol{m}_i^0$ in analogy to eqn. (6.17), are given by:

$$\boldsymbol{m}_i^0 = \boldsymbol{\mu}_i - \alpha \sum_{j \neq i}^{\mathrm{N}} \boldsymbol{T}_{ij} \cdot \boldsymbol{\mu}_j + \alpha^2 \sum_{j \neq i}^{\mathrm{N}} \sum_{k \neq j}^{\mathrm{N}} \boldsymbol{T}_{ij} \cdot \boldsymbol{T}_{jk} \cdot \boldsymbol{\mu}_k + \cdots . \qquad (6.187)$$

When the orientations of the molecules are distributed isotropically, the moment induced in molecule i by the permanent moments of the other molecules is zero in a spherical sample, leading to the approximation $\boldsymbol{m}_i^0 = \boldsymbol{\mu}_i$. Since this means in any case the neglect of the reaction field it is not a very realistic approximation. When the sum of the moments m_i^0 is considered, however, the approximations are compensated to a large extent, and completely when translational fluctuations are neglected. After averaging, all values of i lead to the same result, so that we may write:

$$\langle M^2 \rangle_0 = \mathrm{N} \langle \boldsymbol{\mu}_i \cdot \sum_{j=1}^{\mathrm{N}} \boldsymbol{m}_j^0 \rangle_0 . \qquad (6.188)$$

In this expression the averaging can be performed in two steps: first the permanent moment $\boldsymbol{\mu}_i$ is held fixed and the integration over the positions and orientations of all other molecules is carried out, then the average over molecule i is taken.

Before the calculation is performed, we reduce the number of molecules that have to be taken into account. The total moment $\sum_{j=1}^{\mathrm{N}} \boldsymbol{m}_j^0$ of the N molecules in the sphere is considered to be the sum of the moment $\boldsymbol{M}_i^{0*}$ of a small sphere containing $\mathcal{N}$ molecules and centred on the i-th molecule with fixed orientation, and the moment of the remaining molecules. When $\mathcal{N}$ is sufficiently large, the system consisting of a small sphere containing $\mathcal{N}$ molecules in a large sphere containing N molecules can be treated macroscopically. In section 9g we calculated the moment of a sphere in the field of a charge distribution within it. To be able to apply the equations of section 9g, we have to use the external moment of the small sphere in a dielectric with dielectric constant ε. According to eqn. (4.26), the external moment is given by $\frac{3\varepsilon}{2\varepsilon + 1} \boldsymbol{M}_i^{0*}$. Using eqn. (2.94) with the substitution $\sum_i \boldsymbol{m}_i = \frac{3\varepsilon}{2\varepsilon + 1} \boldsymbol{M}_i^{0*}$, we find for the total moment of the large sphere:

$$\sum_{j=1}^{\mathrm{N}} \boldsymbol{m}_j^0 = \frac{3\varepsilon}{2\varepsilon + 1} \boldsymbol{M}_i^{0*} - \frac{\varepsilon - 1}{\varepsilon + 2} \frac{3\varepsilon}{2\varepsilon + 1} \boldsymbol{M}_i^{0*} = \frac{9\varepsilon}{(\varepsilon + 2)(2\varepsilon + 1)} \boldsymbol{M}_i^{0*} . \qquad (6.189)$$

If we now substitute this result into eqn. (6.188) and the resulting expression into eqn. (6.43), together with eqn. (6.186) and eqn. (6.6), we obtain:

$$\varepsilon - 1 = 4\pi N \frac{\varepsilon + 2}{3} \left\{ \frac{3V}{4\pi \text{N}} \frac{\varepsilon_\infty - 1}{\varepsilon_\infty + 2} + \frac{1}{3kT} \frac{9\varepsilon}{(\varepsilon + 2)(2\varepsilon + 1)} \langle \boldsymbol{\mu}_i \cdot \boldsymbol{M}_i^{0*} \rangle_0 \right\}, \quad (6.190)$$

where $N = \text{N}/V$ is the number density. This expression can be rearranged to obtain:

$$\langle \boldsymbol{\mu}_i \cdot \boldsymbol{M}_i^{0*} \rangle_0 = \frac{9kT}{4\pi N} \frac{(\varepsilon - \varepsilon_\infty)(2\varepsilon + 1)}{3\varepsilon(\varepsilon_\infty + 2)}. \quad (6.191)$$

Expression (6.191) does not depend on the volume V of the large sphere nor on the number of particles contained in it. Therefore, V can be extended to infinity without changing our results. In this way we obtain a system resembling the model used in the derivation of the Kirkwood-Fröhlich equation. A difference is, however, the use of the moments $\boldsymbol{m}_j^0$ in the small sphere instead of the moments $\boldsymbol{\mu}_d$.

The average $\langle \boldsymbol{\mu}_i \cdot \boldsymbol{M}_i^{0*} \rangle_0$ in eqn. (6.191) differs from μ^2 for two reasons: specific molecular interactions lead to preferred orientations around the central molecule, and the moments of the molecules are increased with the moments induced by all other molecules. As a first approximation, we assume that the increased dipole moment is constant during the movements of the molecule, *i.e.* we neglect fluctuations of the moment induced by other molecules. As a further approximation we use the value of the increased moment calculated with the help of the continuum approach.

The latter approximation implies that the specific interactions between the molecules are not taken into account, insofar as they influence the value of the induced moments of the molecules. The effects of the specific interactions now only occur in the preferred orientations of the permanent moments around the central molecule. Since the influence of the specific interactions on the induced moments is much smaller than the influence on the average of the products of permanent moments, we may follow Buckingham[51] in considering the use of the continuum approach for the calculation of the increased moments as a reasonable approximation.

According to eqn. (4.36) with the substitution $n_D^2 = \varepsilon_\infty$, we have:

$$\boldsymbol{m}_j^0 = \frac{2\varepsilon + 1}{2\varepsilon + \varepsilon_\infty} \frac{\varepsilon_\infty + 2}{3} \boldsymbol{\mu}_j. \quad (6.192)$$

In this approximation we may therefore write:

$$\boldsymbol{M}_i^{0*} = \frac{2\varepsilon + 1}{2\varepsilon + \varepsilon_\infty} \frac{\varepsilon_\infty + 2}{3} \boldsymbol{M}_i^*, \quad (6.193)$$

where $\boldsymbol{M}_i^* = \sum_{j=1}^{\mathcal{N}} \boldsymbol{\mu}_j$, is given in eqn. (6.155) or (6.165). Substituting eqn. (6.193) into eqn. (6.191) and using the abbreviation (6.163), we can write:

$$g\mu^2 = \frac{9kT}{4\pi N} \frac{(\varepsilon - \varepsilon_\infty)(2\varepsilon + \varepsilon_\infty)}{\varepsilon(\varepsilon_\infty + 2)^2}. \tag{6.194}$$

This expression is identical to the Kirkwood-Fröhlich equation, eqn. (6.185).

If we compare the two derivations of the Kirkwood-Fröhlich equation given in this section, we see that the first derivation, based on Fröhlich's model, is rigorous as far as it goes, and that the only question is how far the model is applicable to a pure dipole liquid. In the second derivation the molecular distribution can in principle be taken into account, but rather rough approximations are made to avoid the necessity of calculating averages of products of interaction terms of the form $\alpha \boldsymbol{T}_{ij} \cdot \boldsymbol{\mu}_j$. It does not seem worthwhile to refine these approximations, since too little is known about the molecular distribution functions in pure dipole liquids. For crystals, however, the interaction terms can be calculated,[1,52] and the second method of derivation leads to better results (see Volume II).

§40. The Kirkwood correlation factor

The Kirkwood-Fröhlich equation, eqn. (6.185):

$$g\mu^2 = \frac{9kT}{4\pi N} \frac{(\varepsilon - \varepsilon_\infty)(2\varepsilon + \varepsilon_\infty)}{\varepsilon(\varepsilon_\infty + 2)^2},$$

gives the relation between ε, the dielectric constant, ε_∞, the dielectric constant of induced polarization introduced in section 27, the temperature, the density, and the permanent dipole moment, for those cases where the intermolecular interactions are sufficiently well known to calculate the correlation factor g. If there are no specific correlations one has $g = 1$, and the Kirkwood-Fröhlich equation reduces to the Onsager equation, eqn. (5.67):

$$\mu^2 = \frac{9kT}{4\pi N} \frac{(\varepsilon - \varepsilon_\infty)(2\varepsilon + \varepsilon_\infty)}{\varepsilon(\varepsilon_\infty + 2)^2}.$$

If the correlations are not negligible, detailed information about the molecular interactions is required for the calculation of g. For associating compounds, where the occurrence of hydrogen bonds makes relevant the assumption that only certain specific angles between the dipoles of neighbouring molecules are possible, the molecular interactions may be represented by simplified models. For polymers, analogous models can be used if each segment is treated as a separate entity of which the orientation of the dipole moment is correlated with the orientations of the dipole moments of the other segments. In many cases these models will contain undetermined parameters, so that the value of g cannot be calculated explicitly. In these cases g is often calculated as an empirical quantity with the help of eqn. (6.185), and the value of μ determined in some other way. The unknown parameters can then be found from the experimental value of g. For the evaluation of the undetermined parameters in the model it is often useful to determine g as a function of the temperature or as a function of the concentration when the compound is dissolved in a non-polar solvent. For solutions, however, eqn. (6.185) cannot be used as such, because it is only valid for pure dipole liquids. Therefore, it is useful to have available an analogon of eqn. (6.185), valid for solutions of polar compounds in non-polar solvents.

We shall derive this extended equation for those compounds where the molecules or the polar segments form clusters of limited size. This is the case for all compounds at low concentrations, and also at high concentrations for ideal polymers and compounds forming no more than two hydrogen bonds (*e.g.* the mono-alcohols and the mono-carboxylic acids). The chains and rings formed by associating compounds of this kind will be denoted as multimers.

Considering each kind of polymers or multimers as a separate compound and neglecting any specific correlation between the total dipole moments of the polymers or multimers, we can apply eqn. (5.60):

$$\frac{(\varepsilon - 1)(2\varepsilon + 1)}{12\pi\varepsilon} = \frac{N_0\alpha_0}{1 - f_0\alpha_0} + \sum_{n=1}^{\infty} \frac{N_n}{1 - f_n\alpha_n}\left(\alpha_n + \frac{1}{3kT}\,\frac{\overline{\mu_n^2}}{1 - f_n\alpha_n}\right). \qquad (6.195)$$

Here the index 0 refers to the non-polar solvent, and the index n refers to the polymers or multimers containing n polar units. The average $\overline{\mu_n^2}$ is taken over all conformations of the polymer or multimer. The upper limit of the summation can be extended to infinity since N_n, the number of n-mers per cm^3, becomes zero when n becomes large. The polarizability α_0 can be calculated from the Clausius-Mossotti equation for the pure solvent (*cf.* eqns. (5.40) and (5.41)):

$$\alpha_0 = \frac{3}{4\pi}\frac{\varepsilon_0 - 1}{\varepsilon_0 + 2}\frac{M_0}{d_0 N_A}. \tag{6.196}$$

Combining the Clausius-Mossotti equation and the Onsager approximation for the radius of the cavity, we have (*cf.* eqn. (5.64)):

$$\frac{1}{1 - f_0\alpha_0} = \frac{2\varepsilon + 1}{2\varepsilon + \varepsilon_0}\frac{\varepsilon_0 + 2}{3}. \tag{6.197}$$

We assume that the polarizabilities and the molecular volumes of the n-mers are proportional to n, so that:

$$\alpha_n = n\alpha_1, \tag{6.198}$$

where:

$$\alpha_1 = \frac{3}{4\pi}\frac{\varepsilon_\infty - 1}{\varepsilon_\infty + 2}\frac{M_1}{d_1 N_A}. \tag{6.199}$$

Here, d_1 is the density and ε_∞ is the dielectric constant of induced polarization of the polar compound in the pure state. In the same way we find, using Onsager's approximation for the radius of the cavity:

$$\frac{1}{1 - f_n\alpha_n} = \frac{2\varepsilon + 1}{2\varepsilon + \varepsilon_\infty}\frac{\varepsilon_\infty + 2}{3}. \tag{6.200}$$

We now substitute eqns. (6.196) to (6.200) into eqn. (6.195) and divide both members by $(2\varepsilon + 1)$, obtaining:

$$\frac{\varepsilon - 1}{12\pi\varepsilon} = \frac{N_0(\varepsilon_0 - 1)M_0}{4\pi(2\varepsilon + \varepsilon_0)d_0 N_A} + \frac{(\varepsilon_\infty - 1)M_1}{4\pi(2\varepsilon + \varepsilon_\infty)d_1 N_A}\sum_{n=1}^{\infty} nN_n + \\ + \frac{(2\varepsilon + 1)(\varepsilon_\infty + 2)^2}{27kT(2\varepsilon + \varepsilon_\infty)^2}\sum_{n=1}^{\infty} N_n\overline{\mu_n^2}. \tag{6.201}$$

The average $\overline{\mu_n^2}$ can be calculated as:

$$\overline{\mu_n^2} = \overline{\left(\sum_{i=1}^{n} \boldsymbol{\mu}_{n,i}\right)\cdot\left(\sum_{j=1}^{n} \boldsymbol{\mu}_{n,j}\right)}, \tag{6.202}$$

where $\boldsymbol{\mu}_{n,i}$ is the moment of the i-th dipolar unit of the n-mer. By taking the i-th dipolar unit as the representative unit and averaging over all possible positions of the i-th unit in the chain, we may write eqn. (6.202) as follows:

$$\overline{\mu_n^2} = n\,\overline{\boldsymbol{\mu}_{n,i}\cdot\left(\sum_{j=1}^{n} \boldsymbol{\mu}_{n,j}\right)} = ng_n\mu^2, \tag{6.203}$$

where μ has been used to denote the dipole strength of a single unit. The factor g_n represents the average value of the ratio between the component of the moment of the whole n-mer in the direction of the permanent moment of an arbitrary segment and the dipole strength of the segment. Since we assumed that there is no correlation between the different chains, g_n represents the total correlation between the permanent moment of a segment in a n-mer and its surroundings.

To obtain the Kirkwood correlation factor g, we must average g_n over all values of n, with a weight factor equal to the chance that a segment forms part of a n-mer. This chance is given by $nN_n/\sum_{n=1}^{\infty} nN_n$, if N_n is the number of n-mers per cm^3. Thus, we find with the help of eqn. (6.203):

$$g = \frac{\sum_{n=1}^{\infty} nN_n g_n}{\sum_{n=1}^{\infty} n N_n} = \frac{\sum_{n=1}^{\infty} N_n \overline{\mu_n^2}}{\mu^2 \sum_{n=1}^{\infty} n N_n}. \tag{6.204}$$

We now use molar fractions x_0 and x_p for the non-polar and the polar component respectively, regarding each segment as a separate molecule:

$$N_0 = \frac{x_0 N_A}{\varphi}, \tag{6.205}$$

$$\sum_{n=1}^{\infty} n N_n = \frac{x_p N_A}{\varphi}. \tag{6.206}$$

In these equations $\varphi = (x_0 M_0 + x_p M_1)/d$ denotes the molar volume of the mixture. Substituting eqns. (6.204), (6.205), and (6.206) into eqn. (6.201), we find:

$$\frac{\varepsilon - 1}{12\pi\varepsilon} = \frac{x_0(\varepsilon_0 - 1)M_0}{4\pi(2\varepsilon + \varepsilon_0)\varphi d_0} + \frac{x_p(\varepsilon_\infty - 1)M_1}{4\pi(2\varepsilon + \varepsilon_\infty)\varphi d_1} + $$
$$+ \frac{(2\varepsilon + 1)(\varepsilon_\infty + 2)^2 x_p N_A}{27kT(2\varepsilon + \varepsilon_\infty)^2 \varphi} g\mu^2. \tag{6.207}$$

From this it follows:

$$g\mu^2 = \frac{9kT(2\varepsilon + \varepsilon_\infty)^2}{4\pi N_A x_p(\varepsilon_\infty + 2)^2(2\varepsilon + 1)} \cdot \left[\frac{\varphi(\varepsilon - 1)}{\varepsilon} - \frac{3x_0 M_0(\varepsilon_0 - 1)}{(2\varepsilon + \varepsilon_0)d_0} - \frac{3x_p M_1(\varepsilon_\infty - 1)}{(2\varepsilon + \varepsilon_\infty)d_1}\right]. \tag{6.208}$$

This equation makes possible the calculation of the Kirkwood correlation factor g from experimental data for solutions of associating or polymeric compounds in non-polar solvents.

For the pure dipole liquid we have $x_0 = 0$, $x_p = 1$, $\varphi = M_1/d_1$, and eqn. (6.208) reduces to the Kirkwood-Fröhlich equation, eqn. (6.185):

$$g\mu^2 = \frac{9kT}{4\pi N} \frac{(\varepsilon - \varepsilon_\infty)(2\varepsilon + \varepsilon_\infty)}{\varepsilon(\varepsilon_\infty + 2)^2},$$

where $N = d_1 N_A/M_1$ is the number of segments per cm^3. This expression is also obtained for the case that the dielectric constant of induced polarization of the polar component is equal to the dielectric constant of the non-polar solvent provided that the volume of the mixture is an additive quantity.

Eqn. (6.201) reduces for the pure dipole liquid to:

$$\sum_{n=1}^{\infty} N_n \overline{\mu_n^2} = \frac{9kT}{4\pi} \frac{(\varepsilon - \varepsilon_\infty)(2\varepsilon + \varepsilon_\infty)}{\varepsilon(\varepsilon_\infty + 2)^2}. \tag{6.209}$$

Combining (6.209) with the Kirkwood-Fröhlich equation we obtain the expression of g as given in eqn. (6.204) since $N = \sum nN_n$. It can easily be derived from eqn. (5.65) that eqn. (6.209) holds not only for pure associating compounds and polymers but also generally for all mixtures of compounds having the same value of ε_∞.

(a) *Polymers*

A general method for the calculation of g for chain-like polymers based on work of Birshtein and Ptitsyn is given by Volkenstein.[53,54] In this method the average value of μ_n^2 is calculated as:

$$\overline{\mu_n^2} = \overline{\left(\sum_{i=1}^{n} \boldsymbol{\mu}_{n,i}\right) \cdot \left(\sum_{j=1}^{n} \boldsymbol{\mu}_{n,j}\right)} = n\mu^2 + 2 \sum_{i=2}^{n} \sum_{j=1}^{i-1} \overline{\boldsymbol{\mu}_{n,i} \cdot \boldsymbol{\mu}_{n,j}}. \tag{6.210}$$

To calculate $\overline{\boldsymbol{\mu}_{n,i} \cdot \boldsymbol{\mu}_{n,j}}$, we ascribe to each atom in the chain its own right handed coordinate system. The centre of gravity of the atom k is chosen as the origin of the k-th coordinate system, the z-axis is chosen in the direction of the centre of gravity of the $(k+1)$-th atom (or to the centre of a hydrogen atom for $k = n$). The xz-plane is chosen through the centre of the atom $k - 1$, with the positive x-axis pointing away from this atom (see Fig. 33). In this way, for a certain conformation of the chain, each atom is connected with a uniquely defined coordinate system, and the components

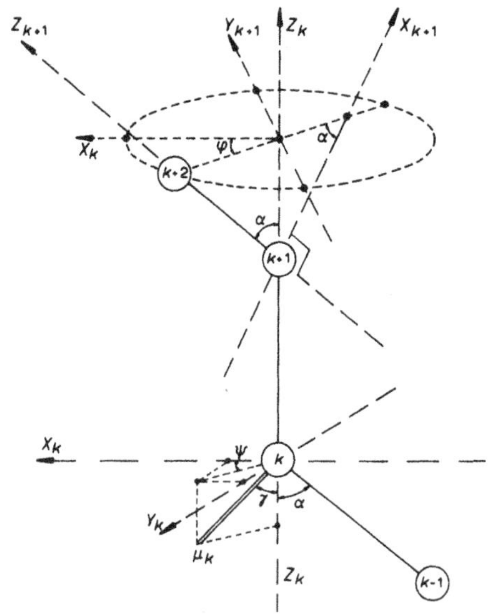

Fig. 33
Coordinate systems for subsequent atoms in a polymer chain.

of each dipole vector $\boldsymbol{\mu}_{n,i}$ can be given in the coordinate system of the k-th atom:

$$\boldsymbol{\mu}_{n,i} = (\mu_{ix}^{(k)}, \mu_{iy}^{(k)}, \mu_{iz}^{(k)}). \tag{6.211}$$

In the components of $\boldsymbol{\mu}_{n,i}$, the index n has been suppressed for the sake of brevity.

For the calculation of the product $\boldsymbol{\mu}_{n,i} \cdot \boldsymbol{\mu}_{n,j}$, one must express both dipole moments in the same coordinate system. This is arranged by using the orthogonal transformation matrix C_k, which transforms the $(k+1)$-th coordinate system into the k-th coordinate system (see Appendix I, section 6). Using the bond angle α_k between the z-axes of the coordinate systems $k+1$ and k, and the angle of rotation φ_k about the k-th bond (the angle between the xz-plane of the $(k+1)$-th coordinate system and the x-axis of the k-th coordinate system), we find:

$$C_k = \begin{pmatrix} -\cos\alpha_k \cos\varphi_k & \sin\varphi_k & \sin\alpha_k \cos\varphi_k \\ -\cos\alpha_k \sin\varphi_k & -\cos\varphi_k & \sin\alpha_k \sin\varphi_k \\ \sin\alpha_k & 0 & \cos\alpha_k \end{pmatrix}. \tag{6.212}$$

Using this matrix, we can write $\boldsymbol{\mu}_i \cdot \boldsymbol{\mu}_j$ as:

$$\boldsymbol{\mu}_i \cdot \boldsymbol{\mu}_j = \boldsymbol{\mu}_j \cdot \boldsymbol{\mu}_i = (\mu_{jx}^{(j)}, \mu_{jy}^{(j)}, \mu_{jz}^{(j)}) \prod_{k=j}^{i-1} C_k \begin{pmatrix} \mu_{ix}^{(i)} \\ \mu_{iy}^{(i)} \\ \mu_{iz}^{(i)} \end{pmatrix}. \tag{6.213}$$

Although this expression is valid generally, we shall use it only for the case that each atom carries a polar group. The other cases will be treated separately.

We now introduce in our model the assumptions that the bond angles α_k are equal to a constant value α and that the rotations about the bonds are independent of each other. Then the average $\overline{\boldsymbol{\mu}_i \cdot \boldsymbol{\mu}_j}$ can be written as:

$$\begin{aligned} \overline{\boldsymbol{\mu}_i \cdot \boldsymbol{\mu}_j} &= \overline{(\mu_{jx}^{(j)}, \mu_{jy}^{(j)}, \mu_{jz}^{(j)}) \prod_{k=j}^{i-1} C_k \begin{pmatrix} \mu_{ix}^{(i)} \\ \mu_{iy}^{(i)} \\ \mu_{iz}^{(i)} \end{pmatrix}} \\ &= \overline{(\mu_{jx}^{(j)}, \mu_{jy}^{(j)}, \mu_{jz}^{(j)})} \bar{C}^{i-j} \overline{\begin{pmatrix} \mu_{ix}^{(i)} \\ \mu_{iy}^{(i)} \\ \mu_{iz}^{(i)} \end{pmatrix}}. \end{aligned} \tag{6.214}$$

$\bar{C}$ is found from eqn. (6.212), using the abbreviations $\eta = \overline{\cos\varphi}$ and $\varepsilon = \overline{\sin\varphi}$:

$$\bar{C} = \begin{pmatrix} -\eta\cos\alpha & \varepsilon & \eta\sin\alpha \\ -\varepsilon\cos\alpha & -\eta & \varepsilon\sin\alpha \\ \sin\alpha & 0 & \cos\alpha \end{pmatrix}. \tag{6.215}$$

In many cases the potential for the rotation about a bond is symmetrical. Since $\sin\varphi$ is an odd function of φ, we then have $\varepsilon = \overline{\sin\varphi} = 0$, and:

$$\bar{C} = \begin{pmatrix} -\eta\cos\alpha & 0 & \eta\sin\alpha \\ 0 & -\eta & 0 \\ \sin\alpha & 0 & \cos\alpha \end{pmatrix}. \tag{6.216}$$

In some cases it is useful to suppose the internal rotation to be free, so that one has $\varepsilon = \eta = 0$, leading to:

$$\bar{C} = \begin{pmatrix} 0 & 0 & 0 \\ 0 & 0 & 0 \\ \sin\alpha & 0 & \cos\alpha \end{pmatrix}. \tag{6.217}$$

The direction of the dipole $\boldsymbol{\mu}_i$ in the i-th coordinate system is characterized by the angle γ with respect to the negative z-axis, and by the angle of rotation ψ (see Fig. 33). So we may write:

$$\overline{(\mu_{ix}^{(i)}, \mu_{iy}^{(i)}, \mu_{iz}^{(i)})} = \mu(\overline{\sin\gamma\cos\psi}, \overline{\sin\gamma\sin\psi}, \overline{-\cos\gamma})\cdot \tag{6.218}$$

Substituting expression (6.218) into (6.214) and then (6.214) into (6.210), we find:

$$\overline{\mu_n^2} = \mu^2\left[n + 2(\overline{\sin\gamma\cos\psi}, \overline{\sin\gamma\sin\psi}, \overline{-\cos\gamma})\cdot \left\{\sum_{i=2}^{n}\sum_{j=1}^{i-1} \bar{C}^{i-j}\right\}\begin{pmatrix} \overline{\sin\gamma\cos\psi} \\ \overline{\sin\gamma\sin\psi} \\ -\overline{\cos\gamma} \end{pmatrix}\right]. \tag{6.219}$$

The factor $\left\{\sum_{i=2}^{n}\sum_{j=1}^{i-1} \bar{C}^{i-j}\right\}$ can be calculated as the summation of a double geometric series:

$$\sum_{i=2}^{n}\sum_{j=1}^{i-1} \bar{C}^{i-j} = (n-1)\bar{C}[I-\bar{C}]^{-1} - \bar{C}^2[I-\bar{C}^{n-1}][I-\bar{C}]^{-2}. \tag{6.220}$$

Substituting this result into eqn. (6.219) and substituting the resulting expression into eqn. (6.204) for g, we find:

$$g = 1 + \frac{2}{\sum_{n=1}^{\infty} nN_n}\sum_{n=1}^{\infty} N_n(\overline{\sin\gamma\cos\psi}, \overline{\sin\gamma\sin\psi}, -\overline{\cos\gamma})\cdot \{(n-1)\bar{C}[I-\bar{C}]^{-1} - \bar{C}^2[I-\bar{C}^{n-1}][I-\bar{C}]^{-2}\}\begin{pmatrix} \overline{\sin\gamma\cos\psi} \\ \overline{\sin\gamma\sin\psi} \\ -\overline{\cos\gamma} \end{pmatrix}. \tag{6.221}$$

In the limit of infinitely long chains this leads to:

$$g = 1 + 2(\overline{\sin\gamma\cos\psi}, \overline{\sin\gamma\sin\psi}, -\overline{\cos\gamma})\bar{C}[I - \bar{C}]^{-1}\begin{pmatrix}\overline{\sin\gamma\cos\psi}\\ \overline{\sin\gamma\sin\psi}\\ -\overline{\cos\gamma}\end{pmatrix}. \qquad (6.222)$$

In these expressions the matrix $\bar{C}$ is given by eqn. (6.215); the matrix $[I - \bar{C}]^{-1}$ is then:

$$[I - \bar{C}]^{-1} = \frac{1}{(1-\eta^2-\varepsilon^2)(1-\cos\alpha)} \cdot \begin{pmatrix}(1+\eta)(1-\cos\alpha) & -\varepsilon(\cos\alpha-1) & \{\varepsilon^2+\eta(1+\eta)\}\sin\alpha\\ -\varepsilon(\cos\alpha-1) & (1-\eta)(1-\cos\alpha) & \varepsilon\sin\alpha\\ (1+\eta)\sin\alpha & \varepsilon\sin\alpha & 1+\eta+\{\varepsilon^2+\eta(1+\eta)\}\cos\alpha\end{pmatrix}. \qquad (6.223)$$

For polymers of the type $[-CX_2-]_n$, such as polytetrafluoroethylene, the direction of the dipole cannot change independently of the conformation of the chain. For tetrahedric bond angles, the direction of the dipole is characterized by the fixed values $\psi = 0$, $\cos\gamma = \frac{1}{3}\sqrt{3}$, and α is given by $\cos\alpha = \frac{1}{3}$. Using these figures in the evaluation of eqn. (6.222), one arrives at:

$$g = 0. \qquad (6.224)$$

This result was to be expected, because it is known that due to the symmetry a CF_4 molecule has no permanent dipole moment, and that a polytetrafluoroethylene molecule can be derived from a CF_4 molecule by successively replacing F atoms by CF_3 groups, which does not change the permanent dipole moment.

For polymers of the type $[-CHX-]_n$, the stereoisomerism of the compounds must be taken into account. In the simplest case, all dipole moments have the same components in the coordinate system of the corresponding atoms. This holds for syndiotactic chains, *i.e.* chains for which in the trans-conformation the substituents also lie trans with respect to each other:

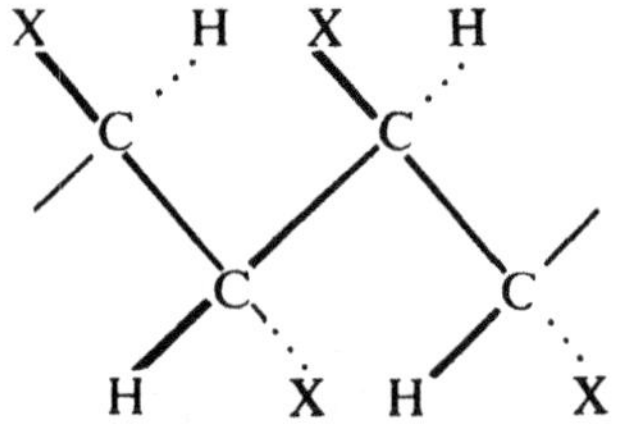

The potential for the rotation around the C–C bonds is then symmetrical, so that $\varepsilon = 0$. For tetrahedric bond angles, $\cos\alpha = \cos\gamma = \frac{1}{3}$ and $\psi = 60°$. Using these values, we find from eqn. (6.222):

$$g = \tfrac{2}{3}\,\frac{1-\eta}{1+\eta}. \tag{6.225}$$

For free rotation $\eta = 0$, so that:

$$g = \tfrac{2}{3}. \tag{6.226}$$

When the chain is held in the trans-conformation exclusively, $\eta = 1$, which leads to:

$$g = 0. \tag{6.227}$$

The latter result was to be expected, because for the trans-chain the dipole moments of two neighbouring groups compensate each other.

Not only syndiotactic but also isotactic chains can be treated in this way. For these chains, the polar substituents are in the cis-position with respect to each other when the polymer chain is in the trans-conformation:

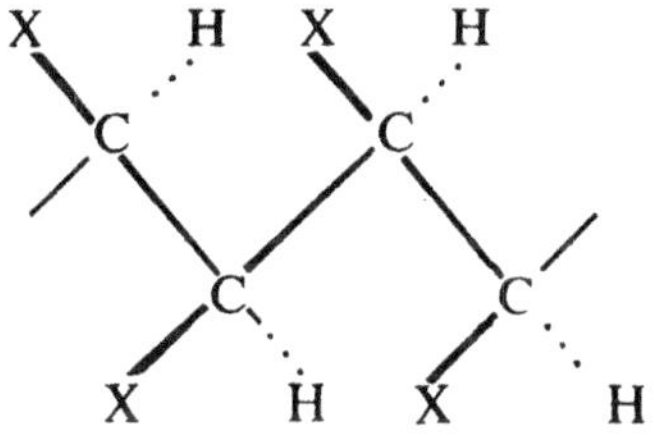

If in this case we ascribe to each atom a righthanded coordinate system, the components of neighbouring dipole moments in their own coordinate system are not identical, but we would have:

$$\psi_k = -\psi_{k+1}. \tag{6.228}$$

For this reason, we now attribute coordinate systems to the atoms that are alternately righthanded and lefthanded. In this way each dipole moment has the same components in its own coordinate system. The matrix of transformation is then indicated by C'_k and given by:

$$C'_k = \begin{pmatrix} -\cos\alpha\cos\varphi_k & -\sin\varphi_k & \sin\alpha\cos\varphi_k \\ -\cos\alpha\sin\varphi_k & \cos\varphi_k & \sin\alpha\sin\varphi_k \\ \sin\alpha & 0 & \cos\alpha \end{pmatrix}. \tag{6.229}$$

The correlation factor g for infinitely long chains with independent rotations about the bonds is given by:

$$g = 1 + 2(\overline{\sin\gamma\cos\psi},\ \overline{\sin\gamma\sin\psi},\ -\overline{\cos\gamma})\overline{C'}[I - \overline{C'}]^{-1}\begin{pmatrix} \overline{\sin\gamma\cos\psi} \\ \overline{\sin\gamma\sin\psi} \\ -\overline{\cos\gamma} \end{pmatrix}, \qquad (6.230)$$

where $\overline{C'}$ is derived from eqn. (6.229) as:

$$\overline{C'} = \begin{pmatrix} -\eta\cos\alpha & -\varepsilon & \eta\sin\alpha \\ -\varepsilon\cos\alpha & \eta & \varepsilon\sin\alpha \\ \sin\alpha & 0 & \cos\alpha \end{pmatrix}. \qquad (6.231)$$

For tetrahedric bond angles this leads to:

$$g = \tfrac{2}{3}\,\frac{1 - \eta^2 - \varepsilon^2}{(1 - \eta)^2 + \varepsilon^2}. \qquad (6.232)$$

For free rotation $\eta = \varepsilon = 0$, and g obtains the same value as for the syndiotactic chain (*cf.* eqn. (6.226)). For the trans-chain, however, it leads to $g = \infty$, owing to the fact that the correlation between the i-th and j-th dipole moment does not disappear for large values of $(i - j)$. For rigid chains the number of segments in the chain is not proportional to the square of the dipole moment of the chain (compare eqn. (6.203)), but to the dipole moment itself. This proportionality can be calculated in a simple way, leading to:

$$\mu_n = \tfrac{1}{3}n\mu\sqrt{6}. \qquad (6.233)$$

It is also possible to give expressions for the g-factor of polymers where not every atom in the chain carries a polar group. For chains of the type $[-CH_2-CX_2]_n$, such as polyvinylidene chloride, the product $\boldsymbol{\mu}_i \cdot \boldsymbol{\mu}_j$ is given by:

$$\boldsymbol{\mu}_i \cdot \boldsymbol{\mu}_j = \boldsymbol{\mu}_j \cdot \boldsymbol{\mu}_i = (\mu_{jx}^{(j)}, \mu_{jy}^{(j)}, \mu_{jz}^{(j)}) \prod_{k=2j}^{2i-1} C_k \begin{pmatrix} \mu_{ix}^{(i)} \\ \mu_{iy}^{(i)} \\ \mu_{iz}^{(i)} \end{pmatrix}. \qquad (6.234)$$

Here a coordinate system is ascribed to the atoms in the chain, which do not carry a polar group, in the same way as for the other atoms. Substituting (6.234) into (6.210) and the result into (6.204), we find in the limit of infinitely long chains:

$$g = 1 + 2(\overline{\sin\gamma\cos\psi}, \overline{\sin\gamma\sin\psi}, -\overline{\cos\gamma})\bar{C}^2[I - \bar{C}^2]^{-1}\begin{pmatrix}\overline{\sin\gamma\cos\psi}\\ \overline{\sin\gamma\sin\psi}\\ -\overline{\cos\gamma}\end{pmatrix}. \quad (6.235)$$

Assuming tetrahedric bond angles, we obtain, since $\varepsilon = 0$ due to the symmetry of the rotation around the C–C bonds:

$$g = \tfrac{3}{4}\,\frac{1+\eta}{1-\eta}. \quad (6.236)$$

It can easily be seen that this expression for the g-factor holds also for compounds like $[-CH_2-O]_n$ if the C–O–C bond angle is tetrahedric, because in such chains the dipole is directed in the same way as in polyvinylidene chloride.

For chains of the type $[-CH_2-CHX]_n$, such as polyvinylchloride and the polyhalostyrenes, isotactic and syndiotactic chains must again be distinguished. For isotactic chains of this type, all dipoles have the same components in their own coordinate system when all coordinate systems are chosen righthanded. In that case the values of $\overline{\sin\varphi_k}$ have alternating signs, however. This can be redressed by ascribing a lefthanded coordinate system to the atoms not connected with a dipolar group. Then the matrix of transformation is given by eqn. (6.229) and for infinitely long chains g is given by:

$$g = 1 + 2(\overline{\sin\gamma\cos\psi}, \overline{\sin\gamma\sin\psi}, -\overline{\cos\gamma})\bar{C}'^2[I - \bar{C}'^2]^{-1}\begin{pmatrix}\overline{\sin\gamma\cos\psi}\\ \overline{\sin\gamma\sin\psi}\\ -\overline{\cos\gamma}\end{pmatrix}. \quad (6.237)$$

For tetrahedric bond angles we find:

$$g = \frac{1}{1-\eta^2-\varepsilon^2}\left\{\tfrac{11}{12}(1+\eta^2+\varepsilon^2) + \tfrac{1}{2}\eta - \tfrac{4}{3}\,\frac{\varepsilon^2}{(1-\eta)^2+\varepsilon^2} + \tfrac{2}{3}\varepsilon\sqrt{3}\right\}. \quad (6.238)$$

For free rotation $\eta = \varepsilon = 0$, and this reduces to:

$$g = \tfrac{11}{12}, \quad (6.239)$$

a value first calculated by Debye and Bueche.[55]

For syndiotactic chains,

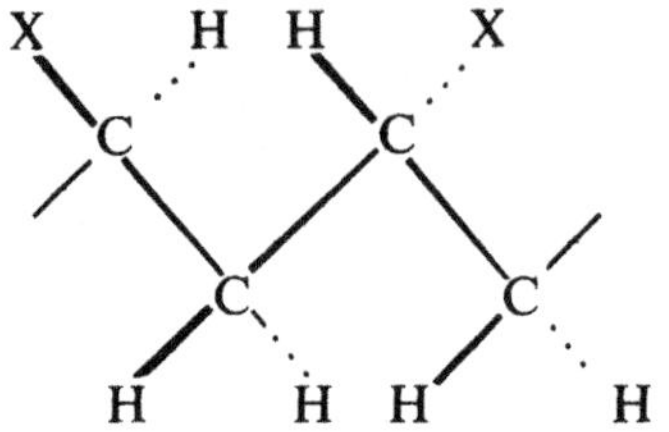

we find, if we ascribe to each atom a righthanded coordinate system:

$$\psi_{i+1} = -\psi_i, \tag{6.240}$$

and:

$$\overline{\sin \varphi_k} = \overline{\sin \varphi_{k+1}} = -\overline{\sin \varphi_{k+2}} = -\overline{\sin \varphi_{k+3}} = \overline{\sin \varphi_{k+4}} = \cdots, \tag{6.241}$$

when k is the number of an atom carrying a group X. We now ascribe to the first couple of atoms righthanded coordinate systems, to the second couple of atoms lefthanded ones, and so on. The matrices of rotation are then alternately equal to C (eqn. (6.212)) and C' (eqn. (6.229)). Thus for infinitely long chains, we find the following expression for g:

$$g = 1 + 2(\overline{\sin \gamma \cos \psi}, \overline{\sin \gamma \sin \psi}, -\overline{\cos \gamma})\overline{C}\overline{C'}[I - \overline{C}\overline{C'}]^{-1} \begin{pmatrix} \overline{\sin \gamma \cos \psi} \\ \overline{\sin \gamma \sin \psi} \\ -\overline{\cos \gamma} \end{pmatrix}. \tag{6.242}$$

For tetrahedric bond angles, this leads to:

$$g = \frac{1 - \eta^2 - \varepsilon^2}{(1 - \eta)^2 + (\eta - \eta^2 - \varepsilon^2)^2} \left\{ \tfrac{11}{12}(1 + \eta^2 + \varepsilon^2) - \tfrac{4}{3}\eta + \tfrac{2}{3}\varepsilon\sqrt{3} \right\}. \tag{6.243}$$

Again, for free rotation this reduces to $g = \frac{11}{12}$.

When there is correlation between the rotations about the bonds, the calculations become very intricate. For polyvinylidene- and polyvinylchloride, Volkenstein gives a method (developed by Ptitsyn and Sharonov) for the incorporation of this correlation to the extent that it is caused by the interaction between neighbouring substituents. If k numbers a carbon atom carrying a chlorine atom, it is assumed that the rotations about the bonds $k - 1$ and k are independent and the rotations about the bonds k and $k + 1$ are correlated. It is then possible to use a matrix B which is the average of the product of the transformation matrices of the coordinate systems k and $k + 1$ and the coordinate systems $k + 1$ and $k + 2$. In the case of polyvinylidenechloride we have:

$$B = \overline{C_k C_{k+1}}, \tag{6.244}$$

while for polyvinylchloride analogous expressions can be given for the syndiotactic and the isotactic chains.

The elements of the matrix B contain not only terms in $\overline{\cos \phi_k}$, $\overline{\cos \varphi_{k+1}}$, $\overline{\sin \varphi_k}$, and $\overline{\sin \varphi_{k+1}}$, but also correlation terms, *e.g.* terms in $\overline{\cos \varphi_k \cos \varphi_{k+1}}$. With the help of the matrix B, for infinitely long chains g can be written as:

$$g = 1 + 2(\overline{\sin\gamma\cos\psi}, \overline{\sin\gamma\sin\psi}, -\overline{\cos\gamma})B[I - B]^{-1}\begin{pmatrix}\overline{\sin\gamma\cos\psi} \\ \overline{\sin\gamma\sin\psi} \\ -\overline{\cos\gamma}\end{pmatrix}. \tag{6.245}$$

The evaluation of this expression is very cumbersome for most cases and will not be treated here.

Suzuki[56] has extended the calculation of the g-factor to the case of atactic chains. Introducing B_+ for the matrix $\overline{C'_k C'_{k+1}}$, which is used in the case of isotactic chains, and B_- for the matrix $\overline{C_k C'_{k+1}}$, which is used in the case of syndiotactic chains, we can apply eqn. (6.245), with B given by:

$$B = p_+ B_+ + p_- B_-, \tag{6.246}$$

where p_+ denotes the fraction of chain elements which are isotactic and $p_- = 1 - p_+$ the fraction of chain elements which are syndiotactic. Suzuki has calculated values of the elements of B and the corresponding values of g, for various cases, limiting the rotation about the bonds to trans-gauche isomerism.

Read[57] has extended eqn. (6.236), which was found to hold for polyoxymethylene, for some other polyethers. In his calculations it is assumed that the potentials of internal rotation around the C–C and the C–O bonds are equal. Using tetrahedric bond angles, he finds for polyethylene oxide:

$$g = \frac{12(1 - \eta^2)}{13 + \eta + 13\eta^2}, \tag{6.247}$$

for polytrimethylene oxide:

$$g = \frac{1 + \eta}{1 - \eta}\,\frac{39 - 24\eta + 39\eta^2}{40 + 64\eta + 40\eta^2}, \tag{6.248}$$

for polytetramethylene oxide:

$$g = \frac{8(1 - \eta^2)(10 + 7\eta + 10\eta^2)}{81(1 + \eta + \eta^2 + \eta^3 + \eta^4) - 135\eta^2 - 15\eta(1 - \eta)^2 - \frac{1}{3}(1 - \eta)^4}. \tag{6.249}$$

The calculations of Read were criticized by Mark[58] on the ground that the interdependence of bond rotational states is ignored and only one quantity is employed to characterize the rotation about three different kinds of bonds. Mark has given calculations of g for polyethylene oxide[59] and for polytetramethylene oxide, which incorporate both refinements but in which the rotation about the bonds is limited to trans-gauche isomerism; g-factors of dimethylsiloxane chains were calculated on the same basis.[60]

For the determination of the g-factor from the experimental data with the help of eqn. (6.208), the value of ε_∞ and μ must be estimated. For the dielectric constant of induced polarization we recommend the estimate $\varepsilon_\infty = 1.05n^2$ (*cf.* section 29). Sometimes, however, $\varepsilon_\infty = n^2$, $\varepsilon_\infty = 1.1n^2$, or estimates from the dielectric relaxation behaviour are used; thus, different values of g may be calculated on the basis of the same experimental data. The other unknown quantity is μ, the dipole strength of one segment. Usually, μ is taken equal to the dipole strength of a comparable monomeric compound. Not in all cases, however, is there general agreement on the question with which compound the segment should be compared. For instance, in the case of poly-p-chlorostyrene Debye and Bueche[55] and Kotera[61] considered p-chlorotoluene to be the comparable monomeric compound, whereas Fattakhov[62] took chlorobenzene; this leads to differences of more than 30% in the values of g determined with the help of eqn. (6.208).

The values of g found in this way may depend on the physical state of the compound. Although for a number of esters of polymethacrylic acid Volkenstein[62] quotes g-factors (derived from Mikhailov and Burshtein) which are almost equal for the amorphous solid state and for the dilute solution, in the case of poly-p-chlorostyrene, however, the value $g = 0.72$ found for the solid state[63] deviates markedly from the value $g = 0.42$ found for dilute solutions.[55] A comparable difference is found for polystyrene.[64] Moreover, in some cases the experimental g-factor is strongly dependent on the solvent. In the case of polyhexane-1-sulphone and poly-2-methylpentane-1-sulphone Bates *et al.*[65] found that the g-factor increases as the solvent becomes poorer. For polyvinylbromide a dependence of g on the concentration was found.[66]

We shall now give some cases in which the results of dielectric measurements on polymer systems are interpreted in terms of the conformations of the chains. Unfortunately, there are few possibilities for a direct application of the general theory, since most of the above mentioned halo-polymers are insoluble. In the case of the polyhalostyrenes a comparison between theoretical and experimental values for the g-factor is complicated by the lack of knowledge about the tacticity.[55,61] The same problem arises for polymethylmethacrylate, H–[CH_2–C(CH_3)($COOCH_3$)–]$_n$H, which has been studied frequently.[67–76] Moreover, for this compound the polar group is not rigidly linked to the polymer chains, and the average values of the components of the moment of the polar group have to be used (*cf.* eqn. (6.214)). If the polar group moved as though it was not connected with the chain, one would have $\overline{\mu_{ix}^{(i)}} = \overline{\mu_{iy}^{(i)}} = \overline{\mu_{iz}^{(i)}} = 0$ and the value $g = 1$ would be found. The fact that this value is not found experimentally, indicates that there is a net resultant moment. The experimental value of g generally proves to be dependent on the tacticity, but the reports concerning this dependence do not agree.[71,73,76] Roig and Horta[76] explain this on the basis of the supposition that the details of the stereoregularity play an important role.

The best possibilities for the application of Volkenstein's general method are given by the polyethers. Here, however, the drawing of definite conclusions from static dielectric measurements about the conformation of the polymer molecules is sometimes impossible due to the conflicting reports of different authors concerning the values of $\overline{\mu_n^2}$ determined under the same circumstances.

For instance, this is the case for the investigations which have been done on polyethylene oxide. Measurements on this compound in benzene were performed by Marchal and Bénoit,[77,78] both in the form of a polyglycol, H–[O–CH_2–CH_2–]$_m$OH, and as a polyether, C_2H_5–[O–CH_2–CH_2–]$_m$-OC_2H_5, with m varying from $m = 2$ to $m = 227$ in the first case, and in the second case m taking the values $m = 2$ and $m = 6$. As was to be expected, $\overline{\mu_n^2}/n$ has a limiting value for large values of n; applying eqn. (6.247), one finds $\eta = 0.3$. The values of $\overline{\mu_n^2}$ found for the oligomers are in agreement with this value of η. In terms of trans-gauche isomerism the value $\eta = 0.3$ indicates that there is a preference for the trans-positions in the chain. The same conclusion was reached, although on the basis of somewhat different values of $\overline{\mu_n^2}$, for samples of oligomers of HOC_2H_4OH in dioxane by Uchida *et al.*[79] and for samples of these oligomers in benzene by Kotera *et al.*[80] The measurements of Bak *et al.*[81] on polyethylene oxide in benzene also indicate a predominance of the trans-positions. These authors performed measurements at various temperatures and found a decrease of the predominance of the trans-position with increasing temperature. Kimura and Fujishiro,[82] however, found higher values of $\overline{\mu_n^2}$ for samples of CH_3–[O–CH_2–CH_2–$]_m$O–CH_3 in benzene and hexane than the foregoing authors had found for the corresponding systems; from these values it was concluded that there is free rotation about the bonds. For H–[O–CH_2–CH_2–$]_m$-OH in benzene, Rossi and Magnasco[83] found values of $\overline{\mu_n^2}$ so high that these authors concluded that the C–O bonds prefer the gauche position. It appears that new careful investigations are necessary to elucidate this situation. The occurrence of intramolecular hydrogen bonds in solutions of H–[O–CH_2–CH_2–$]_m$OH, which also influences the conformation of the chain for moderate values of m, must also be taken into account.[79]

For the comparable compound H–[NH–CH_2–CH_2–$]_m$$NH_2$, Kimura *et al.*[84] concluded from measurements on the oligomers in benzene that the rotation is free, as Kimura and Fujishiro had concluded for CH_3–[O–CH_2–CH_2–$]_m$O–CH_3.

The other polyethers received less study. Uchida *et al.*[85] investigated some oligomers H–[O–CH_2–$]_m$OH in hexane. They found from the dipole moment a preference for the gauche-positions leading to a coiling up of the chain. For the comparable oligomers of dimethylsiloxane, $(CH_3)_3$Si–[O–Si$(CH_3)_2$–$]_m$O–Si$(CH_3)_3$, however, Mark[60] concluded from the measurements of Dasgupta and Smyth[86] a preference for the trans-positions.

Polytetramethylene oxide was studied in non-polar solvents by Bak *et al.*,[81] and in the solid state by Wetton and Williams.[87] The value of g found by Wetton and Williams is somewhat higher than the values obtained by Bak. Both conclude that there is a preference for the trans-positions. Wetton and Williams applied eqn. (6.249) and found $\eta = 0.6$.

Williams also studied polyacetaldehyde, CH_3–[O–CH(CH_3)–$]_m$OCH_3,[88] and polypropylene oxide, CH_3–[O–CH(CH_3)–CH_2–$]_m$OCH$(CH_3)_2$,[89] in the solid state. In these cases he also applied eqns. (6.236) and (6.247), although the tacticity is wholly neglected in this way.

(b) *Associating compounds*

We shall now treat a number of associating compounds, for which explanations of the behaviour of the experimental g-factor have been proposed in terms of the specific influence of the hydrogen bonds. These compounds are water, the mono-alcohols, the carboxylic acids, the amides, and the cyanides.

Water

For water, it is found that the experimental g-factor decreases with increasing temperature from $g = 2.75$ at 0°C to $g = 2.49$ at 83°C. Theoretical values for g were calculated by Oster and Kirkwood,[90] who regarded only the first coordination shell around the representative molecule. Assuming free rotation around the hydrogen bond, eqn. (6.167) leads to:

$$g = 1 + z \cos^2 \tfrac{1}{2}\theta, \tag{6.250}$$

where z is the average number of neighbours in the first coordination shell, and θ is the H–O–H bond angle in water. The coordination number z was calculated from X-ray data and θ was assumed to be $\theta = 105°$. In spite of the crude approximations introduced in this way, Oster and Kirkwood calculated values of g which did not differ more than 13% from the experimental values in the temperature range 0°–83°C. It was pointed out by Hill,[91,92] however, that the experimental value of g depends strongly on the dielectric constant of induced polarization ε_∞. If the value $\varepsilon_\infty = 4.5$ is chosen, a value which is quite compatible with the dielectric relaxation data, one finds approximately $g = 1$. Therefore, for this assumption concerning the dielectric constant of induced polarization, the association has no effect on the dielectric constant.

Mono-alcohols

For the mono-alcohols, Middelhoek and Böttcher[93] demonstrated that, in contrast to water, an adaptation of ε_∞ such that the Kirkwood correlation factor becomes equal to unity, is incompatible with the dielectric relaxation behaviour. The static dielectric behaviour of these compounds has been investigated by many authors, some of whom restricted themselves to the temperature dependence of the dielectric constant of the pure alcohols,[90,93–104] whereas others also investigated the dielectric behaviour of solutions of alcohols in non-polar solvents[105–121] or studied the influence of changes in the pressure.[122–125]

Fig. 34 shows the temperature dependence of the Kirkwood correlation factor for some pure primary, secondary, and tertiary alcohols. Values $g > 1$ and $g < 1$ both occur, indicating the predominance of multimers with a large and a small dipole moment, respectively. It appears that the influence

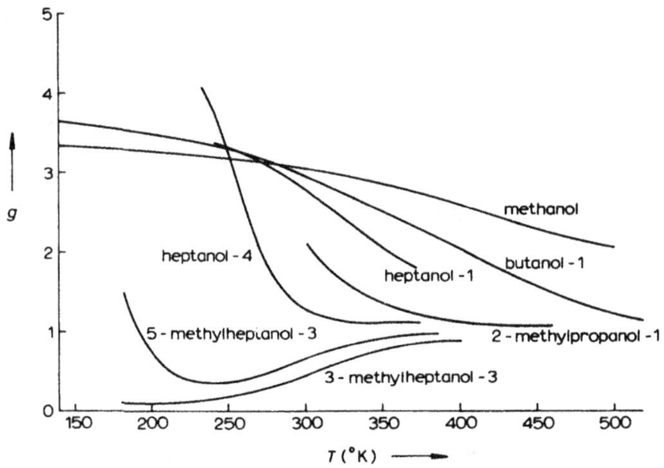

Fig. 34
Kirkwood correlation factors for some mono-alcohols in dependence on temperature.

of the multimers with a large dipole moment decreases with increasing temperature, increasing chain length, and an increasingly central position of the hydroxyl group in the molecule. At high temperatures a limiting value of approximately $g = 1$ is reached, as must be expected from the consideration that at high temperature all hydrogen bonds will be destroyed, so that Onsager's equation will then be valid. It may not be concluded, however, from the occurrence of a value $g = 1$, that no hydrogen bonding occurs, because the effects of both kinds of multimers may cancel each other out.[102,103] For the primary alcohols, Fig. 34 indicates that a limiting value of the g-factor also occurs at low temperatures. This limiting value increases with increasing chain length. The g-factor appears not to be influenced by the occurrence of double or triple bonds in the carbon chain.[104]

Figs. 35, 36, and 37 give the experimental g-factor as a function of the concentration, for various alcohols, solvents, and temperatures, respectively. Except for the solvent benzene in Fig. 36, all these graphs show a minimum in g, indicating that for lower concentrations the multimers with a small dipole moment dominate, even when the multimers with a high dipole moment dominate in the pure state of the alcohol. This implies that in general the multimers with a small dipole moment will contain fewer molecules than the multimers with a large dipole moment.

Ibbitson and Moore[115,117] investigated very diluted solutions (less than

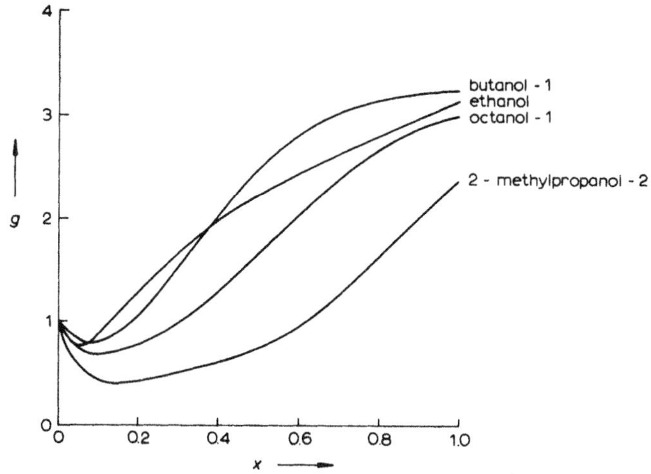

Fig. 35
Kirkwood correlation factors for some mono-alcohols in cyclohexane at 25°C in dependence on concentration.

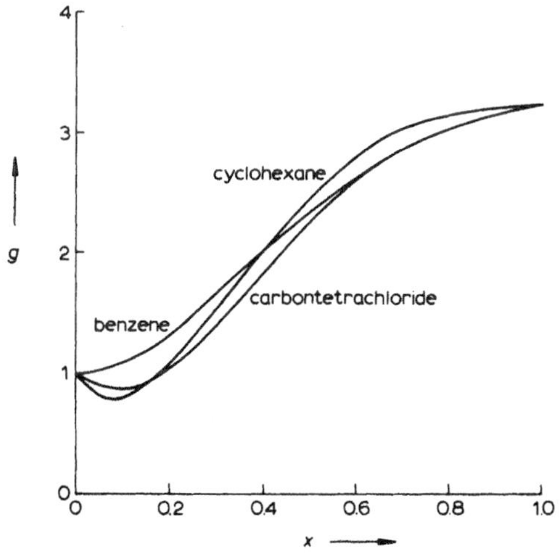

Fig. 36
Kirkwood correlation factors for butanol-1 in different solvents at 25°C in dependence on concentration.

2 mole per cent) of alcohols in non-polar solvents. From their observations it must be concluded that at very low concentrations g shows a maximum

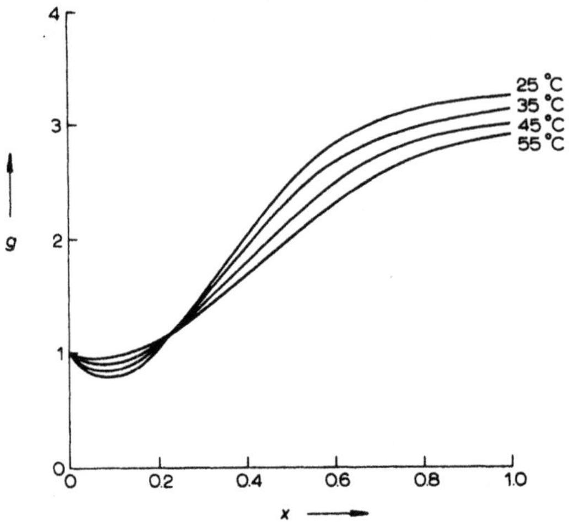

Fig. 37
Kirkwood correlation factors for butanol-1 in cyclohexane at different temperatures in dependence on concentration.

(due to the scale this maximum is not discernible in Figs. 35, 36, and 37). This conclusion was recently confirmed by other measurements.[127] It implies the existence of a kind of small multimer with a relatively high dipole moment and containing fewer molecules than the multimers with a low dipole moment dominating in the concentration range where g has a minimum.

From Fig. 36 it is evident that over the whole concentration range the experimental value of g depends on the solvent to a high degree. This implies that the equilibria between various kinds of multimers depend on the medium. Consequently it is not possible to use concentration-independent equilibrium constants, since the properties of the medium depend on the concentration of the alcohol in the mixture.

Fig. 38 shows the pressure dependence of the g-factor of some alcohols. Both increasing and decreasing values of g are found with increasing pressure. Increasing values occur when the value of g at normal pressure is relatively low, *i.e.* when the multimers with a small dipole moment play an important role. This is in accordance with the observation that the g-factor of butanol-2 at high temperatures increases with increasing pressure, whereas at lower temperatures, where multimers with a high dipole moment predominate, it shows a decrease.

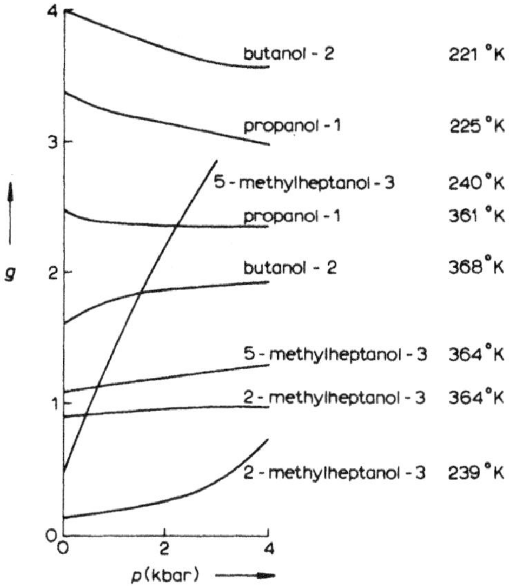

Fig. 38
Kirkwood correlation factors for some mono-alcohols in dependence on pressure.

Both kinds of pressure dependence of the g-factor can be explained if it is assumed that the hydrogen bonds are destroyed under the influence of the pressure, *i.e.*, the multimers take a larger volume per molecule than the monomers. This is in agreement with the results of the scarce high-pressure infra-red studies on alcohols, which indicate that, as far as differences may be noted, the increase of the pressure leads to an increase of the number of hydroxyl groups which do not act as hydrogen bond donors.[128] It must then further be assumed that the multimers with a low dipole moment take a higher volume per molecule than the multimers with a high dipole moment.

The multimers with a small dipole are generally considered to be cyclic, with the hydroxyl groups in one plane. In view of the maximum in g found at very low concentrations as mentioned above, these cyclic multimers must be built up of at least three molecules. A number of authors assume the multimers with a high dipole moment to be linear. The decrease of the g-factor with increasing temperature is then explained from the decreasing average multimer size; the increase of the low-temperature limit of the g-factor with increasing length of the carbon chain is explained from the increasingly restricted rotation about the hydrogen bonds. Calculations of the g-factor based on this model[90,94,96,97,129] have been made with methods analogous to those of Volkenstein for polymers. In this way, however, the results of infra-red[130] and dielectric relaxation[102,131] measurements cannot be explained. For the interpretation of the latter results it must be assumed that the large multimers with a high dipole moment are cyclic and are all built up from the same number of molecules. From the high dipole moment it then follows that in this type of multimers the oxygen and hydrogen atoms do not lie in one plane.

Carboxylic acids

As in the case of the alcohols, for the carboxylic acids[132–135] both values $g < 1$ and $g > 1$ are found (see Fig. 39). The low values of g are explained from the occurrence of closed dimers 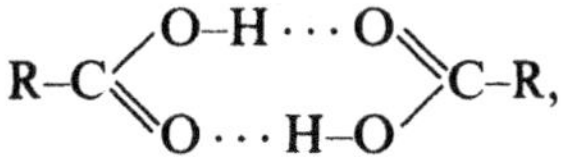 , in which

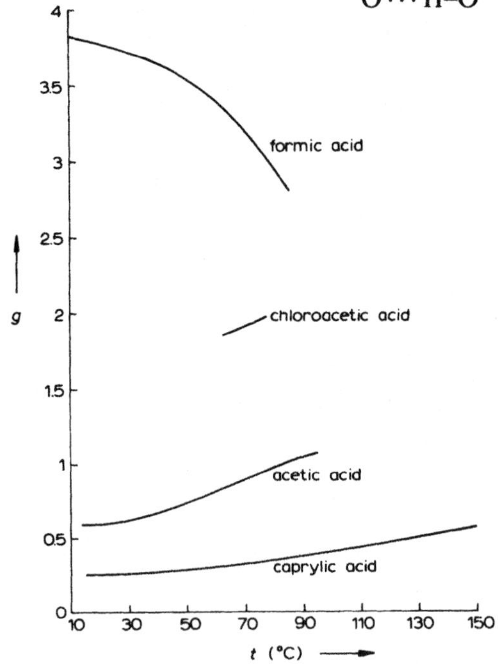

Fig. 39
Kirkwood correlation factors for some carboxylic acids in dependence on temperature.

the dipole moments of both molecules compensate each other. It appears from Fig. 39 that for the stronger acids these closed dimers are less stable, and multimers with a high dipole moment dominate. These multimers are assumed to be linear.[133]

Amides

For the amides,[136–141] 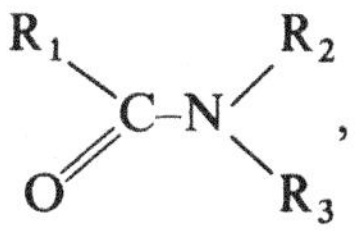

, the formation of hydrogen bonds when $R_3 = H$ can lead to exceptionally high values of ε, up to $\varepsilon = 348$ for N-methylpropionamide at –40°C. In Fig. 40 the temperature dependence of the g-factor is given for a number of different amides. For N,N-dimethylformamide one finds $g \approx 1$, as was to be expected, because the N,N-dialkylamides cannot act as strong proton donors in the hydrogen bond formation.

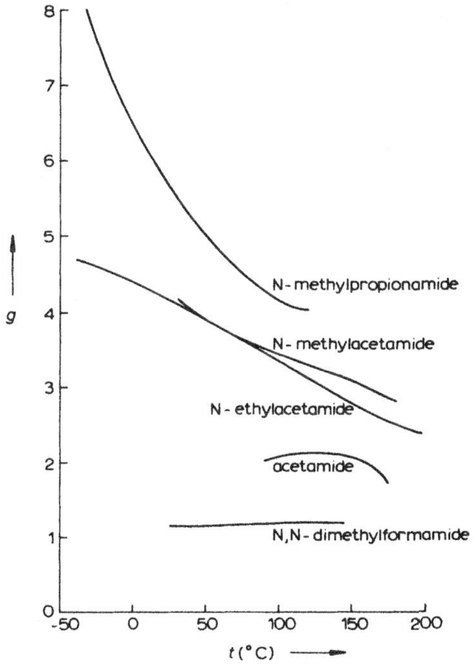

Fig. 40
Kirkwood correlation factors for some amides in dependence on temperature.

It follows from measurements on higher N,N-dialkylamides, however, that in these cases the CH_3– or –CH_2– groups can act as hydrogen bond donors to some extent, leading at 25°C to correlation factors up to $g = 1.3$ for N,N-diethylacetamide.[142] This indicates that the g-factor found for N,N-dimethylformamide is due to multimers with high and low dipole moments compensating each other.

For the N-alkylamides higher values of g are generally found than for the amides in which the nitrogen is not substituted. This can be explained from stereochemical considerations as follows. The amide molecules are planar, with for the N-alkylamides the trans-conformation 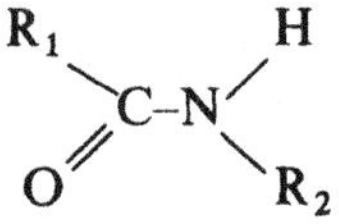 as the dominating species. This implies that the N-alkylamides cannot form closed dimers as is the case for those amides where the nitrogen atom is not substituted at all: 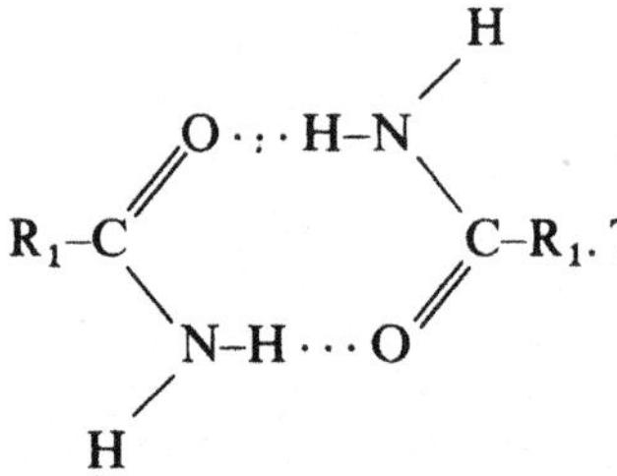. These dimers have a vanishing dipole moment, in contrast to the linear multimers formed by the N-alkylamides. In this way the different concentration dependences of the dielectric constant of solutions of N-alkylamides[143,144] and other amides[145] in benzene can also be explained.

Bass *et al.*[140] have given an explanation for the difference between the g-factors of N-methylpropionamide and the isomeric N-ethylacetamide. They assume the N-alkylamides to be associated to linear chains in which the rotation about the hydrogen bonds is restricted due to the interaction of the R_1-groups. It appears from steric models that the interaction between the R_2-groups and between the R_1- and R_2-groups is of far less importance. When the rotation about the hydrogen bonds is more restricted, the correlation between the dipoles in the chain increases, leading for this type of molecules to higher values of g. Thus, the transition from N-methylacetamide ($R_1 = CH_3$) to N-methylpropionamide ($R_1 = C_2H_5$) leads to a much higher value of g, whereas the transition from the former compound to N-ethylacetamide does not change the g-factor to any important degree. Bass' model thus gives a good explanation of the static dielectric behaviour of the amides. It must be mentioned, however, that so far no explanation of the dielectric relaxation behaviour has been given on the basis of this model.

Nitriles

For hydrogen cyanide it is found that the Kirkwood correlation factor decreases from $g = 4.04$ at the melting point (−13.3°C) to $g = 2.38$ at the boiling point (25.7°C). Cole[146] has given a quantitative interpretation of the temperature dependence of the g-factor for this compound. Following him, we assume linear association, H–C≡N···H–C≡N···H–C≡N, and find for the dipole moment of a multimer:

$$\mu_n = n\mu. \tag{6.251}$$

Substituting this into eqn. (6.204), we obtain:

$$g = \frac{\sum_{n=1}^{\infty} n^2 N_n}{\sum_{n=1}^{\infty} n N_n}. \tag{6.252}$$

We now introduce the assumption that all equilibria

$$(HCN)_n + HCN \leftrightarrows (HCN)_{n+1},$$

are characterized by the same equilibrium constant K so that:

$$N_n = (KN_1/N_A)^{n-1} N_1. \tag{6.253}$$

Substituting this into eqn. (6.252), we find after execution of the summations:

$$g = \frac{1 + KN_1/N_A}{1 - KN_1/N_A}. \tag{6.254}$$

The formal concentration $N = \sum_{n=1}^{\infty} nN_n$ can be calculated in the same way, which leads to:

$$N = \frac{N_1}{(1 - KN_1/N_A)^2}. \tag{6.255}$$

Combining (6.254) and (6.255), we have:

$$g = \left(1 + \frac{4KN}{N_A}\right)^{\frac{1}{2}}. \tag{6.256}$$

With the help of this relation the equilibrium constant K at different temperatures can be calculated from the experimental values of g. The model can then be tested by a rate plot of $\log K$ versus $1/T$; a linear plot is found, from which follows a heat of formation of the hydrogen bond of 4.6 kcal/mole.

In the same way Dannhauser and Flueckinger[147] calculated for cyanoethyne(H–C≡C–C≡N) a heat of formation of the hydrogen bond of 2.8 kcal/mole from the experimental values of g, indicating that for this compound the hydrogen bond is weaker than for hydrogen cyanide. These authors also investigated the dielectric behaviour of a number of other nitriles.[148] They find values $g \approx 1$, which indicates that the hydrogen bonding can be neglected for these compounds. The deviations from Onsager's equation for these compounds result from deviations from the spherical shape of the molecules, and can therefore better be treated with the help of the continuum approach, as shown in section 30.

Mixtures

For mixtures of non-associating compounds eqn. (5.60) shows that the dielectric constant is a monotonous function of the concentration. For mixtures of associating compounds, however, this need not be true. For instance a minimum in the dielectric constant is found for solutions of water in some mono-alcohols,[149–152] whereas for mixtures of heptanol-1 and -4 below 20°C a maximum is found.[153] The interpretation of these phenomena in terms of the Kirkwood correlation factor is not straightforward, however, because for these mixtures both kinds of molecules have values of g different from 1.

From the examples of the behaviour of associating compounds given above, as well as from the behaviour of polymers treated in part (a) of this section, it is clear that a description of the dielectric behaviour in terms of the Kirkwood correlation factor is often very useful. In a number of cases the application of the theory is hampered by lack of knowledge of the molecular configuration or of the mode of association.

References

1. M. Mandel and P. Mazur, *Physica* **24** (1958) 116.
2. J. G. Kirkwood, *J. Chem. Phys.* **3** (1935) 300.
3. *e.g.* J. S. Rowlinson, *Liquids and Liquid Mixtures*, Butterworth, London 1969, Ch. 8.
4. J. G. Kirkwood, *J. Chem. Phys.* **4** (1936) 592.
5. J. de Boer, F. van der Maesen, and C. A. ten Seldam, *Physica* **19** (1953) 265.
6. J. Yvon, *Recherches sur la théorie cinétique des liquides : I, Fluctuations en densité* (*Actualités Scientifiques et Industrielles No. 542*), Hermann et Cie., Paris 1937.
7. W. F. Brown Jr., *J. Chem. Phys.* **18** (1950) 1193.
8. D. R. Johnston, G. J. Oudemans, and R. H. Cole, *J. Chem. Phys.* **33** (1960) 1310.
9. G. J. Oudemans, *Thesis* Amsterdam 1967.
10. F. I. Mopsik, *J. Chem. Phys.* **50** (1969) 2559.
11. A. Michels, C. A. ten Seldam, and S. D. J. Overdijk, *Physica* **17** (1951) 781.
12. Ref. 9, p. 73.
13. A. Michels, C. Michels, and A. Bijl, *Nature* **138** (1936) 509; see also A. Michels, J. de Boer, and A. Bijl, *Physica* **4** (1937) 981.
14. S. R. de Groot and C. A. ten Seldam, *Physica* **13** (1947) 47; and C. A. ten Seldam and S. R. de Groot, *Physica* **18** (1952) 905, 910.
15. L. Jansen and P. Mazur, *Physica* **21** (1955), 193, 208.
16. L. Jansen and A. D. Solem, *Phys. Rev.* **104** (1956) 1291.
17. L. Jansen, *Phys. Rev.* **112** (1958) 434.
18. R. H. Orcutt and R. H. Cole, *J. Chem. Phys.* **46** (1967) 697.
19. J. Heinrichs, *Chem. Phys. Letters* **1** (1967) 467.
20. B. Lindner and R. A. Kromhout, *J. Chem. Phys.* **52** (1970) 1615; and T. K. Lim, B. Lindner, and R. A. Kromhout, *J. Chem. Phys.* **52** (1970) 3831.
21. A. D. Buckingham and J. A. Pople, *Trans. Faraday Soc.* **51** (1955) 1029; and A. D. Buckingham and J. A. Pople, *J. Chem. Phys.* **27** (1957) 820.
22. R. W. Zwanzig, *J. Chem. Phys.* **25** (1956) 211; and R. W. Zwanzig, *J. Chem. Phys.* **27** (1957) 821.
23. D. R. Johnston and R. H. Cole, *J. Chem. Phys.* **36** (1962) 2026.
24. J. E. Mayer and M. G. Mayer, *Statistical Mechanics*, Wiley, New York 1957, Ch. 13.
25. R. H. Fowler, *Statistical Mechanics*, University Press, Cambridge 1936, Ch. IX, X.
26. W. H. Stockmayer, *J. Chem. Phys.* **9** (1941) 398.
27. A. D. Buckingham and J. A. Pople, *Trans. Faraday Soc.* **51** (1955) 1173.
28. A. D. Buckingham and J. A. Pople, *Trans. Faraday Soc.* **51** (1955) 1179.
29. A. D. Buckingham and J. A. Pople, *Trans. Faraday Soc.* **51** (1955) 1029.
30. R. H. Orcutt and R. H. Cole, *Physica* **31** (1965) 1779.
31. Ref. 9, Table 2-I, II, III, and IV, and p. 39.
32. Ref. 9, pp. 117 and 132.
33. D. E. Stogryn and A. P. Stogryn, *Mol. Phys.* **11** (1966) 371.
34. Ref. 9, p. 133.
35. H. Sutter and R. H. Cole, *J. Chem. Phys.* **52** (1970) 132.
36. G. Briegleb, *Fortschr. Chem. Forschung* **1** (1950) 642.
37. H. A. Stuart, *Molekülstruktur*, Springer, Berlin 1967.
38. L. Pauling, *Nature of the Chemical Bond*, Cornell University Press, Ithaca N.Y. 1960.
39. J. R. Sweet and W. A. Steele, *J. Chem. Phys.* **50** (1969) 668.

40. Chang Lyoul Kong, *J. Chem. Phys.* **53** (1970) 1522.
41. A. D. Buckingham and R. E. Raab, *Trans. Faraday Soc.* **54** (1958) 623.
42. T. K. Bose and R. H. Cole, *J. Chem. Phys.* **52** (1970) 140.
43. S. Kielich, *Mol. Phys.* **9** (1965) 549.
44. D. A. McQuarrie and H. B. Levine, *Physica* **31** (1965) 749; and H. B. Levine and D. A. McQuarrie, *J. Chem. Phys.* **44** (1966) 3500.
45. T. L. Hill, *J. Chem. Phys.* **28** (1958) 61.
46. A. N. Kaufmann and K. M. Watson, *Phys. Fluids* **4** (1961) 931.
47. A. Isihara and R. V. Hanks, *J. Chem. Phys.* **36** (1962) 433; and A. Isihara, *J. Chem. Phys.* **38** (1963) 2437.
48. J. G. Kirkwood, *J. Chem. Phys.* **7** (1939) 911.
49. H. Fröhlich, *Theory of Dielectrics*, Oxford University Press, London 1958.
50. R. H. Cole, *J. Chem. Phys.* **27** (1957) 33.
51. A. D. Buckingham, *Proc. Roy. Soc.* **A238** (1957) 235.
52. R. H. Cole, *J. Chem. Phys.* **39** (1963) 2602.
53. M. V. Volkenstein, *Configurational Statistics of Polymeric Chains*, Interscience, New York 1963.
54. M. V. Volkenstein, *J. Pol. Sci.* **29** (1958) 441.
55. P. Debye and F. Bueche, *J. Chem. Phys.* **19** (1951) 589.
56. Y. Suzuki, *J. Chem. Phys.* **34** (1961) 79.
57. B. E. Read, *Trans. Faraday Soc.* **61** (1965) 2140.
58. J. E. Mark, *J. Am. Chem. Soc.* **88** (1966) 3708.
59. J. E. Mark and P. Flory, *J. Am. Chem. Soc.* **88** (1966) 3702.
60. J. E. Mark, *J. Chem. Phys.* **49** (1968) 1398.
61. Ref. 53, p. 347.
62. Ref. 53, p. 351.
63. R. N. Work and Y. M. Tréhu, *J. Appl. Phys.* **27** (1956) 1003.
64. W. R. Krigbaum and A. Roig, *J. Chem. Phys.* **31** (1959) 544.
65. T. W. Bates, K. J. Ivin, and G. Williams, *Trans. Faraday Soc.* **63** (1967) 1976.
66. M. Kryszewski and J. Marchal, *J. Pol. Sci.* **29** (1958) 103.
67. L. de Brouckère, D. Buess, J. de Bock, and J. Versluys, *Bull. Soc. Chim. Belg.* **64** (1955) 669.
68. L. de Brouckère and L. K. H. van Beek, *Rec. Trav. Chim.* **75** (1956) 355.
69. L. de Brouckère, D. Buess, and L. K. H. van Beek, *J. Pol. Sci.* **23** (1957) 233.
70. L. de Brouckère and M. Mandel, *Adv. Chem. Phys.* **1** (1958) 77.
71. H. A. Pohl, R. Bacskai, and W. P. Purcell, *J. Phys. Chem.* **64** (1960) 1701.
72. L. de Brouckère and A. Lecocq-Robert, *Bull. Soc. Chim. Belg.* **70** (1961) 549.
73. R. Salovey, *J. Pol. Sci.* **50** (1961) 57.
74. R. W. J. le Fèvre and K. M. S. Sundoram, *J. Chem. Soc.* **1963** 1880.
75. M. Mandel, *Pol. Lett.* **1** (1963) 265.
76. A. Roig and A. Horta, *J. Pol. Sci.* **C16** (1968) 3501.
77. J. Marchal and H. Bénoit, *J. Chim. Phys.* **52** (1955) 818.
78. J. Marchal and H. Bénoit, *J. Pol. Sci.* **23** (1957) 223.
79. T. Uchida, Y. Kurita, N. Koizumi, and M. Kubo, *J. Pol. Sci.* **21** (1956) 313.
80. A. Kotera, K. Suzuki, K. Matsumura, T. Nakano, and T. Oyama, *Bull. Chem. Soc. Jap.* **35** (1962) 797.
81. K. Bak, G. Elefante, and J. E. Mark, *J. Phys. Chem.* **71** (1967) 4007.
82. K. Kimura and R. Fujishiro, *Bull. Chem. Soc. Jap.* **39** (1966) 608.
83. C. Rossi and V. Magnasco, *J. Pol. Sci.* **58** (1962) 977.

84. K. Kimura, Y. Toshiyasu, and R. Fujishiro, *Bull. Chem. Soc. Jap.* **39** (1966) 1681.
85. T. Uchida, Y. Kurita, and M. Kubo, *J. Pol. Sci.* **19** (1956) 365.
86. S. Dasgupta and C. P. Smyth, *J. Chem. Phys.* **47** (1967) 2911.
87. R. Wetton and G. Williams, *Trans. Faraday Soc.* **61** (1965) 2132.
88. G. Williams, *Trans. Faraday Soc.* **59** (1963) 1397.
89. G. Williams, *Trans. Faraday Soc.* **61** (1965) 1564.
90. G. Oster and J. G. Kirkwood, *J. Chem. Phys.* **11** (1943) 175.
91. N. E. Hill, *Trans. Faraday Soc.* **59** (1963) 344.
92. N. E. Hill, *J. Phys.* **C3** (1970) 238.
93. J. Middelhoek and C. J. F. Böttcher, in *Molecular Relaxation Processes, Chem. Soc. Spec. Publ.* **20** (1966) 69.
94. W. Dannhauser and R. H. Cole, *J. Chem. Phys.* **23** (1955) 1762.
95. R. Philippe and A. M. Piette, *Bull. Soc. Chim. Belg.* **64** (1955) 600.
96. P. C. Brot, *Ann. de Phys.* **13–2** (1957) 714.
97. W. Dannhauser and L. W. Bahe, *J. Chem. Phys.* **40** (1964) 3058.
98. W. Dannhauser, L. W. Bahe, R. Y. Lin, and A. F. Flueckinger, *J. Chem. Phys.* **43** (1965) 257.
99. J. Middelhoek, *Thesis* Leiden 1967.
100. W. Dannhauser, *J. Chem. Phys.* **48** (1968) 1911.
101. G. P. Johari and W. Dannhauser, *J. Phys. Chem.* **72** (1968) 3273.
102. P. Bordewijk, *Thesis* Leiden 1968.
103. P. Bordewijk, F. Gransch, and C. J. F. Böttcher, *Trans. Faraday Soc.* **66** (1970) 293.
104. W. Dannhauser, R. Guering, and A. F. Flueckinger, *J. Chem. Phys.* **52** (1970) 6447.
105. G. Oster, *J. Am. Chem. Soc.* **68** (1946) 2036.
106. R. Mecke and A. Reuter, *Z. Naturforschung* **4a** (1949) 368.
107. C. Brot and A. Soulard, *Arch. des. Sci.* **12** fasc.spéc. (1959) 9.
108. R. J. W. le Fèvre and A. J. Williams, *J. Chem. Soc.* **1960** 108.
109. P. Huyskens and A. Cracco, *Bull. Soc. Chim. Belg.* **69** (1960) 422.
110. R. Liébaert, A. Lebrun, and Y. Leroy, *C.R.* **253** (1961) 2496.
111. P. Huyskens, R. Henry, and G. Gillerot, *Bull. Soc. Chim. F.* **1962** 720.
112. R. Liébaert, *Thesis* Lille 1962.
113. R. Liébaert and Y. Leroy, in J. Smidt (ed.), *Magnetic and Electric Resonance and Relaxation*, Proc. of the XIth Coll. Ampère, North-Holland, Amsterdam 1963, p. 316.
114. L. Raczy, E. Constant, and M. Moriamez, ibidem p. 321.
115. D. A. Ibbitson and L. F. Moore, *Chem. Comm.* **15** (1965) 339.
116. Th. Clerbaux and Th. Zeegers-Huyskens, *Bull. Soc. Chim. Belg.* **69** (1960) 422.
117. D. A. Ibbitson and L. F. Moore, *J. Chem. Soc.* **B1967**, 76, 80.
118. A. Weisbecker, *J. Chim. Phys.* **64** (1967) 297.
119. L. Raczy, E. Constant, and A. Lebrun, *J. Chim. Phys.* **64** (1967) 1180.
120. P. J. Gold and R. Perrine, *J. Phys. Chem.* **71** (1967) 4218.
121. R. van Loon, J. P. Dauchot, and A. Bellemans. *Bull. Soc. Chim. Belg.* **77** (1968) 397.
122. A. Gilchrist, J. E. Early, and R. H. Cole, *J. Chem. Phys.* **26** (1957) 196.
123. G. P. Johari and W. Dannhauser, *J. Chem. Phys.* **48** (1968) 5114.
124. T. Chen, W. Dannhauser, and G. P. Johari, *J. Chem. Phys.* **50** (1969) 2046.
125. H. Hartmann, R. Engelmann, and A. Neumann, *Z. Phys. Chem. N.F.* **66** (1969) 268.
126. Y. A. Atanov and M. I. Shakparonov, *Russ. J. Phys. Chem.* **43** (1969) 1232.
127. P. Bordewijk, M. Kunst and A. Rip, *J. Phys. Chem.*, to be published.
128. R. J. Jakobson, Y. Mikawa, and J. W. Brasch, *Applied Spectr.* **24** (1970) 333.
129. Ref. 99, section 4.

130. L. P. Kuhn and R. E. Bowman, *Spectrochim. Acta* **17** (1961) 650.
131. P. Bordewijk, F. Gransch, and C. J. F. Böttcher, *J. Phys. Chem.* **73** (1969) 3255.
132. J. F. Johnson and R. H. Cole, *J. Am. Chem. Soc.* **73** (1951) 4536.
133. A. E. Lutskii and S. A. Mikhailenko. *J. Struct. Chem.* **4** (1963) 12.
134. J. Lafontaine, *Bull. Soc. Chim. Belg.* **67** (1958) 153.
135. J. Liszi, *Acta Chim. Acad. Sci. Hung.* **62** (1969) 263.
136. G. R. Leader and J. F. Gormley, *J. Am. Chem. Soc.* **73** (1951) 5731.
137. J. W. Vaugh and P. G. Sears, *J. Phys. Chem.* **62** (1958) 183.
138. A. E. Lutskii and S. A. Mikhailenko, *J. Struct. Chem.* **4** (1963) 323.
139. R. Y. Lin and W. Dannhauser, *J. Phys. Chem.* **67** (1963) 1805.
140. S. J. Bass, W. I. Nathan, R. M. Meighan, and R. H. Cole, *J. Phys. Chem.* **68** (1964) 509.
141. W. Dannhauser and G. P. Johari, *Can. J. Chem.* **46** (1968) 3143.
142. M. Steffen, *Ber. Bunsenges.* **74** (1970) 505.
143. J. E. Worsham Jr. and M. E. Hobbs, *J. Am. Chem. Soc.* **76** (1954) 206.
144. R. M. Meighan and R. H. Cole, *J. Phys. Chem.* **68** (1964) 503.
145. M. E. Hobbs and W. W. Bates, *J. Am. Chem. Soc.* **74** (1952) 746.
146. R. H. Cole, *J. Am. Chem. Soc.* **77** (1954) 2012.
147. W. Dannhauser and A. F. Flueckinger, *J. Chem. Phys.* **38** (1963) 69.
148. W. Dannhauser and A. F. Flueckinger, *J. Phys. Chem.* **68** (1964) 114.
149. A. C. Brown and D. J. G. Ives, *J. Chem. Soc.* **1962** 1608.
150. F. Franks and D. J. G. Ives, *Quart. Rev.* **20** (1966) 1.
151. A. S. C. Lawrence, M. D. McDonald, and J. V. Stevens, *Trans. Faraday Soc.* **65** (1969) 3231.
152. T. H. Tjia, P. Bordewijk, and C. J. F. Böttcher, *J. Phys. Chem.* **74** (1970) 2851.
153. Ref. 102, section 4.3.

CHAPTER VII

NON-LINEAR EFFECTS

§41. Introduction

So far, we have considered the polarization to be proportional with the field strength, in accordance with eqn. (2.45):

$$\boldsymbol{P} = \chi \boldsymbol{E}.$$

This simple relation is only valid at moderate field intensities, however. At field intensities of about 10^4 volts/cm or higher, deviations of relation (2.45) become noticeable. In that case we need the next term of the series development of $\boldsymbol{P}$ with respect to $\boldsymbol{E}$, and find (eqn. (2.47)):

$$\boldsymbol{P} = \chi \boldsymbol{E} + \xi E^2 \boldsymbol{E}.$$

It must be noted that the series development of $\boldsymbol{P}$ contains only odd powers of $\boldsymbol{E}$, because a reversal of the direction of $\boldsymbol{E}$ leads to a reversal of the direction of $\boldsymbol{P}$:

$$\boldsymbol{P}(\boldsymbol{E}) = -\boldsymbol{P}(-\boldsymbol{E}). \tag{7.1}$$

Eqn. (2.47) of course holds only for isotropic systems. For anisotropic systems we have:

$$\boldsymbol{P} = \boldsymbol{\chi} \cdot \boldsymbol{E} + \boldsymbol{\xi} \vdots \boldsymbol{EEE}, \tag{7.2}$$

where $\boldsymbol{\xi}$ is a tensor of the fourth degree.

If the polarization is non-linear in the field strength, the dependence of the dielectric displacement $\boldsymbol{D}$ on the field strength will also be non-linear:

$$\boldsymbol{D} = \boldsymbol{E} + 4\pi \boldsymbol{P} = \varepsilon_0 \boldsymbol{E} + 4\pi \xi E^2 \boldsymbol{E}, \tag{7.3}$$

where ε_0 is written instead of ε for the dielectric constant at low field intensity to emphasize the difference between the two cases. When measurements of the non-linear effect are made by superposing a low intensity alternating field on a static field of high intensity, the quantity measured is the field-dependent incremental dielectric constant ε_E:

$$\varepsilon_E = \frac{\partial D}{\partial E} = \varepsilon_0 + 12\pi\xi E^2. \tag{7.4}$$

The non-linear effect is then characterized by the quantity $\Delta\varepsilon/E^2$:

$$\frac{\Delta\varepsilon}{E^2} = \frac{\varepsilon_E - \varepsilon_0}{E^2} = 12\pi\xi. \tag{7.5}$$

The non-linearity of the polarization can be caused by a number of effects:

1. At large field intensities the higher terms of the Langevin function:

$$L(a) = \tfrac{1}{3}a - \tfrac{1}{45}a^3 + \tfrac{2}{945}a^5 - \tfrac{2}{9450}a^7 + \cdots,$$

must be taken into account (see section 26). The contribution of this effect to $\Delta\varepsilon/E^2$ is negative; it is called normal saturation.

2. If there is an equilibrium between entities with higher and lower dipole moments, the equilibrium is shifted under the influence of the field in favour of the more polar entities. The contribution of this phenomenon to $\Delta\varepsilon/E^2$ is positive; it is called anomalous saturation.

3. If the particles have an anisotropic polarizability, they are directed with their axis of highest polarizability in the direction of the field (*cf.* section 15). The contribution of this phenomenon to $\Delta\varepsilon/E^2$ is positive; it is related to the Kerr effect.

4. Theoretically, the hyperpolarizabilities of the molecule also play a role; then the total moment of a molecule in a homogenous external field is given by (see section 44):

$$\boldsymbol{m} = \boldsymbol{\mu} + \boldsymbol{\alpha} \cdot \boldsymbol{E}_0 + \tfrac{1}{2}\boldsymbol{\beta} : \boldsymbol{E}_0\boldsymbol{E}_0 + \tfrac{1}{6}\gamma \vdots \boldsymbol{E}_0\boldsymbol{E}_0\boldsymbol{E}_0 + \cdots, \tag{7.6}$$

where $\boldsymbol{\beta}$ and γ are tensors of the third and fourth degree, respectively. The sign of the contribution of this effect to $\Delta\varepsilon/E^2$ can be positive or negative, depending on the components of $\boldsymbol{\beta}$ and γ.

5. Finally, if an electric field is applied at constant pressure, the density is increased by the electric field. This phenomenon is called electrostriction. It gives rise to a positive contribution to $\Delta\varepsilon/E^2$.

The first and second effect occur only for polar systems; the other effects can be found for polar and non-polar systems. If the molecules have a permanent dipole moment, the contribution of the normal saturation to $\Delta\varepsilon/E^2$ is larger in absolute value than the contribution of the anisotropy of the polarizability, and the electrostriction.

For non-linear dielectrics, Laplace's equation, eqn. (1.16):

$$\Delta\phi = 0,$$

which was found to be valid for linear dielectrics in all charge-free parts of the dielectric (p. 72), does not hold for inhomogeneous fields. This can be shown as follows.

For the divergence of $\boldsymbol{E}$ we may write:

$$\begin{aligned}\operatorname{div}\boldsymbol{E} &= \operatorname{div}\boldsymbol{D} - 4\pi \operatorname{div}\boldsymbol{P}\\ &= 0 - 4\pi\chi \operatorname{div}\boldsymbol{E} - 4\pi\xi \operatorname{div} E^2\boldsymbol{E}\\ &= -4\pi\chi \operatorname{div}\boldsymbol{E} - 4\pi\xi E^2 \operatorname{div}\boldsymbol{E} - 4\pi\xi\boldsymbol{E}\cdot\operatorname{grad} E^2. \end{aligned} \tag{7.7}$$

Introducing a quantity ε', given by:*

$$\varepsilon' = \frac{D}{E} = 1 + 4\pi\chi + 4\pi\xi E^2, \tag{7.8}$$

we rewrite eqn. (7.7):

$$\operatorname{div}\boldsymbol{E} = -\frac{4\pi\xi}{\varepsilon'}\boldsymbol{E}\cdot\operatorname{grad} E^2. \tag{7.9}$$

Since by definition $\boldsymbol{E} = -\operatorname{grad}\phi$ (eqn. (1.14)) we have:

$$\Delta\phi = \frac{4\pi\xi}{\varepsilon'}\boldsymbol{E}\cdot\operatorname{grad} E^2. \tag{7.10}$$

For inhomogeneous fields this differs from zero.

In the case of a spherical cavity in a dielectric, we found (section 9b), for a homogeneous external field $\boldsymbol{E}_0$, the field in the dielectric to be inhomogeneous in the neighbourhood of the cavity. Hence, if the dielectric is non-linear, Laplace's equation may not be used for the dielectric in the calculation of the cavity field. This makes an exact calculation of the cavity field for non-linear dielectrics extremely intricate. Because the calculations of the directing and the internal field are based on the calculation of the cavity field, a rigorous calculation of these fields is also very intricate, even for the continuum model. To find reasonable approximations for these fields, in the calculation of the directing and the internal field in the following sections we shall assume the ratio $\varepsilon' = D/E$ to be constant over the dielectric and given by:

* The ratio $\varepsilon' = D/E$ must be distinguished from the incremental dielectric constant ε_E as defined by eqn. (7.4). Their relation is given by $\varepsilon' - \varepsilon_0 = \frac{1}{3}(\varepsilon_E - \varepsilon_0)$.

$$\varepsilon' = \frac{D}{E} = \varepsilon_0 + 4\pi\xi E_0^2, \tag{7.11}$$

where E_0 is the homogeneous part of the field in the dielectric (compare eqn. (2.63)).

§42. Normal saturation

In Chapter V we found the polarization to be built up of the induced polarization $\boldsymbol{P}_\alpha$ and the dipole polarization $\boldsymbol{P}_\mu$:

$$\boldsymbol{P} = \boldsymbol{P}_\alpha + \boldsymbol{P}_\mu = N(\alpha \boldsymbol{E}_\mathrm{i} + \bar{\boldsymbol{\mu}}), \tag{7.12}$$

where N is the number of particles per cm^3. In that chapter we considered $\bar{\boldsymbol{\mu}}$ to be given by eqn. (5.14):

$$\bar{\boldsymbol{\mu}} = \frac{\mu^2}{3kT}\boldsymbol{E}_\mathrm{d}.$$

In this section we shall include the second term of the series development of the Langevin function in the expression for the average dipole moment in the direction of the field:

$$\overline{(\boldsymbol{\mu})_E} = \mu\,\overline{\cos\theta} = \mu\left\{\tfrac{1}{3}\frac{\mu E_\mathrm{d}}{kT} - \tfrac{1}{45}\left(\frac{\mu E_\mathrm{d}}{kT}\right)^3\right\}. \tag{7.13}$$

Hence:

$$\bar{\boldsymbol{\mu}} = \left(\frac{\mu^2}{3kT} - \frac{\mu^4 E_\mathrm{d}^2}{45(kT)^3}\right)\boldsymbol{E}_\mathrm{d}, \tag{7.14}$$

and thus:

$$\boldsymbol{P} = N\left\{\alpha \boldsymbol{E}_\mathrm{i} + \left(\frac{\mu^2}{3kT} - \frac{\mu^4 E_\mathrm{d}^2}{45(kT)^3}\right)\boldsymbol{E}_\mathrm{d}\right\}. \tag{7.15}$$

In the case of a gas at moderate pressure we can take $\boldsymbol{E}_\mathrm{i} = \boldsymbol{E}_\mathrm{d} = \boldsymbol{E}$ (compare section 31). Eqn. (7.15) then gives:

$$\boldsymbol{P} = N\left(\alpha + \frac{\mu^2}{3kT}\right)\boldsymbol{E} - \frac{N\mu^4}{45(kT)^3}E^2\boldsymbol{E}. \tag{7.16}$$

Comparing this with eqn. (2.47):

$$P = \chi E + \xi E^2 E,$$

we find:

$$\xi = -\frac{N\mu^4}{45(kT)^3}. \tag{7.17}$$

Substituting this result into eqn. (7.5), we have:

$$\Delta\varepsilon = -\frac{4\pi N\mu^4}{15(kT)^3}E^2. \tag{7.18}$$

Hence for a gas with $\mu = 2D$ at 300°K and 1 atm, we have:

$$\Delta\varepsilon = -0.62\ 10^{-16}E^2,$$

if E is expressed in volts/cm. Therefore, for a field intensity of 10^5 volts/cm, which is about the highest value realizable, the decrease of ε would be only $0.62\ 10^{-6}$. Consequently, the experimental proof of electric saturation in the case of gaseous dielectrics is extremely difficult.

The expected change of ε is larger in the case of liquid dielectrics, where N is much larger. The theoretical treatment of the problem is much more intricate, however, since for liquids we cannot use the approximation $\boldsymbol{E}_i = \boldsymbol{E}_d = \boldsymbol{E}$. The righthand member of eqn. (7.15):

$$\boldsymbol{P} = N\left\{\alpha \boldsymbol{E}_i + \left(\frac{\mu^2}{3kT} - \frac{\mu^4 E_d^2}{45(kT)^3}\right)\boldsymbol{E}_d\right\},$$

depends on E^3 in two ways: by the term $-\mu^4 E_d^2/45(kT)^3$, and by the influence of the saturation on the values of $\boldsymbol{E}_i$ and $\boldsymbol{E}_d$.

In the calculation of $\boldsymbol{E}_i$ and $\boldsymbol{E}_d$ we attribute to the dielectric a linear dielectric behaviour, given by eqn. (7.11):

$$\frac{D}{E} = \varepsilon' = \varepsilon_0 + 4\pi\xi E_0^2,$$

where E_0 is equal to the homogeneous part of the Maxwell field in the dielectric.

According to eqns. (5.54) and (5.55), and using eqn. (7.14) for $\bar{\boldsymbol{\mu}}$, we have:

$$\boldsymbol{P} = N\{\alpha \boldsymbol{E}_i(\varepsilon') + \bar{\boldsymbol{\mu}}\} =$$

$$= N\left\{\alpha \boldsymbol{E}_{\rm d}(\varepsilon') + \frac{\alpha f(\varepsilon')\bar{\boldsymbol{\mu}}}{1 - f(\varepsilon')\alpha} + \bar{\boldsymbol{\mu}}\right\}$$

$$= N\left\{\alpha \boldsymbol{E}_{\rm d}(\varepsilon') + \frac{1}{1 - f(\varepsilon')\alpha}\frac{\mu^2}{3kT}\boldsymbol{E}_{\rm d}(\varepsilon') - \frac{1}{1 - f(\varepsilon_0)\alpha}\frac{\mu^4\{E_{\rm d}(\varepsilon_0)\}^2}{45(kT)^3}\boldsymbol{E}_{\rm d}(\varepsilon_0)\right\}. \quad (7.19)$$

In the last term in the last member of this equation ε' is replaced by ε_0 because terms in powers of the field higher than the third can be neglected.

Developing $\boldsymbol{E}_{\rm d}(\varepsilon')$ in a Taylor series around ε_0 and neglecting terms in E^4 and higher powers of E we have, after substitution of eqn. (7.11):

$$\boldsymbol{E}_{\rm d}(\varepsilon') = \boldsymbol{E}_{\rm d}(\varepsilon_0) + (\varepsilon' - \varepsilon_0)\left(\frac{\partial \boldsymbol{E}_{\rm d}(\varepsilon)}{\partial \varepsilon}\right)_{\varepsilon=\varepsilon_0}$$

$$= \boldsymbol{E}_{\rm d}(\varepsilon_0) + 4\pi\xi E_0^2\left(\frac{\partial \boldsymbol{E}_{\rm d}(\varepsilon)}{\partial \varepsilon}\right)_{\varepsilon=\varepsilon_0}. \quad (7.20)$$

In this expression the argument of $\boldsymbol{E}_{\rm d}$ in the derivative has been denoted by ε to avoid confusion with ε'. In the same way, we have:

$$\frac{\boldsymbol{E}_{\rm d}(\varepsilon')}{1 - f(\varepsilon')\alpha} = \frac{\boldsymbol{E}_{\rm d}(\varepsilon_0)}{1 - f(\varepsilon_0)\alpha} + 4\pi\xi E_0^2\left(\frac{\partial}{\partial \varepsilon}\frac{\boldsymbol{E}_{\rm d}(\varepsilon)}{1 - f(\varepsilon)\alpha}\right)_{\varepsilon=\varepsilon_0}. \quad (7.21)$$

Substituting these expressions into eqn. (7.19) and using:

$$\boldsymbol{P} = \chi\boldsymbol{E}_0 + \xi E_0^2\boldsymbol{E}_0, \quad (7.22)$$

we find, equating terms in the third power of the field:

$$\xi E_0^2\boldsymbol{E}_0 = N\left\{4\pi\alpha\xi E_0^2\left(\frac{\partial \boldsymbol{E}_{\rm d}(\varepsilon)}{\partial \varepsilon}\right)_{\varepsilon=\varepsilon_0} + 4\pi\frac{\mu^2}{3kT}\xi E_0^2\left(\frac{\partial}{\partial \varepsilon}\frac{\boldsymbol{E}_{\rm d}(\varepsilon)}{1 - f(\varepsilon)\alpha}\right)_{\varepsilon=\varepsilon_0} - \right.$$

$$\left. - \frac{\mu^4\{E_{\rm d}(\varepsilon_0)\}^2}{45(kT)^3}\frac{\boldsymbol{E}_{\rm d}(\varepsilon_0)}{1 - f(\varepsilon_0)\alpha}\right\}. \quad (7.23)$$

Solving this equation for ξ, and using for the directing field the expression given in eqn. (5.52):

$$\boldsymbol{E}_{\rm d}(\varepsilon) = \frac{3\varepsilon\boldsymbol{E}_0}{(2\varepsilon + 1)(1 - f(\varepsilon)\alpha)},$$

we find, since $\boldsymbol{E}_{\rm d}$ and $\boldsymbol{E}_0$ are in the same direction:

$$\xi = \frac{-\dfrac{N\mu^4\{E_d(\varepsilon_0)\}^3}{45(kT)^3(1-f(\varepsilon_0)\alpha)}}{E_0^2\left\{E_0 - 4\pi N\alpha\left(\dfrac{\partial E_d(\varepsilon)}{\partial\varepsilon}\right)_{\varepsilon=\varepsilon_0} - 4\pi N\dfrac{\mu^2}{3kT}\left(\dfrac{\partial}{\partial\varepsilon}\dfrac{E_d(\varepsilon)}{1-f(\varepsilon)\alpha}\right)_{\varepsilon=\varepsilon_0}\right\}}$$

$$= \frac{-3N\mu^4\varepsilon_0^3/5(kT)^3(2\varepsilon_0+1)^3(1-f(\varepsilon_0)\alpha)^4}{1-4\pi N\alpha\left(\dfrac{\partial}{\partial\varepsilon}\dfrac{1}{1-f(\varepsilon)\alpha}\dfrac{3\varepsilon}{2\varepsilon+1}\right)_{\varepsilon=\varepsilon_0} - 4\pi N\dfrac{\mu^2}{3kT}\left(\dfrac{\partial}{\partial\varepsilon}\dfrac{1}{(1-f(\varepsilon)\alpha)^2}\dfrac{3\varepsilon}{2\varepsilon+1}\right)_{\varepsilon=\varepsilon_0}} \tag{7.24}$$

Taking $(1 - f(\varepsilon)\alpha)^{-1} = (\varepsilon_\infty + 2)(2\varepsilon + 1)/3(2\varepsilon + \varepsilon_\infty)$ (*cf.* eqn. (5.64)) the differentiations can be performed. Substituting α as given by the Lorenz-Lorentz relation (eqn. (5.43)):

$$\alpha = \frac{3}{4\pi N}\frac{\varepsilon_\infty - 1}{\varepsilon_\infty + 2},$$

and in the denominator μ^2 as given by the Onsager equation (eqn. (5.67)):

$$\mu^2 = \frac{9kT}{4\pi N}\frac{(\varepsilon_0 - \varepsilon_\infty)(2\varepsilon_0 + \varepsilon_\infty)}{\varepsilon_0(\varepsilon_\infty + 2)^2},$$

we arrive at:

$$\xi = -\frac{N\mu^4}{135(kT)^3}\frac{\varepsilon_0^4(\varepsilon_\infty + 2)^4}{(2\varepsilon_0 + \varepsilon_\infty)^2(2\varepsilon_0^2 + \varepsilon_\infty^2)}, \tag{7.25}$$

or:

$$\Delta\varepsilon = -\frac{4\pi N\mu^4}{45(kT)^3}\frac{\varepsilon_0^4(\varepsilon_\infty + 2)^4}{(2\varepsilon_0 + \varepsilon_\infty)^2(2\varepsilon_0^2 + \varepsilon_\infty^2)}E^2. \tag{7.26}$$

This equation was first derived by Thiébaut.[1,2]

If we neglect the polarizability of the molecules, taking $\varepsilon_\infty = 1$, eqn. (7.26) is reduced to:

$$\Delta\varepsilon = -\frac{4\pi N\mu^4}{15(kT)^3}\frac{27\varepsilon_0^4}{(2\varepsilon_0 + 1)^2(2\varepsilon_0^2 + 1)}E^2, \tag{7.27}$$

an expression* first derived by Van Vleck.[3]

For high values of ε_0, eqn. (7.26) reduces to:

* Van Vleck defines $\Delta\varepsilon = \varepsilon' - \varepsilon_0$ instead of the more usual definition $\Delta\varepsilon = \varepsilon_E - \varepsilon_0$; hence our eqn. (7.24) and his eqn. (12) differ by a factor 3.

TABLE 28

$\varepsilon_0^4(\varepsilon_\infty + 2)^4/3(2\varepsilon_0 + \varepsilon_\infty)^2(2\varepsilon_0^2 + \varepsilon_\infty^2)$

ε_0	$\varepsilon_\infty = 1$	$\varepsilon_\infty = 1.5$	$\varepsilon_\infty = 2$	$\varepsilon_\infty = 2.5$
1	1.000	—	—	—
4	2.568	4.143	6.068	8.298
10	3.046	5.350	8.643	13.091
20	3.208	5.792	9.627	15.018
80	3.333	6.136	10.403	16.556
∞	3.375	6.253	10.667	17.086

$$\Delta\varepsilon = \frac{\pi N \mu^4(\varepsilon_\infty + 2)^4}{90(kT)^3} E^2. \tag{7.28}$$

Hence the saturation term is then independent of ε.

The saturation term for the case of liquids as given by eqn. (7.26) differs by a factor $\varepsilon_0^4(\varepsilon_\infty + 2)^4/3(2\varepsilon_0 + \varepsilon_\infty)^2(2\varepsilon_0^2 + \varepsilon_\infty^2)$ from the saturation term for gases (eqn. (7.18)). In Table 28 this factor is given for some values of ε_0 and ε_∞.

In Table 29 we give values of $\Delta\varepsilon/E^2$ at 20°C computed from eqn. (7.26), which are compared with experimental values of $\Delta\varepsilon/E^2$ given by Thiébaut.[2] The theoretical and experimental values differ at most by a factor 2. This indicates that if the experimental values of $\Delta\varepsilon/E^2$ (which are proportional to μ^4) are used for the calculation of values of μ, the divergence from the values determined from low field dielectric measurements in the gas phase will not exceed 20%. It must be expected that if a correction for the deviations of the spherical shape were applied, the agreement between the experimental and theoretical values would be better.

TABLE 29

VALUES OF $\Delta\varepsilon/E^2$ FOR SOME PURE POLAR LIQUIDS (IN $(\mathrm{V/cm})^{-2}$)

Compound	eqn. (7.26)	exp.
Chloroform	$-8.2\ 10^{-15}$	$-16\ 10^{-15}$
Ethoxyethane	$-8.4\ 10^{-15}$	$-15\ 10^{-15}$
1,1,1-Trichloroethane	$-70.0\ 10^{-15}$	$-78\ 10^{-15}$

§43. Anomalous saturation

In 1936, Piekara and Piekara[4] found that for very pure nitrobenzene the dielectric constant increases with the field strength, in contrast to the normal saturation effect discussed in the last section. This anomalous saturation effect was also found for the related compounds[5] o-nitrotoluene and m-nitrotoluene and for the vicinal-dihalogenides[6,7] 1,2-dichloroethane and 1,2-dibromoethane. Furthermore, this was found for some tertiary alcohols,[8,9] which have a relatively low value of the Kirkwood correlation factor, and for solutions of other alcohols in non-polar solvents.[8,11] For alcohols which have a high value of the Kirkwood correlation factor in the pure state, like the normal alcohols, the anomalous saturation effect does not occur in the pure compounds or in mixtures with a high alcohol content.[8–11] In contrast, nitrobenzene and related compounds do not show the anomalous saturation effect in non-polar solvents, whereas the vicinal-dihalogenides show

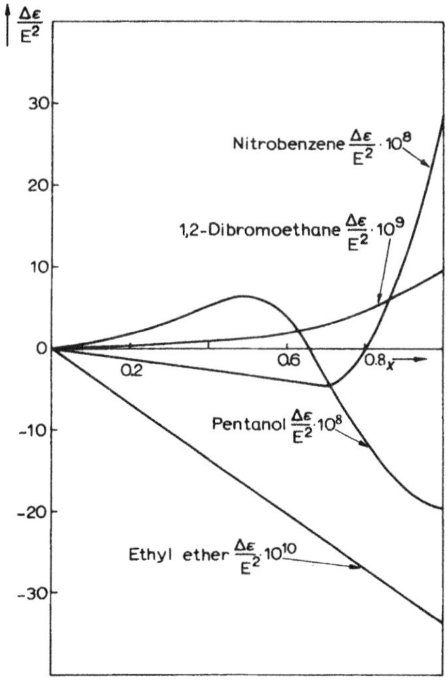

Fig. 41

$\Delta\varepsilon/E^2$ for different kinds of polar molecules dissolved in a non-polar solvent.[19]

anomalous saturation behaviour in the whole concentration range (see Fig. 41).

A qualitative explanation of the positive saturation effect for nitrobenzene was given by Piekara[12] on the basis of the formation of molecular pairs in the liquid. The molecular pairs have a net dipole moment, depending on the angle between the molecules constituting the pair. The field influences the average angle between the molecules constituting a pair in such a way, that the high average dipole moment of the molecular pair is increased at higher field strengths. If the effect of this phenomenon on $\Delta\varepsilon$ is larger than the normal saturation effect, $\Delta\varepsilon$ will become positive.

For a quantitative treatment of the anomalous saturation effect, we use, following Schellman[13] and Barriol and Greffe,[14] the method of Fröhlich (see section 39, p. 251). We then consider the orientation polarization $\boldsymbol{P}_{\text{or}}$ of a dielectric sphere of volume V, containing $\mathcal{N}$ point dipoles with a moment $\boldsymbol{\mu}_{\text{d}}$ embedded in a continuum with dielectric constant ε_∞. The sphere is surrounded by an infinite dielectric with the same macroscopic properties as the sphere. The average component in the direction of the field of the dipole moment due to the dipoles in the sphere, $\langle \boldsymbol{M}_{\text{d}} \cdot \boldsymbol{e} \rangle = VP_{\text{or}}$ (*cf.* eqn. (6.172)), is given by eqn. (6.174):

$$\langle \boldsymbol{M}_{\text{d}} \cdot \boldsymbol{e} \rangle = \frac{\int \mathrm{d}X^{\mathcal{N}} \boldsymbol{M}_{\text{d}} \cdot \boldsymbol{e} \exp(-U/kT)}{\int \mathrm{d}X^{\mathcal{N}} \exp(-U/kT)}.$$

In section 39 we calculated the dielectric constant at low field intensities (denoted by ε_0 in this chapter) by retaining only the linear term in the series development of the average moment in powers of the field (compare eqn. (6.176)):

$$\varepsilon_0 - \varepsilon_\infty = \frac{4\pi}{V}\left(\frac{\partial \langle \boldsymbol{M}_{\text{d}} \cdot \boldsymbol{e} \rangle}{\partial E}\right)_{E=0} = \frac{4\pi}{V}\,\frac{3\varepsilon_0}{2\varepsilon_0 + \varepsilon_\infty}\left(\frac{\partial \langle \boldsymbol{M}_{\text{d}} \cdot \boldsymbol{e} \rangle}{\partial E_{\text{F}}}\right)_{E_{\text{F}}=0}. \tag{7.29}$$

In this equation expression (6.175) for the Fröhlich field $\boldsymbol{E}_{\text{F}}$ has been used. In this way we derived the Kirkwood-Fröhlich equation, eqn. (6.185):

$$\frac{4\pi N g \mu^2}{9kT} = \frac{(\varepsilon_0 - \varepsilon_\infty)(2\varepsilon_0 + \varepsilon_\infty)}{\varepsilon_0(\varepsilon_\infty + 2)^2}.$$

The macroscopic saturation behaviour of the dielectric is characterized by ξ, the coefficient of the third power of E in expression (2.47) for the polarization $\boldsymbol{P}$. Therefore, to express the macroscopic saturation behaviour in microscopic quantities, besides the first derivative higher derivatives of the average moment have to be taken into account. Obviously the series development of $\langle \boldsymbol{M}_d \cdot \boldsymbol{e} \rangle$ with respect to E_F contains only odd powers of E_F:

$$\langle \boldsymbol{M}_d \cdot \boldsymbol{e} \rangle = \left(\frac{\partial \langle \boldsymbol{M}_d \cdot \boldsymbol{e} \rangle}{\partial E_F} \right)_{E_F = 0} E_F + \tfrac{1}{6} \left(\frac{\partial^3 \langle \boldsymbol{M}_d \cdot \boldsymbol{e} \rangle}{\partial E_F^3} \right)_{E_F = 0} E_F^3 + \cdots, \quad (7.30)$$

where E_F now is given by (compare eqn. (7.20)):

$$E_F = \frac{3\varepsilon_0}{2\varepsilon_0 + \varepsilon_\infty} E + 4\pi\xi \left(\frac{\partial}{\partial \varepsilon} \frac{3\varepsilon}{2\varepsilon + \varepsilon_\infty} \right)_{\varepsilon = \varepsilon_0} E^3 + \cdots$$

$$= \frac{3\varepsilon_0}{2\varepsilon_0 + \varepsilon_\infty} E + \frac{12\pi\xi\varepsilon_\infty}{(2\varepsilon_0 + \varepsilon_\infty)^2} E^3 + \cdots. \quad (7.31)$$

The third derivative of $\langle \boldsymbol{M}_d \cdot \boldsymbol{e} \rangle$ with respect to E_F can be found in a way analogous to the procedure in section 36 (*cf.* eqns. (6.35)–(6.43)). Differentiating the above expression for $\langle \boldsymbol{M}_d \cdot \boldsymbol{e} \rangle$ (eqn. (6.174)) and taking into account the fact that $\partial U / \partial E_F = \boldsymbol{M}_d \cdot \boldsymbol{e}$ we obtain:

$$\frac{\partial^3 \langle \boldsymbol{M}_d \cdot \boldsymbol{e} \rangle}{\partial E_F^3} = \frac{1}{(kT)^3} \{ \langle (\boldsymbol{M}_d \cdot \boldsymbol{e})^4 \rangle - 4 \langle (\boldsymbol{M}_d \cdot \boldsymbol{e})^3 \rangle \langle \boldsymbol{M}_d \cdot \boldsymbol{e} \rangle +$$

$$+ 12 \langle (\boldsymbol{M}_d \cdot \boldsymbol{e})^2 \rangle \langle \boldsymbol{M}_d \cdot \boldsymbol{e} \rangle^2 - 3 \langle (\boldsymbol{M}_d \cdot \boldsymbol{e})^2 \rangle^2 - 6 \langle \boldsymbol{M}_d \cdot \boldsymbol{e} \rangle^4 \}. \quad (7.32)$$

Since $\langle \boldsymbol{M}_d \cdot \boldsymbol{e} \rangle_0 = 0$ (where as usual the subscript 0 indicates the absence of the field), we find:

$$\left(\frac{\partial^3 \langle \boldsymbol{M}_d \cdot \boldsymbol{e} \rangle}{\partial E_F^3} \right)_{E_F = 0} = \frac{1}{(kT)^3} \{ \langle (\boldsymbol{M}_d \cdot \boldsymbol{e})^4 \rangle_0 - 3 \langle (\boldsymbol{M}_d \cdot \boldsymbol{e})^2 \rangle_0^2 \}. \quad (7.33)$$

Because for zero field all directions of $\boldsymbol{M}_d$ are equally probable, the directions of $\boldsymbol{M}_d$ and $\boldsymbol{e}$ are independent and we find, writing θ for the angle between $\boldsymbol{M}_d$ and $\boldsymbol{e}$:

$$\langle (\boldsymbol{M}_d \cdot \boldsymbol{e})^4 \rangle_0 = \langle M_d^4 \cos^4 \theta \rangle_0 = \langle M_d^4 \rangle_0 \langle \cos^4 \theta \rangle_0 = \tfrac{1}{5} \langle M_d^4 \rangle_0, \quad (7.34)$$

whereas an expression for $\langle (\boldsymbol{M}_d \cdot \boldsymbol{e})^2 \rangle_0$ follows from eqn. (6.41):

$$\langle (\boldsymbol{M}_d \cdot \boldsymbol{e})^2 \rangle_0 = \tfrac{1}{3} \langle M_d^2 \rangle_0.$$

Hence:

$$\left(\frac{\partial^3 \langle \boldsymbol{M}_d \cdot \boldsymbol{e} \rangle}{\partial E_F^3} \right)_{E_F = 0} = \frac{1}{15 (kT)^3} \{ 3 \langle M_d^4 \rangle_0 - 5 \langle M_d^2 \rangle_0^2 \}. \quad (7.35)$$

By combining eqns. (7.30), (7.31), and (7.35) $\langle \boldsymbol{M}_d \cdot \boldsymbol{e} \rangle$ is found as a series in powers of E:

$$\langle \boldsymbol{M}_d \cdot \boldsymbol{e} \rangle = \frac{\varepsilon_0}{2\varepsilon_0 + \varepsilon_\infty} \frac{\langle M_d^2 \rangle_0}{kT} E + \frac{4\pi\xi\varepsilon_\infty}{(2\varepsilon_0 + \varepsilon_\infty)^2} \frac{\langle M_d^2 \rangle_0}{kT} E^3 +$$

$$+ \frac{27\varepsilon_0^3}{(2\varepsilon_0 + \varepsilon_\infty)^3} \frac{3\langle M_d^4 \rangle_0 - 5\langle M_d^2 \rangle_0^2}{90(kT)^3} E^3 + O(E^5). \quad (7.36)$$

Comparing this with the general relation:

$$\langle \boldsymbol{M}_d \cdot \boldsymbol{e} \rangle = VP_{or} = \frac{\varepsilon_0 - \varepsilon_\infty}{4\pi} VE + \xi VE^3 + \cdots, \quad (7.37)$$

we get, equating the terms in E^3:

$$\xi V = \frac{4\pi\xi\varepsilon_\infty}{(2\varepsilon_0 + \varepsilon_\infty)^2} \frac{\langle M_d^2 \rangle_0}{kT} + \frac{27\varepsilon_0^3}{(2\varepsilon_0 + \varepsilon_\infty)^3} \frac{3\langle M_d^4 \rangle_0 - 5\langle M_d^2 \rangle_0^2}{90(kT)^3}. \quad (7.38)$$

By equating the terms in E in eqns. (7.36) and (7.37) we obtain (compare eqn. (6.179)):

$$\langle M_d^2 \rangle_0 = \frac{kTV}{4\pi} \frac{(\varepsilon_0 - \varepsilon_\infty)(2\varepsilon_0 + \varepsilon_\infty)}{\varepsilon_0}.$$

Substituting this expression into the first term of the righthand member of eqn. (7.38) and solving for ξ, we find:

$$\xi = \frac{3\varepsilon_0^4}{10(2\varepsilon_0 + \varepsilon_\infty)^2(2\varepsilon_0^2 + \varepsilon_\infty^2)} \frac{3\langle M_d^4 \rangle_0 - 5\langle M_d^2 \rangle_0^2}{(kT)^3 V}. \quad (7.39)$$

For the calculation of the statistical-mechanical averages we use eqn. (6.173):

$$\boldsymbol{M}_d = \sum_{i=1}^{\mathcal{N}} (\boldsymbol{\mu}_d)_i,$$

where $\boldsymbol{\mu}_d$ is given by eqn. (6.182):

$$\boldsymbol{\mu}_d = \boldsymbol{\mu} \frac{\varepsilon_\infty + 2}{3}.$$

In section 39 the average squared moment was written in terms of the angles between the dipoles (eqn. (6.159)):

$$\langle M^2\rangle_0 = \mathscr{N}\mu^2 \sum_{j=1}^{\mathscr{N}} \langle\cos\theta_{ij}\rangle,$$

where θ_{ij} is the angle between the moments of the molecules i and j. In the case of dipoles embedded in a dielectric with dielectric constant ε_∞ we have analogously:

$$\langle M_d^2\rangle_0 = \mathscr{N}\mu_d^2 \sum_{j=1}^{\mathscr{N}} \langle\cos\theta_{ij}\rangle = \mathscr{N}\mu^2\left(\frac{\varepsilon_\infty + 2}{3}\right)^2 \sum_{j=1}^{\mathscr{N}} \langle\cos\theta_{ij}\rangle.$$

In the same way we derive:

$$\langle M_d^4\rangle_0 = \mathscr{N}\mu^4\left(\frac{\varepsilon_\infty + 2}{3}\right)^4 \sum_{j=1}^{\mathscr{N}}\left\langle\cos\theta_{ij}\sum_{k=1}^{\mathscr{N}}\sum_{l=1}^{\mathscr{N}}\cos\theta_{kl}\right\rangle.$$

Substituting the expressions for $\langle M_d^2\rangle_0$ and $\langle M_d^4\rangle_0$ into eqn. (7.39), we obtain:

$$\xi = \frac{N\mu^4\varepsilon_0^4(\varepsilon_\infty + 2)^4}{270(kT)^3(2\varepsilon_0 + \varepsilon_\infty)^2(2\varepsilon_0^2 + \varepsilon_\infty^2)}\cdot$$

$$\cdot\left[3\sum_{j=1}^{\mathscr{N}}\left\langle\cos\theta_{ij}\sum_{k=1}^{\mathscr{N}}\sum_{l=1}^{\mathscr{N}}\cos\theta_{kl}\right\rangle - 5\mathscr{N}\left\{\sum_{j=1}^{\mathscr{N}}\langle\cos\theta_{ij}\rangle\right\}^2\right]. \tag{7.40}$$

Expression (7.40) shows that it is possible for ξ to have negative as well as positive values; the sign of ξ depends on the last factor, in which the correlations between the orientations of the molecules are incorporated. Generally, the calculation of this factor is too complicated. In the following we shall treat some simple cases.

First, we assume that there is no correlation between the orientations of the molecules. We then have:

$$\sum_{j=1}^{\mathscr{N}} \langle\cos\theta_{ij}\rangle = \sum_{j=1}^{\mathscr{N}} \delta_{ij} = 1,$$

and:

$$\sum_{j=1}^{\mathscr{N}}\left\langle\cos\theta_{ij}\sum_{k=1}^{\mathscr{N}}\sum_{l=1}^{\mathscr{N}}\cos\theta_{kl}\right\rangle = \sum_{j=1}^{\mathscr{N}}\sum_{k=1}^{\mathscr{N}}\sum_{l=1}^{\mathscr{N}}\langle\cos\theta_{ij}\cos\theta_{kl}\rangle$$

$$= \sum_{k=1}^{\mathscr{N}}\langle\cos\theta_{ii}\cos\theta_{kk}\rangle + \sum_{\substack{j=1\\ j\neq i}}^{\mathscr{N}}\sum_{k=1}^{\mathscr{N}}\sum_{\substack{l=1\\ l\neq k}}^{\mathscr{N}}\langle\cos\theta_{ij}\cos\theta_{kl}\rangle$$

$$= \mathcal{N} + \sum_{\substack{j=1\\j\neq i}}^{\mathcal{N}} \sum_{k=1}^{\mathcal{N}} \sum_{\substack{l=1\\l\neq k}}^{\mathcal{N}} \tfrac{1}{3}(\delta_{ik}\delta_{jl} + \delta_{il}\delta_{jk})$$

$$= \mathcal{N} + \tfrac{2}{3}(\mathcal{N} - 1) = \tfrac{1}{3}(5\mathcal{N} - 2).$$

In this derivation we have used the fact that $\langle \cos\theta_{ij} \cos\theta_{kl} \rangle$ only differs from zero, when either $i = j$ and $k = l$, or $i = k$ and $j = l$, or $i = l$ and $j = k$. After substitution of the above expressions into eqn. (7.40), we arrive at the same expression for ξ as the one given in eqn. (7.25) for the case of normal saturation behaviour:

$$\xi = -\frac{N\mu^4}{135(kT)^3} \frac{\varepsilon_0^4(\varepsilon_\infty + 2)^4}{(2\varepsilon_0 + \varepsilon_\infty)^2(2\varepsilon_0^2 + \varepsilon_\infty^2)}.$$

We now consider the case of compounds having several conformations with different dipole moments, as for instance the vicinal-dihalogenides. The anomalous saturation behaviour will be dominated by the influence of the field on the equilibrium between the conformations so that the orientational correlations between the molecules can be neglected. In the derivation above the neglect of orientational correlations led to the same result as the restricting of the sphere to one molecule. Therefore it appears reasonable to restrict the sphere to one molecule in this calculation also. We then have $\mathcal{N} = 1$ and:

$$\boldsymbol{M}_\mathrm{d} = \boldsymbol{\mu}\frac{\varepsilon_\infty + 2}{3}. \tag{7.41}$$

Eqn. (7.39) now leads to:

$$\xi = \frac{\varepsilon_0^4(\varepsilon_\infty + 2)^4}{270(2\varepsilon_0 + \varepsilon_\infty)^2(2\varepsilon_0^2 + \varepsilon_\infty^2)} \frac{N}{(kT)^3}\{3\overline{\mu^4} - 5(\overline{\mu^2})^2\} \tag{7.42}$$

or, with the help of eqn. (7.5) expressed in terms of $\Delta\varepsilon$:

$$\Delta\varepsilon = \frac{\varepsilon_0^4(\varepsilon_\infty + 2)^4}{(2\varepsilon_0 + \varepsilon_\infty)^2(2\varepsilon_0^2 + \varepsilon_\infty^2)} \frac{2\pi N}{45(kT)^3}\{3\overline{\mu^4} - 5(\overline{\mu^2})^2\}E^2. \tag{7.43}$$

In eqns. (7.42) and (7.43) the bar denotes an average over the conformations with the appropriate weight factor. This average over the conformations replaces the general statistical-mechanical average in eqn. (7.39).

The factor in curly brackets in eqns. (7.42) and (7.43), $\{3\overline{\mu^4} - 5(\overline{\mu^2})^2\}$, can be split up in two parts, one equal to $-2(\overline{\mu^2})^2$ and representing normal saturation (*cf.* eqns. (7.25) and (7.26)), and one equal to $3\{\overline{\mu^4} - (\overline{\mu^2})^2\}$ and representing anomalous saturation. Denoting the contribution to $\Delta\varepsilon$ due to anomalous saturation by $\Delta\varepsilon_a$, we find by subtracting eqn. (7.26) from eqn. (7.43):

$$\Delta\varepsilon_a = \frac{\varepsilon_0^4(\varepsilon_\infty + 2)^4}{(2\varepsilon_0 + \varepsilon_\infty)^2(2\varepsilon_0^2 + \varepsilon_\infty^2)} \frac{2\pi N}{15(kT)^3}\{\overline{\mu^4} - (\overline{\mu^2})^2\}E^2. \tag{7.44}$$

For the vicinal-dihalogenides we have an equilibrium between two conformations:[15] a fraction x_t of the molecules is in the trans-conformation (dipole moment μ_t), and a fraction $1 - x_t$ of the molecules is in the gauche-conformation (dipole moment μ_g). Accordingly, we rewrite the factor $\{\overline{\mu^4} - (\overline{\mu^2})^2\}$ in eqn. (7.44) as:

$$\begin{aligned}\overline{\mu^4} - (\overline{\mu^2})^2 &= x_t\mu_t^4 + (1 - x_t)\mu_g^4 - \{x_t\mu_t^2 + (1 - x_t)\mu_g^2\}^2 \\ &= x_t(1 - x_t)(\mu_t^2 - \mu_g^2)^2. \end{aligned} \tag{7.45}$$

Substituting this expression into eqn. (7.44) we obtain:

$$\Delta\varepsilon_a = \frac{\varepsilon_0^4(\varepsilon_\infty + 2)^4}{(2\varepsilon_0 + \varepsilon_\infty)^2(2\varepsilon_0^2 + \varepsilon_\infty^2)} \frac{2\pi N}{15(kT)^3} x_t(1 - x_t)(\mu_t^2 - \mu_g^2)^2 E^2. \tag{7.46}$$

From this expression we see that the anomalous saturation term is always positive, when the anomalous saturation effect is caused by the existence of an equilibrium between two entities with different dipole moments. This was to be expected, since the application of an electric field will always favour the entities with a high dipole moment, thus shifting the equilibrium in the corresponding direction and giving rise to an increase in the dielectric constant.

For 1,2-dibromoethane, we can calculate $\Delta\varepsilon_a/E^2$ and $\Delta\varepsilon/E^2$, using the following estimates given by Altona and Hageman:[16] $\mu_t = 0.5\,D$, $\mu_g = 3.0\,D$, $x_t = 0.91$. We then find $\Delta\varepsilon_a/E^2 = +86\ 10^{-15}$ (V/cm)$^{-2}$. The normal saturation contributes only $-9\ 10^{-15}$ (V/cm)$^{-2}$, so that we obtain $\Delta\varepsilon/E^2 = +77\ 10^{-15}$ (V/cm)$^{-2}$. The experimental value[7] of $\Delta\varepsilon/E^2$ is $+123\ 10^{-15}$ (V/cm)$^{-2}$.

The agreement between theoretical and experimental values is only in order of magnitude, but it should be noted that it is much better than would have been obtained if only the normal saturation had been taken into account. It should also be realized that in the derivation of eqn. (7.43) the molecules are assumed to be spherical; deviations from the spherical shape may therefore account for the difference between theoretical and experimental values.

In the case of solutions of vicinal-dihalogenides in non-polar solvents anomalous saturation behaviour is found over the whole concentration range. This can be explained with the help of eqn. (7.46). The sign of $\Delta\varepsilon_a$ remains positive whereas the relative magnitudes of the normal and anomalous saturation terms will not change. Therefore, $\Delta\varepsilon$ will be positive over the whole concentration range.*

Anomalous saturation not only occurs as a result of a shift in equilibrium between conformations with different dipole moments, it can also be the result of strong intermolecular orientational correlations, as for instance with nitrobenzene and the alcohols. In such a case, one has to fall back on the general expression (7.40). The factor

$$\left[3\sum_{j=1}^{\mathcal{N}}\left\langle\cos\theta_{ij}\sum_{k=1}^{\mathcal{N}}\sum_{l=1}^{\mathcal{N}}\cos\theta_{kl}\right\rangle - 5\mathcal{N}\left\{\sum_{j=1}^{\mathcal{N}}\langle\cos\theta_{ij}\rangle\right\}^2\right]$$

in eqn. (7.40) will then have a value different from -2, the value found if the sphere is restricted to one molecule (compare eqn. (7.25)).

After Piekara,[17] we shall introduce a reduction factor of saturation, denoted by R_s, which represents the influence of interaction, in analogy with the introduction of the Kirkwood correlation factor (see section 39). Thus we write:

$$\xi = -\frac{NR_s\mu^4}{135(kT)^3}\frac{\varepsilon_0^4(\varepsilon_\infty+2)^4}{(2\varepsilon_0+\varepsilon_\infty)^2(2\varepsilon_0^2+\varepsilon_\infty^2)}. \qquad (7.47)$$

Comparing eqns. (7.47) and (7.40), we find for R_s:

$$R_s = -\tfrac{1}{2}\left[3\sum_{j=1}^{\mathcal{N}}\left\langle\cos\theta_{ij}\sum_{k=1}^{\mathcal{N}}\sum_{l=1}^{\mathcal{N}}\cos\theta_{kl}\right\rangle - 5\mathcal{N}\left\{\sum_{j=1}^{\mathcal{N}}\langle\cos\theta_{ij}\rangle\right\}^2\right]. \qquad (7.48)$$

This expression for R_s was first given by Kielich.[18]

Clearly, for fully independent orientations of the molecules we have:

$$R_s = 1. \qquad (7.49)$$

In contrast to the Kirkwood correlation factor, the reduction factor of

* Since for different solvents the fraction x_t may vary widely, the anomalous saturation depends on the solvent used.

saturation can have both positive and negative values. Negative values of R_s indicate that the anomalous saturation effect dominates the normal saturation effect.

For solutions of nitrobenzene and the alcohols in non-polar solvents the reduction factor of saturation as a function of the concentration changes sign. For nitrobenzene, at high concentrations the reduction factor of saturation is negative (corresponding to positive values of $\Delta\varepsilon$), whereas it is positive at lower values of the concentration. This can be explained by assuming that at low concentrations there is no orientational correlation between the nitrobenzene molecules.

For the normal alcohols such an explanation cannot be given, because we find positive values of R_s at high concentrations (as is also found for the pure compounds). To explain this behaviour, we assume that for the normal alcohols at high concentrations and in the pure state there exists an equilibrium between various kinds of multimers that is dominated by a kind of multimer that is rigid and has a high dipole moment.* Then the effect of a shift in equilibrium under the influence of the field is negligible and only a normal saturation effect is observed. At lower concentrations, other kinds of multimers with different dipole moments become more important, and the shift in the equilibrium has a more noticeable effect. When there is steric hindrance of the association, the shift in equilibrium is also more important. This explains the occurrence of negative values of R_s for moderate concentrations of all alcohols and for tertiary alcohols in the pure state.**

We have seen that the anomalous saturation effect can often be explained as a result of a shift in a chemical equilibrium under the influence of a strong field. Such a shift can also be calculated with the help of thermodynamics.[20] From this calculation, we shall derive an expression for $\Delta\varepsilon_a$, the anomalous saturation term, and compare it with the expressions derived previously.

When temperature and volume are kept constant, the change of the free energy is given by eqn. (3.51):

$$(\mathrm{d}F)_{T,V} = \mathrm{d}W. \tag{7.50}$$

* This assumption has already been used in section 40 to explain the dependence on concentration of the Kirkwood correlation factor for the alcohols.

** The explanation given here for the saturation behaviour of alcohols is not generally accepted. See for a different explanation Piekara.[19]

In the case under consideration, $\mathrm{d}W$ consists of two parts, the chemical work $\mathrm{d}W_{\mathrm{chem}}$, and the electrical work $\mathrm{d}W_{\mathrm{el}}$. The chemical work given by:

$$\mathrm{d}W_{\mathrm{chem}} = \iiint\limits_V A\,\mathrm{d}y\,\mathrm{d}v. \tag{7.51}$$

Here A is the affinity:

$$A = \sum_k \nu_k g_k, \tag{7.52}$$

where ν_k is the stoechiometric coefficient ($\nu_k < 0$ for the reactants, $\nu_k > 0$ for the reaction products) and g_k is the chemical potential of compound k. The differential of the reaction coordinate y is given by:

$$\mathrm{d}y = \frac{\mathrm{d}n_k}{\nu_k} \qquad \text{for any } k, \tag{7.53}$$

where n_k is the number of moles of compound k per unit of volume. The electrical work is given by eqn. (3.40):

$$\mathrm{d}W_{\mathrm{el}} = \frac{1}{4\pi} \iiint\limits_\infty \boldsymbol{E} \cdot \boldsymbol{dD}\,\mathrm{d}v.$$

Hence we find:

$$(\mathrm{d}F)_{T,V} = \iiint\limits_V A\,\mathrm{d}y\,\mathrm{d}v + \frac{1}{4\pi} \iiint\limits_\infty \boldsymbol{E} \cdot \boldsymbol{dD}\,\mathrm{d}v. \tag{7.54}$$

We now assume, that the electric field is zero outside the sample, and is homogeneous within the sample; this condition is fulfilled for instance in the case of an ideal capacitor completely filled with the sample. The composition of the reaction mixture will then also be the same throughout the sample, and the integrations in eqn. (7.54) can be performed. We then obtain:

$$(\mathrm{d}F)_{T,V} = VA\,\mathrm{d}y + \frac{V}{4\pi} \boldsymbol{E} \cdot \boldsymbol{dD}. \tag{7.55}$$

Since we want to use E instead of D as independent variable, we define a transformed free energy:

$$\tilde{F} = F - \frac{V}{4\pi} \boldsymbol{E} \cdot \boldsymbol{D}, \tag{7.56}$$

so that we have:

$$(\mathrm{d}\tilde{F})_{T,V} = (\mathrm{d}F)_{T,V} - \frac{V}{4\pi}\boldsymbol{E}\cdot\boldsymbol{dD} - \frac{V}{4\pi}\boldsymbol{D}\cdot\boldsymbol{dE}$$

$$= VA\,\mathrm{d}y - \frac{V}{4\pi}\boldsymbol{D}\cdot\boldsymbol{dE}. \tag{7.57}$$

Thus, with the help of the transformed free energy $\tilde{F}$, we may express all quantities of interest at constant temperature and volume as a function of y and E.

The quantity measured in saturation experiments is the incremental dielectric constant ε_E at chemical equilibrium. Since temperature and volume are held constant and the affinity A is constant at chemical equilibrium (it has the value zero), we may write:

$$\varepsilon_E = \left(\frac{\partial D}{\partial E}\right)_{T,V,A}. \tag{7.58}$$

Since we consider all quantities as functions of y and E, we write D as a function of y and E and obtain from eqn. (7.58):

$$\varepsilon_E = \left(\frac{\partial D}{\partial E}\right)_{T,V,y} + \left(\frac{\partial D}{\partial y}\right)_{T,V,E}\left(\frac{\partial y}{\partial E}\right)_{T,V,A}. \tag{7.59}$$

The term $(\partial D/\partial E)_{T,V,y}$ in this expression for ε_E is the incremental dielectric constant at constant reaction coordinate, *i.e.* when there are no effects due to a shift in the equilibrium. Therefore, it represents the normal saturation effect. The second term represents the contribution of the anomalous saturation. Comparing (7.5) with (7.59), we see that it is equal to the anomalous saturation term $\Delta\varepsilon_a$ introduced in eqn. (7.44). Using $D = \varepsilon E$, we find for $\Delta\varepsilon_a$:

$$\Delta\varepsilon_a = \left(\frac{\partial D}{\partial y}\right)_{T,V,E}\left(\frac{\partial y}{\partial E}\right)_{T,V,A}$$

$$= E\left(\frac{\partial \varepsilon}{\partial y}\right)_{T,V,E}\left(\frac{\partial y}{\partial E}\right)_{T,V,A}. \tag{7.60}$$

The factor $(\partial y/\partial E)_{T,V,A}$ can be calculated with the help of a general thermodynamic formula[21] as:

$$\left(\frac{\partial y}{\partial E}\right)_{T,V,A} = -\frac{(\partial A/\partial E)_{T,V,y}}{(\partial A/\partial y)_{T,V,E}}. \tag{7.61}$$

The numerator of this expression can be derived with the help of eqn. (7.57), by using the cross-differentiation identity of a perfect differential:

$$\left(\frac{\partial A}{\partial E}\right)_{T,V,y} = -\frac{1}{4\pi}\left(\frac{\partial D}{\partial y}\right)_{T,V,E} = -\frac{1}{4\pi}E\left(\frac{\partial \varepsilon}{\partial y}\right)_{T,V,E}. \tag{7.62}$$

To calculate the denominator $(\partial A/\partial y)_{T,V,E}$ we use the definitions (7.52) and (7.53) to write:

$$\left(\frac{\partial A}{\partial y}\right)_{T,V,E} = \frac{\partial}{\partial y}\left(\sum_k v_k g_k\right)_{T,V,E} = \sum_k v_k\left(\frac{\partial g_k}{\partial y}\right)_{T,V,E}$$
$$= \sum_k v_k^2\left(\frac{\partial g_k}{\partial n_k}\right)_{T,V,E}.$$

For a further evaluation we need an expression for the chemical potential g_k. As an approximation we use the expression for the chemical potential in an ideal mixture:

$$g_k = g_k^* + RT\ln n_k, \tag{7.63}$$

where g_k^* is the chemical potential in the reference state of a concentration of one mole per unit of volume. Then we find:

$$\left(\frac{\partial A}{\partial y}\right)_{T,V,E} = RT\sum_k \frac{v_k^2}{n_k}. \tag{7.64}$$

Substituting (7.62) and (7.64) into (7.61) and the result into eqn. (7.60), we obtain:

$$\Delta\varepsilon_a = \frac{1}{4\pi}E^2\left(\frac{\partial \varepsilon}{\partial y}\right)^2_{T,V,E}\frac{1}{RT}\left(\sum_k \frac{v_k^2}{n_k}\right)^{-1}. \tag{7.65}$$

The value of the derivative $(\partial\varepsilon/\partial y)_{T,V,E}$ can be derived from the dependence of ε on the composition of the system. As an expression for ε we may use eqn. (5.60):

$$\frac{(\varepsilon - 1)(2\varepsilon + 1)}{12\pi\varepsilon} = \sum_k N_k \frac{1}{1 - f_k\alpha_k}\left(\alpha_k + \frac{1}{3kT}\frac{\mu_k^2}{1 - f_k\alpha_k}\right).$$

To simplify this relation, we assume that the dielectric constant characteristic for the induced polarization is the same for all particles. This leads to an equation analogous to eqn. (6.209):

$$\frac{(\varepsilon - \varepsilon_\infty)(2\varepsilon + \varepsilon_\infty)}{\varepsilon(\varepsilon_\infty + 2)^2} = \frac{4\pi}{9kT}\sum_k N_k\mu_k^2.$$

The derivative $(\partial\varepsilon/\partial y)_{T,V,E}$ can be conveniently calculated by differentiating both sides of this expression with respect to y, considering ε as a function of y. Using (7.53), this leads to:

$$\left(\frac{\partial\varepsilon}{\partial y}\right)_{T,V,E} = \frac{\dfrac{\partial}{\partial y}\dfrac{4\pi}{9kT}\sum_k N_k\mu_k^2}{\dfrac{\partial}{\partial\varepsilon}\dfrac{(\varepsilon-\varepsilon_\infty)(2\varepsilon+\varepsilon_\infty)}{\varepsilon(\varepsilon_\infty+2)^2}} = \frac{4\pi}{9kT}\frac{\varepsilon^2(\varepsilon_\infty+2)^2}{2\varepsilon^2+\varepsilon_\infty^2}N_A\sum_k \nu_k\mu_k^2. \quad (7.66)$$

Substituting this result into eqn. (7.65), we arrive at:

$$\Delta\varepsilon_a = 4\pi E^2\frac{N_A}{81(kT)^3}\frac{\varepsilon^4(\varepsilon_\infty+2)^4}{(2\varepsilon^2+\varepsilon_\infty^2)^2}\left(\sum_k\frac{\nu_k^2}{n_k}\right)^{-1}\left(\sum_k \nu_k\mu_k^2\right)^2. \quad (7.67)$$

For the case of an equilibrium between trans- and gauche-isomers, eqn. (7.67) reduces to:

$$\Delta\varepsilon_a = 4\pi E^2\frac{N}{81(kT)^3}\frac{\varepsilon^4(\varepsilon_\infty+2)^4}{(2\varepsilon^2+\varepsilon_\infty^2)^2}x_t(1-x_t)(\mu_t^2-\mu_g^2)^2. \quad (7.68)$$

Comparing this expression with eqn. (7.46), we see that the expressions are similar. The difference is a factor $10(2\varepsilon+\varepsilon_\infty)^2/27(2\varepsilon^2+\varepsilon_\infty^2)$; this factor can have values between 10/9 (when $\varepsilon=\varepsilon_\infty$) and 20/27 (when $\varepsilon=\infty$). This difference is caused by the neglect of the contribution of the intermolecular interactions in expression (7.63) for g_k. It does not appear possible to take the intermolecular interactions into account in the thermodynamic derivation in a simple way.

§44. Non-linear effects due to the anisotropy of the polarizability and to the hyperpolarizabilities

In the calculations of the non-linear part of the polarization we have so far considered the effects due to the permanent moments of the molecules only. The induced polarization plays also a role, however. In section 15 it was mentioned that a molecule with an anisotropic polarizability is directed by an external field with its axis of greatest polarizability in the direction of the field, which leads to a non-linear term in the polarization. Furthermore, deviations of the linearity of the induced moment as a function of the applied field must be considered. We shall investigate the consequences of these effects on the dielectric constant. We start with the case of a gas at normal pressure and follow Buckingham's derivation.[22]

The moment of a molecule in a homogeneous external field can be written[23] as:

$$\boldsymbol{m} = \boldsymbol{\mu} + \boldsymbol{\alpha} \cdot \boldsymbol{E}_0 + \tfrac{1}{2}\boldsymbol{\beta} : \boldsymbol{E}_0 \boldsymbol{E}_0 + \tfrac{1}{6}\boldsymbol{\gamma} \vdots \boldsymbol{E}_0 \boldsymbol{E}_0 \boldsymbol{E}_0. \tag{7.69}$$

Here, the hyperpolarizabilities $\boldsymbol{\beta}$ and $\boldsymbol{\gamma}$ are tensors of the third and fourth order, respectively. Like the polarizability tensor $\boldsymbol{\alpha}$, the hyperpolarizability tensors are symmetric in all indices. For molecules with a centre of symmetry, we have $\boldsymbol{m}(\boldsymbol{E}) = -\boldsymbol{m}(-\boldsymbol{E})$, so that $\boldsymbol{\mu} = \boldsymbol{0}$ and $\boldsymbol{\beta} = \boldsymbol{0}$. If there is an axis of symmetry, which is then taken as the z-axis of the molecular co-ordinate system, the only non-vanishing elements are μ_3, $\alpha_{11} = \alpha_{22}$, α_{33}, $\beta_{113} = \beta_{223}$, β_{333}, $\gamma_{1111} = \gamma_{2222} = 3\gamma_{1122}$, $\gamma_{1133} = \gamma_{2233}$, and γ_{3333}. Furthermore, for spherical symmetry, we have for the non-zero elements $\alpha_{11} = \alpha_{22} = \alpha_{33} = \alpha$, $\gamma_{1111} = \gamma_{2222} = \gamma_{3333} = 3\gamma_{1122} = 3\gamma_{2233} = 3\gamma_{3311} = \gamma$, so that in that case:

$$\boldsymbol{m} = \alpha \boldsymbol{E}_0 + \gamma E_0^2 \boldsymbol{E}_0. \tag{7.70}$$

From eqn. (7.69) we can calculate the energy of a molecule in a strong uniform external field, in analogy with the calculation of expression (3.101) for the energy of a molecule in an external field of moderate intensity:

$$W = -\boldsymbol{\mu} \cdot \boldsymbol{E}_0 - \tfrac{1}{2}\boldsymbol{E}_0 \cdot \boldsymbol{\alpha} \cdot \boldsymbol{E}_0.$$

In this way we find:

$$\begin{aligned}
W &= V_{\mathrm{m}} + W_{\mathrm{pol}} \\
&= -\boldsymbol{m} \cdot \boldsymbol{E}_0 + \int_0^{\boldsymbol{E}_0} \boldsymbol{E} \cdot d\boldsymbol{p} \\
&= -\boldsymbol{\mu} \cdot \boldsymbol{E}_0 - \boldsymbol{\alpha} : \boldsymbol{E}_0 \boldsymbol{E}_0 - \tfrac{1}{2}\boldsymbol{\beta} \vdots \boldsymbol{E}_0 \boldsymbol{E}_0 \boldsymbol{E}_0 - \tfrac{1}{6}\boldsymbol{\gamma} \vdots \boldsymbol{E}_0 \boldsymbol{E}_0 \boldsymbol{E}_0 \boldsymbol{E}_0 + \\
&\quad + \int_0^{\boldsymbol{E}_0} \boldsymbol{E} \cdot (\boldsymbol{\alpha} + \boldsymbol{\beta} \cdot \boldsymbol{E} + \tfrac{1}{2}\boldsymbol{\gamma} : \boldsymbol{E}\boldsymbol{E}) \cdot d\boldsymbol{E} \\
&= -\boldsymbol{\mu} \cdot \boldsymbol{E}_0 - \tfrac{1}{2}\boldsymbol{\alpha} : \boldsymbol{E}_0 \boldsymbol{E}_0 - \tfrac{1}{6}\boldsymbol{\beta} \vdots \boldsymbol{E}_0 \boldsymbol{E}_0 \boldsymbol{E}_0 - \tfrac{1}{24}\boldsymbol{\gamma} \vdots \boldsymbol{E}_0 \boldsymbol{E}_0 \boldsymbol{E}_0 \boldsymbol{E}_0.
\end{aligned} \tag{7.71}$$

If we use the fact that the components e_1, e_2, and e_3 of the unit vector $\boldsymbol{e}$ in the direction of the external field are the direction cosines for the angles between the direction of the field and the axes of the molecular coordinate system, eqn. (7.71) changes into:

$$W = -E_0 \sum_{k=1}^{3} \mu_k e_k - \tfrac{1}{2}E_0^2 \sum_{k=1}^{3}\sum_{l=1}^{3} \alpha_{kl} e_k e_l - \tfrac{1}{6}E_0^3 \sum_{k=1}^{3}\sum_{l=1}^{3}\sum_{m=1}^{3} \beta_{klm} e_k e_l e_m - \\ - \tfrac{1}{24}E_0^4 \sum_{k=1}^{3}\sum_{l=1}^{3}\sum_{m=1}^{3}\sum_{n=1}^{3} \gamma_{klmn} e_k e_l e_m e_n. \tag{7.72}$$

In the same way, the component of $\boldsymbol{m}$ in the direction of $\boldsymbol{E}_0$ is given by:

$$\boldsymbol{m}\cdot\boldsymbol{e} = \sum_{k=1}^{3} \mu_k e_k + E_0 \sum_{k=1}^{3}\sum_{l=1}^{3} \alpha_{kl} e_k e_l + \tfrac{1}{2}E_0^2 \sum_{k=1}^{3}\sum_{l=1}^{3}\sum_{m=1}^{3} \beta_{klm} e_k e_l e_m + \\ + \tfrac{1}{6}E_0^3 \sum_{k=1}^{3}\sum_{l=1}^{3}\sum_{m=1}^{3}\sum_{n=1}^{3} \gamma_{klmn} e_k e_l e_m e_n. \tag{7.73}$$

For a gas at normal pressure, we may take the field which acts on the molecule equal to the Maxwell field $\boldsymbol{E}$. Thus, in the following we may use eqns. (7.72) and (7.73) with $\boldsymbol{E}_0$ replaced by $\boldsymbol{E}$. The polarization is then given by:

$$P = N\langle \boldsymbol{m}\cdot\boldsymbol{e}\rangle = \chi E + \xi E^3. \tag{7.74}$$

From this, we find:

$$\chi = N\left(\frac{\partial}{\partial E}\langle \boldsymbol{m}\cdot\boldsymbol{e}\rangle\right)_{E=0} \\ = N\left(\frac{\partial}{\partial E}\frac{\int \boldsymbol{m}\cdot\boldsymbol{e}\, e^{-W/kT}\,\mathrm{d}\tau}{\int e^{-W/kT}\,\mathrm{d}\tau}\right)_{E=0}, \tag{7.75}$$

where the integrations are performed over all orientations of the molecule. Eqn. (7.75) leads to (compare eqns. (6.35) and (6.36)):

$$\chi = N\left\{\left\langle\frac{\partial(\boldsymbol{m}\cdot\boldsymbol{e})}{\partial E}\right\rangle_0 - \frac{1}{kT}\left\langle \boldsymbol{m}\cdot\boldsymbol{e}\frac{\partial W}{\partial E}\right\rangle_0\right\} \\ = N\left[\sum_{k=1}^{3}\sum_{l=1}^{3}\left\{\langle\alpha_{kl}e_k e_l\rangle_0 + \frac{1}{kT}\langle m_k m_l e_k e_l\rangle_0\right\}\right]. \tag{7.76}$$

Using $\langle e_k e_l\rangle_0 = \tfrac{1}{3}\delta_{kl}$ (compare eqn. (6.42)), we find:

$$\chi = N\left(\bar{\alpha} + \frac{\mu^2}{3kT}\right), \tag{7.77}$$

in agreement with eqn. (6.44).

In the same way, we find from eqn. (7.74) for ξ:

$$\xi = \tfrac{1}{6}N\left(\frac{\partial^3}{\partial E^3}\langle \boldsymbol{m}\cdot\boldsymbol{e}\rangle\right)_{E=0}$$

$$= \tfrac{1}{6}N\left[\left\langle\frac{\partial^3 \boldsymbol{m}\cdot\boldsymbol{e}}{\partial E^3}\right\rangle_0 - \frac{3}{kT}\left\langle\frac{\partial^2 \boldsymbol{m}\cdot\boldsymbol{e}}{\partial E^2}\frac{\partial W}{\partial E} + \frac{\partial \boldsymbol{m}\cdot\boldsymbol{e}}{\partial E}\left\{\frac{\partial^2 W}{\partial E^2} - \left\langle\frac{\partial^2 W}{\partial E^2}\right\rangle_0 - \right.\right.\right.$$

$$\left.\left. - \frac{1}{kT}\left(\frac{\partial W}{\partial E}\right)^2 + \frac{1}{kT}\left\langle\left(\frac{\partial W}{\partial E}\right)^2\right\rangle_0\right\}\right\rangle_0 -$$

$$- \frac{1}{kT}\left\langle \boldsymbol{m}\cdot\boldsymbol{e}\left\{\frac{\partial^3 W}{\partial E^3} - \frac{3}{kT}\frac{\partial^2 W}{\partial E^2}\frac{\partial W}{\partial E} + \frac{3}{kT}\frac{\partial W}{\partial E}\left\langle\frac{\partial^2 W}{\partial E^2}\right\rangle_0 + \right.\right.$$

$$\left.\left.\left. + \frac{1}{(kT)^2}\left(\frac{\partial W}{\partial E}\right)^3 - \frac{3}{(kT)^2}\frac{\partial W}{\partial E}\left\langle\left(\frac{\partial W}{\partial E}\right)^2\right\rangle_0\right\}\right\rangle_0\right]$$

$$= \tfrac{1}{6}N\left[\left\langle\sum_{k=1}^{3}\sum_{l=1}^{3}\sum_{m=1}^{3}\sum_{n=1}^{3}\left\{\gamma_{klmn} + \frac{4}{kT}\mu_n\beta_{klm} + \frac{3}{kT}\alpha_{kl}\alpha_{mn} + \right.\right.\right.$$

$$\left.\left. + \frac{6}{(kT)^2}\alpha_{kl}\mu_m\mu_n + \frac{1}{(kT)^3}\mu_k\mu_l\mu_m\mu_n\right\}e_k e_l e_m e_n\right\rangle_0 -$$

$$\left. - \frac{3}{kT}\left\langle\sum_{k=1}^{3}\sum_{l=1}^{3}\left\{\alpha_{kl} + \frac{1}{kT}\mu_k\mu_l\right\}e_k e_l\right\rangle_0^2\right]. \tag{7.78}$$

Using:

$$\langle e_k e_l\rangle_0 = \tfrac{1}{3}\delta_{kl},$$

and:

$$\langle e_k e_l e_m e_n\rangle_0 = \tfrac{1}{15}\{\delta_{kl}\delta_{mn} + \delta_{km}\delta_{ln} + \delta_{kn}\delta_{lm}\},$$

and the fact that $\boldsymbol{\alpha}$, $\boldsymbol{\beta}$, and γ are symmetric in all indices, we find:

$$\xi = \tfrac{1}{90}N\left[\sum_{k=1}^{3}\sum_{l=1}^{3}\left\{3\gamma_{kkll} + \frac{12}{kT}\mu_k\beta_{kll}\right\} + \frac{6}{kT}(\alpha_1^2 + \alpha_2^2 + \alpha_3^2 - 3\bar{\alpha}^2) + \right.$$

$$\left. + \frac{12}{(kT)^2}(\alpha_1\mu_1^2 + \alpha_2\mu_2^2 + \alpha_3\mu_3^2 - \bar{\alpha}\mu^2) - \frac{2}{(kT)^3}\mu^4\right]. \tag{7.79}$$

For molecules with hyperpolarizabilities $\boldsymbol{\beta} = \mathbf{0}$ and $\gamma = \mathbf{0}$ and with an isotropic polarizability: $\bar{\alpha} = \alpha_1 = \alpha_2 = \alpha_3$, eqn. (7.79) reduces to eqn. (7.17):

$$\xi = -\frac{N\mu^4}{45(kT)^3}.$$

For molecules with both an axis and a centre of symmetry, we have $\mu = 0$, $\alpha_1 = \alpha_2 = \alpha_b$, $\alpha_3 = \alpha_a$, $\boldsymbol{\beta} = \mathbf{0}$, $\gamma_{1111} = \gamma_{2222} = 3\gamma_{1122} = \gamma_b$, $\gamma_{1133} = \gamma_{2233} = \gamma_{ab}$, $\gamma_{3333} = \gamma_a$, and eqn. (7.79) reduces to:

$$\xi = \tfrac{1}{90} N \left\{ 3\gamma_a + 12\gamma_{ab} + 8\gamma_b + \frac{4}{kT}(\alpha_a - \alpha_b)^2 \right\}. \tag{7.80}$$

For molecules with spherical symmetry we have $\alpha_a = \alpha_b$, $\gamma_a = \gamma_b = 3\gamma_{ab} = \gamma$, and we find:

$$\xi = \tfrac{1}{6} N\gamma. \tag{7.81}$$

Buckingham[22] uses the Lorentz field (eqn. (5.22)) instead of the Maxwell field as the field that acts on the molecules. Thus his expressions differ by a factor $\{(\varepsilon_0 + 2)/3\}^4$. In the case of gases at normal pressure, however, this factor can be taken equal to one (compare section 31).

The expressions (7.79), (7.80) and (7.81) for ξ in the gaseous state have only theoretical significance, because the corresponding values of $\Delta\varepsilon$ are too small to be measured. This can be inferred from data concerning the hyperpolarizabilities and the anisotropy of the polarizability which can be obtained from measurements of the electro-optical Kerr constant (see Vol. II).

For condensed systems the contribution of the anisotropy of the polarizability to the non-linear part of the polarization is larger than that for gases. For the calculation of the anisotropy contribution for condensed non-polar systems we shall use the continuum theory.

Assuming the shape of the molecule to be ellipsoidal, with axes along the principal axes of the molecular polarizability tensor, we find for the component of the internal field in the direction of the axis k (compare eqns. (5.68) and (5.69)):

$$(E_i)_k = \frac{1}{1 - f_k\alpha_k}(E_c)_k = \frac{1}{1 - f_k\alpha_k}\,\frac{\varepsilon}{\varepsilon + (1 - \varepsilon)A_k} E e_k, \tag{7.82}$$

since $(E)_k = Ee_k$. Hence, the component of the moment of the molecule in the direction of the external field is given by:

$$\boldsymbol{m}\cdot\boldsymbol{e} = \sum_{k=1}^{3} \alpha_k(E_i)_k e_k = E\varepsilon \sum_{k=1}^{3} \frac{\alpha_k e_k^2}{(1-f_k\alpha_k)\{\varepsilon+(1-\varepsilon)A_k\}}. \tag{7.83}$$

For a zero field, we have the following expressions independent of the orientation of the molecule:

$$(\boldsymbol{m}\cdot\boldsymbol{e})_{E=0} = 0, \tag{7.84}$$

$$\left(\frac{\partial^2 \boldsymbol{m}\cdot\boldsymbol{e}}{\partial E^2}\right)_{E=0} = 0, \tag{7.85}$$

$$\left(\frac{\partial^3 \boldsymbol{m}\cdot\boldsymbol{e}}{\partial E^3}\right)_{E=0} = 0, \tag{7.86}$$

if we neglect the dependence of $\boldsymbol{m}\cdot\boldsymbol{e}$ on E via ε. The susceptibility at low field intensities can be found with the help of eqn. (7.76). Using (7.84) in the second member of eqn. (7.76), we find:

$$\begin{aligned} \chi &= N\left\langle \frac{\partial \boldsymbol{m}\cdot\boldsymbol{e}}{\partial E} \right\rangle_0 \\ &= N\varepsilon \left\langle \sum_{k=1}^{3} \frac{\alpha_k e_k^2}{(1-f_k\alpha_k)\{\varepsilon+(1-\varepsilon)A_k\}} \right\rangle_0 \\ &= \tfrac{1}{3}N\varepsilon \sum_{k=1}^{3} \frac{\alpha_k}{(1-f_k\alpha_k)\{\varepsilon+(1-\varepsilon)A_k\}}, \end{aligned} \tag{7.87}$$

in agreement with eqn. (5.112). Here it is taken that for $E = 0$ all directions are equally probable, so that $\langle e_k^2 \rangle_0 = \frac{1}{3}$.

In the same way, we find an expression for ξ by substituting eqns. (7.84), (7.85) and (7.86) into the second member of eqn. (7.78):

$$\xi = -\frac{N}{2kT}\left\langle \frac{\partial \boldsymbol{m}\cdot\boldsymbol{e}}{\partial E}\left[\frac{\partial^2 W}{\partial E^2} - \left\langle \frac{\partial^2 W}{\partial E^2} \right\rangle_0 - \frac{1}{kT}\left(\frac{\partial W}{\partial E}\right)^2 + \right.\right.$$

$$\left.\left. + \frac{1}{kT}\left\langle \left(\frac{\partial W}{\partial E}\right)^2 \right\rangle_0 \right] \right\rangle_0. \tag{7.88}$$

We now need an expression for W, the electrostatic energy of the particle. This energy consists of four terms:

W_1: the potential energy of the polarized particle in the internal field,

W_2: the work of polarization of the particle,

W_3: the work of polarization of the environment of the particle, which gives rise to the reaction field, and

W_4: the work of formation of the cavity in which the particle is situated.
The potential energy W_1 is given by:

$$W_1 = -\boldsymbol{m} \cdot \boldsymbol{E}_i = -E^2\varepsilon^2 \sum_{k=1}^{3} \frac{\alpha_k e_k^2}{(1 - f_k\alpha_k)^2\{\varepsilon + (1 - \varepsilon)A_k\}^2}. \tag{7.89}$$

The work of polarization W_2 of the particle is given by:

$$W_2 = \int_0^{\boldsymbol{m}} \boldsymbol{E} \cdot d\boldsymbol{p} = \tfrac{1}{2}E^2\varepsilon^2 \sum_{k=1}^{3} \frac{\alpha_k e_k^2}{(1 - f_k\alpha_k)^2\{\varepsilon + (1 - \varepsilon)A_k\}^2}. \tag{7.90}$$

The work of polarization W_3 of the environment of the particle can be found by considering the energy necessary for the formation of a dipole in a cavity. This energy is given by:

$$W = -\int_{\lambda=0}^{1} \lambda \boldsymbol{R} \cdot \boldsymbol{m} \mathrm{d}\lambda = -\tfrac{1}{2}\boldsymbol{m} \cdot \boldsymbol{F} \cdot \boldsymbol{m}. \tag{7.91}$$

This energy consists of the electric energy of the dipole in the reaction field, which is equal to $-\boldsymbol{m} \cdot \boldsymbol{F} \cdot \boldsymbol{m}$, and of the work of polarization of the environment, which is therefore given by $\frac{1}{2}\boldsymbol{m} \cdot \boldsymbol{F} \cdot \boldsymbol{m}$, so that:

$$W_3 = \tfrac{1}{2}E^2\varepsilon^2 \sum_{k=1}^{3} \frac{f_k\alpha_k^2 e_k^2}{(1 - f_k\alpha_k)^2\{\varepsilon + (1 - \varepsilon)A_k\}^2}. \tag{7.92}$$

The work of formation of the cavity, W_4, is the difference between the energies of the dielectric with and without the cavity. This difference is given by the energy of the apparent charges induced on the boundary of the cavity, in the field in the dielectric due to the true charges, which is equal to $\varepsilon\boldsymbol{E}$. This energy is given by:

$$W_4 = -\tfrac{1}{2}\boldsymbol{m}' \cdot \varepsilon\boldsymbol{E}, \tag{7.93}$$

where $\boldsymbol{m}'$ is the moment of the distribution of the apparent charges. This moment can be found from the field in the cavity due to the apparent charges. This field is given by:

$$\boldsymbol{E}' = \boldsymbol{E}_c - \boldsymbol{E}, \tag{7.94}$$

where the components of the cavity field $\boldsymbol{E}_c$ can be written, according to eqn. (4.43), as:

$$(E_c)_k = \frac{\varepsilon}{\varepsilon + (1 - \varepsilon)A_k} (E)_k. \tag{7.95}$$

The relation between the moment of an apparent charge distribution and the field inducing it can be deduced from the expression for the field due to the apparent charges on the boundary of a homogeneous ellipsoid in a homogeneous external field *in vacuo*. According to eqn. (2.80) and using the notation of section 9, we find for $\boldsymbol{P}_2$, the polarization in the ellipsoid:

$$\boldsymbol{P}_2 = (\boldsymbol{E}_0 - \boldsymbol{E}_2) \cdot \boldsymbol{L}^{-1}. \tag{7.96}$$

Using (2.83) for an ellipsoid *in vacuo* ($\varepsilon_1 = 1$), we find for the moment of the ellipsoid:

$$\boldsymbol{m}_2 = \frac{4\pi}{3} abc\boldsymbol{P}_2 = -\tfrac{1}{3}abc(\boldsymbol{E}_0 - \boldsymbol{E}_2) \cdot \boldsymbol{A}^{-1}, \tag{7.97}$$

where $\boldsymbol{A}$ is a tensor with elements A_{ij} given by:

$$A_{ij} = A_i \delta_{ij}. \tag{7.98}$$

The polarization of the ellipsoid is completely given by the moment of the apparent charges on the boundary. Since the field $(\boldsymbol{E}_0 - \boldsymbol{E}_2)$ corresponding with the apparent charges with moment $\boldsymbol{m}_2$, does not depend on the way the apparent charges have arisen, the relation between field and moment given in eqn. (7.97) can be used generally. An example is the apparent charge distribution on the boundary of an ellipsoid giving rise to a homogeneous field within the ellipsoid. Thus, returning to the problem of a cavity in a dielectric, we find for the moment of the apparent charges that are the source of the field $\boldsymbol{E}'$:

$$\boldsymbol{m}' = -\tfrac{1}{3}abc\boldsymbol{E}' \cdot \boldsymbol{A}^{-1}. \tag{7.99}$$

Substituting this result into eqn. (7.93), we find with (7.94) and (7.95):

$$W_4 = \tfrac{1}{6}abc\, \varepsilon(\varepsilon - 1)E^2 \sum_{k=1}^{3} \frac{1}{\varepsilon + (1 - \varepsilon)A_k}. \tag{7.100}$$

Collecting our results, we obtain:

$$W = W_1 + W_2 + W_3 + W_4$$

$$= \tfrac{1}{2}\varepsilon E^2 \sum_{k=1}^{3} e_k^2 \left[\frac{-\varepsilon\alpha_k}{(1 - f_k\alpha_k)\{\varepsilon + (1 - \varepsilon)A_k\}^2} + \frac{\tfrac{1}{3}abc(\varepsilon - 1)}{\varepsilon + (1 - \varepsilon)A_k} \right]. \tag{7.101}$$

W should be zero in the case of a dielectric without a cavity. For homogeneously polarizable ellipsoids that fill up the whole volume (the generalization of Onsager's approximation, (eqn. (5.86)) the dielectric constant of the ellipsoid is the same as the over-all dielectric constant of the system. Substitution of expressions (2.105) and (4.102) into (7.101) then leads to $W = 0$, as was to be expected.

Substituting eqn. (7.101) into (7.88), we find:

$$\xi = -\frac{N}{2kT}\left\langle\left[\sum_{k=1}^{3}\frac{\varepsilon e_k^2\alpha_k}{(1-f_k\alpha_k)\{\varepsilon+(1-\varepsilon)A_k\}}\right]\cdot\right.$$
$$\cdot\left[\varepsilon\sum_{k=1}^{3}e_k^2\left\{\frac{-\varepsilon\alpha_k}{(1-f_k\alpha_k)\{\varepsilon+(1-\varepsilon)A_k\}^2}+\frac{\frac{1}{3}abc(\varepsilon-1)}{\varepsilon+(1-\varepsilon)A_k}\right\}-\right.$$
$$\left.\left.-\left\langle\varepsilon\sum_{k=1}^{3}e_k^2\left\{\frac{-\varepsilon\alpha_k}{(1-f_k\alpha_k)\{\varepsilon+(1-\varepsilon)A_k\}^2}+\frac{\frac{1}{3}abc(\varepsilon-1)}{\varepsilon+(1-\varepsilon)A_k}\right\}\right\rangle_0\right]\right\rangle_0. \tag{7.102}$$

Changing one of the dummy indices k to l and using $\langle e_k^2\rangle_0 = \frac{1}{3}$ and $\langle e_k^2 e_l^2\rangle_0 = \frac{1}{15} + \frac{2}{15}\delta_{kl}$, we find:

$$\xi = -\frac{N}{2kT}\sum_{k=1}^{3}\sum_{l=1}^{3}\frac{\varepsilon\alpha_k}{(1-f_k\alpha_k)\{\varepsilon+(1-\varepsilon)A_k\}}\left\{\frac{-\varepsilon\alpha_l}{(1-f_l\alpha_l)\{\varepsilon+(1-\varepsilon)A_l\}^2}+\right.$$
$$\left.+\frac{\frac{1}{3}abc(\varepsilon-1)}{\varepsilon+(1-\varepsilon)A_l}\right\}\left(\tfrac{2}{15}\delta_{kl}-\tfrac{2}{45}\right). \tag{7.103}$$

For gases we have $\varepsilon = 1$, and (7.103) reduces to the second term of eqn. (7.79).

For spheroids ($\alpha_2 = \alpha_3$, $A_2 = A_3$), eqn. (7.103) changes into:

$$\xi = \frac{2N}{45kT}\left[\frac{\varepsilon\alpha_1}{(1-f_1\alpha_1)\{\varepsilon+(1-\varepsilon)A_1\}}-\frac{\varepsilon\alpha_2}{(1-f_2\alpha_2)\{\varepsilon+(1-\varepsilon)A_2\}}\right]\cdot$$
$$\cdot\left[\frac{\varepsilon\alpha_1}{(1-f_1\alpha_1)\{\varepsilon+(1-\varepsilon)A_1\}^2}-\frac{\frac{1}{3}abc(\varepsilon-1)}{\varepsilon+(1-\varepsilon)A_1}-\right.$$
$$\left.-\frac{\varepsilon\alpha_2}{(1-f_2\alpha_2)\{\varepsilon+(1-\varepsilon)A_2\}^2}+\frac{\frac{1}{3}abc(\varepsilon-1)}{\varepsilon+(1-\varepsilon)A_2}\right]. \tag{7.104}$$

From eqn. (7.104) we calculate for carbon disulfide, using the same values for the molecular parameters as in section 33, $\Delta\varepsilon/E^2 = 0.6\ 10^{-15}$, whereas experimentally[7] $\Delta\varepsilon/E^2 = 1.2\ 10^{-15}$ was found. The agreement between the theoretical and experimental values is rather good, especially when it is taken into account that no correction for electrostriction has been made.

§45. Electrostriction

In section 15 it was found that an inhomogeneous field $\boldsymbol{E}_0$ exerts on a molecule a translational force, given by eqn. (3.103):

$$\boldsymbol{F} = \boldsymbol{\mu} \cdot \nabla \boldsymbol{E}_0 + \alpha \boldsymbol{E}_0 \cdot \nabla \boldsymbol{E}_0.$$

Thus if the permanent moment points in the direction of $\boldsymbol{E}_0$, the molecule will be displaced towards regions of higher field strengths. In a macroscopic sample the average moment is in the direction of the field, *i.e.* the molecules favour orientations with their permanent moments in the direction of the field. Therefore, an inhomogeneous field acting on a macroscopic sample causes a concentration gradient with high concentrations at high field strengths. Thus, if a sample is situated partially in a strong field and is held at constant pressure, the density of the matter in the field will increase, leading to an increase of the dielectric constant. This effect is called electrostriction. We shall calculate the contribution of the electrostriction to $\Delta\varepsilon/E^2$ with the help of thermodynamics.

The experimental situation is as follows: in a capacitor with volume V_c the electric field strength and the electric displacement are given by $\boldsymbol{E}$ and $\boldsymbol{D}$, respectively, whereas both fields are zero outside the capacitor. The investigated compound is situated both inside and outside the capacitor; its total volume is V and it is held under a constant pressure p. The quantity measured is the incremental dielectric constant (at constant pressure) of the part of the sample in the field, which can be written as:

$$\varepsilon_E = \left(\frac{\partial D}{\partial E}\right)_{T,p} = \left(\frac{\partial D}{\partial E}\right)_{T,d} + \left(\frac{\partial D}{\partial d}\right)_{T,E}\left(\frac{\partial d}{\partial E}\right)_{T,p}, \tag{7.105}$$

where d denotes the density of the part of the sample inside the capacitor. In eqn. (7.105) the term $(\partial D/\partial E)_{T,d}$ represents the incremental dielectric constant including all non-linear effects at constant density. The second term of the last member of eqn. (7.105) is the contribution to ε_E due to the electrostriction; denoting this term by $\Delta\varepsilon_e$, we have:

$$\begin{aligned}\Delta\varepsilon_e &= \left(\frac{\partial D}{\partial d}\right)_{T,E}\left(\frac{\partial d}{\partial E}\right)_{T,p}\\ &= E\left(\frac{\partial \varepsilon}{\partial d}\right)_{T,E}\left(\frac{\partial d}{\partial E}\right)_{T,p}.\end{aligned} \tag{7.106}$$

The differential increase of the density inside the capacitor, $\mathrm{d}d$, corresponds to an increase in mass equal to $V_c\,\mathrm{d}d$. This increase in mass is equal to a decrease in mass outside the capacitor given by $-d_0\,\mathrm{d}(V - V_c) = -d_0\,\mathrm{d}V$, so that $\mathrm{d}d$ is given by $-(d_0/V_c)\,\mathrm{d}V$. Thus, we may rewrite (7.106):

$$\Delta\varepsilon_e = -E\left(\frac{\partial\varepsilon}{\partial d}\right)_{T,E}\frac{d_0}{V_c}\left(\frac{\partial V}{\partial E}\right)_{T,p}. \tag{7.107}$$

The derivative $(\partial V/\partial E)_{T,p}$ can be found from the differential of the free energy, which in this case is given by:

$$\mathrm{d}F = -S\,\mathrm{d}T - p\,\mathrm{d}V + \frac{V_c}{4\pi}E\,\mathrm{d}D. \tag{7.108}$$

Introducing the transformed free enthalpy by:

$$\tilde{G} = F + pV - \frac{V_c}{4\pi}ED, \tag{7.109}$$

we have for the differential of $\tilde{G}$:

$$\mathrm{d}\tilde{G} = -S\,\mathrm{d}T + V\mathrm{d}p - \frac{V_c}{4\pi}D\,\mathrm{d}E. \tag{7.110}$$

From this equation we find:

$$\left(\frac{\partial V}{\partial E}\right)_{T,p} = -\frac{V_c}{4\pi}\left(\frac{\partial D}{\partial p}\right)_{T,E} = -\frac{V_c E}{4\pi}\left(\frac{\partial\varepsilon}{\partial p}\right)_{T,E}. \tag{7.111}$$

After substitution of this result into eqn. (7.107) we have:

$$\Delta\varepsilon_e = \frac{E^2}{4\pi}d_0\left(\frac{\partial\varepsilon}{\partial d}\right)_{T,E}\left(\frac{\partial\varepsilon}{\partial p}\right)_{T,E}. \tag{7.112}$$

We now use:

$$\left(\frac{\partial\varepsilon}{\partial p}\right)_{T,E} = \left(\frac{\partial\varepsilon}{\partial d}\right)_{T,E}\left(\frac{\partial d}{\partial p}\right)_{T,E} = \beta d_0\left(\frac{\partial\varepsilon}{\partial d}\right)_T, \tag{7.113}$$

where $\beta = -(1/V)(\partial V/\partial p)_T$ is the compressibility in the absence of the field. In the last member of (7.113), terms depending on E have been neglected, since they lead to terms in powers of E higher than the second in eqn. (7.112). In the same approximation we obtain after substitution of (7.113) into (7.112):

$$\Delta\varepsilon_e = \frac{E^2}{4\pi}\beta d_0^2\left(\frac{\partial\varepsilon}{\partial d}\right)_T^2, \tag{7.114}$$

or:

$$\Delta\varepsilon_e = \frac{E^2}{4\pi}\frac{1}{\beta}\left(\frac{\partial\varepsilon}{\partial p}\right)_T^2. \tag{7.115}$$

Thus, the magnitude of the contribution to the dielectric constant due to the electrostriction can be calculated from the compressibility and the dependence of the dielectric constant on either the density or the pressure. In this way a correction can be made for the effects of electrostriction when the non-linear part of the dielectric constant is measured at constant pressure as is usually done. Accordingly, in the previous sections values of $\Delta\varepsilon$ corrected for the electrostriction were used wherever possible.

The increase of the density due to electrostriction can be calculated from $(\partial V/\partial E)_{T,p}$:

$$\begin{aligned}\Delta d &= \int_0^E \left(\frac{\partial d}{\partial E}\right)_{T,p} \mathrm{d}E \\ &= \int_0^E -\frac{d_0}{V_c}\left(\frac{\partial V}{\partial E}\right)_{T,p} \mathrm{d}E \\ &= \int_0^E \frac{Ed_0}{4\pi}\left(\frac{\partial\varepsilon}{\partial p}\right)_{T,E} \mathrm{d}E \\ &= \int_0^E \frac{E\beta d_0^2}{4\pi}\left(\frac{\partial\varepsilon}{\partial d}\right)_T \mathrm{d}E. \end{aligned} \tag{7.116}$$

To obtain the last member of this equation, eqn. (7.113) was used, which means that in the expression for Δd terms in powers of E higher than the second have been neglected. Then the integration can easily be performed, leading to:

$$\Delta d = \frac{E^2}{8\pi}\beta d_0^2\left(\frac{\partial\varepsilon}{\partial d}\right)_T. \tag{7.117}$$

For the case of ideal gases, we have $\beta = M/d_0RT$, where M is the molecular weight. For this case $(\partial\varepsilon/\partial d)_T$ can be calculated from eqn. (5.98), leading to:

$$\left(\frac{\partial\varepsilon}{\partial d}\right)_T = \frac{4\pi N_A}{M}\left(\alpha + \frac{\mu^2}{3kT}\right). \tag{7.118}$$

Substitution into eqn. (7.117) gives:

$$\Delta d = \frac{E^2 N_A d_0}{2RT}\left(\alpha + \frac{\mu^2}{3kT}\right). \tag{7.119}$$

This result can also be derived in a different way. According to Boltzmann's distribution law, n, the number of moles per cm^3 of the gas at a point with field strength E, is given by:

$$n = n_0 e^{-W/kT}, \tag{7.120}$$

where W is the average value of the work required to bring a molecule into the field $\boldsymbol{E}$, and n_0 is the number of moles per cm^3 at a point where E is zero. According to section 35, the contribution of the induced moment to W is given by $-\frac{1}{2}\alpha E^2$, if the anisotropy of the polarizability and the hyperpolarizabilities are neglected. The average contribution of the permanent dipole moment to W, W_{dip} is dependent on the mean direction of the dipole vector (*cf.* section 26). To calculate this contribution we let the field increase from 0 to E. At each stage of this process the mean value of the component of the dipole moment in the direction of the field is given by eqn. (5.14):

$$\mu\,\overline{\cos\theta} = \frac{\mu^2}{3kT}E. \tag{7.121}$$

From this, it follows that a change dE in E leads to a change in $\overline{\cos\theta}$, given by:

$$\mathrm{d}\,\overline{\cos\theta} = \frac{\mu}{3kT}\,\mathrm{d}E. \tag{7.122}$$

This change in $\overline{\cos\theta}$ leads to a change in W_{dip}, given by:

$$\mathrm{d}\,W_{\text{dip}} = -\mu E\,\mathrm{d}\,\overline{\cos\theta}. \tag{7.123}$$

Combining (7.122) and (7.123), we have:

$$\mathrm{d}\,W_{\text{dip}} = -\frac{\mu^2}{3kT}E\,\mathrm{d}E. \tag{7.124}$$

Performing the integration we obtain:

$$W_{\text{dip}} = -\int_0^E \frac{\mu^2}{3kT} E \, \mathrm{d}E = -\frac{\mu^2}{6kT} E^2. \tag{7.125}$$

Thus the value for W to be used in eqn. (7.120) is:

$$W = -\tfrac{1}{2}\left(\alpha + \frac{\mu^2}{3kT}\right)E^2, \tag{7.126}$$

and we find:

$$n = n_0 e^{\frac{1}{2}(\alpha + \mu^2/3kT)E^2/kT}. \tag{7.127}$$

Neglecting terms in higher powers of E than the second, we have:

$$\Delta d = M(n - n_0) = Mn_0 \tfrac{1}{2}\left(\alpha + \frac{\mu^2}{3kT}\right)\frac{E^2}{kT} = \frac{E^2 N_A d_0}{2RT}\left(\alpha + \frac{\mu^2}{3kT}\right), \tag{7.128}$$

in agreement with eqn. (7.119).

If the dielectric is held at constant density instead of at constant pressure, the pressure will change under the influence of an applied electric field. The change of the pressure can be calculated from $(\partial p/\partial E)_{T,V}$. This partial derivative can be found from the transformed free energy $\tilde{F} = F - (V_c/4\pi)ED$:

$$\mathrm{d}\tilde{F} = -S\,\mathrm{d}T - p\,\mathrm{d}V - \frac{V_c}{4\pi} D\,\mathrm{d}E, \tag{7.129}$$

from which it follows that:

$$\begin{aligned} \left(\frac{\partial p}{\partial E}\right)_{T,V} &= \frac{V_c}{4\pi}\left(\frac{\partial D}{\partial V}\right)_{T,E} \\ &= \frac{V_c}{4\pi} E \left(\frac{\partial \varepsilon}{\partial V}\right)_{T,E} \\ &= -\frac{Ed}{4\pi}\left(\frac{\partial \varepsilon}{\partial d}\right)_{T,E}, \end{aligned} \tag{7.130}$$

so that:

$$\Delta p = -\frac{E^2}{8\pi} d \left(\frac{\partial \varepsilon}{\partial d}\right)_T. \tag{7.131}$$

Since $(\partial\varepsilon/\partial d)_T > 0$, the pressure will be lowered upon application of an electric field. This decrease of the pressure at constant density following from (7.131) is called the electrostrictive pressure.

In the above the pressure p has been defined as the pressure at a plane not in the field. As a consequence V_c was held constant in our derivation. Other definitions of the pressure in a dielectric in an electric field are possible, however.[24,25] Koenig[24] defines:

1. p_{pl}, the pressure on the capacitor plates,
2. p_{pa}, the pressure on a plane in the capacitor parallel to the capacitor plates,
3. p_{tr}, the pressure on a plane in the capacitor perpendicular to the capacitor plates.

These different pressures can in principle be measured in the following ways.

1. A closed capacitor with movable plates is filled with the dielectric. Upon application of the field, the dielectric will contract and the plates will move towards each other. The decrease of the external pressure on the plates necessary to restore the original volume is now equal to $-\Delta p_{pl}$.
2. A closed rigid capacitor the plates of which are permeable for gases is completely filled with the gaseous dielectric. The pressure of the gas outside the capacitor is made equal to the pressure inside. Upon application of the field, the gas inside will contract and suck in gas through the plates. To prevent this, the pressure of the gas outside must be decreased by an amount $-\Delta p_{pa}$.
3. If in experimental situation 2 the insulating walls rather than the plates are permeable for gases, the decrease of the pressure of the gas necessary to prevent contraction of the gas inside is equal to $-\Delta p_{tr}$.

For the calculation of Δp_{pl} we start from the differential of the free energy, which is given by:

$$\mathrm{d}F = -S\,\mathrm{d}T - p_{pl}\,\mathrm{d}V + \frac{V}{4\pi}E\,\mathrm{d}D. \qquad (7.132)$$

The differential of the transformed free energy $\tilde{F} = F - (V/4\pi)ED$ is then given by:

$$\mathrm{d}\tilde{F} = -S\,\mathrm{d}T - p_{pl}\,\mathrm{d}V - \frac{ED}{4\pi}\mathrm{d}V - \frac{VD}{4\pi}\mathrm{d}E. \qquad (7.133)$$

From this, we find:

$$\left(\frac{\partial}{\partial E}\left(p_{pl} + \frac{ED}{4\pi}\right)\right)_{T,V} = \left(\frac{\partial}{\partial V}\left(\frac{VD}{4\pi}\right)\right)_{T,E}, \qquad (7.134)$$

or:

$$\left(\frac{\partial p_{pl}}{\partial E}\right)_{T,V} + 2\frac{\varepsilon E}{4\pi} = \frac{\varepsilon E}{4\pi} + \frac{VE}{4\pi}\left(\frac{\partial\varepsilon}{\partial V}\right)_{T,E}. \qquad (7.135)$$

Thus:

$$\left(\frac{\partial p_{pl}}{\partial E}\right)_{T,V} = \frac{VE}{4\pi}\left(\frac{\partial\varepsilon}{\partial V}\right)_{T,E} - \frac{\varepsilon E}{4\pi}$$

$$= -\frac{Ed}{4\pi}\left(\frac{\partial\varepsilon}{\partial d}\right)_{T,E} - \frac{\varepsilon E}{4\pi}, \qquad (7.136)$$

so that after integration we have:

$$\Delta p_{\text{pl}} = -\frac{E^2}{8\pi} d\left(\frac{\partial \varepsilon}{\partial d}\right)_T - \frac{\varepsilon E^2}{8\pi}. \tag{7.137}$$

By using eqn. (7.131), this can be written as:

$$\Delta p_{\text{pl}} = \Delta p - \frac{\varepsilon E^2}{8\pi}. \tag{7.138}$$

The term $-\varepsilon E^2/8\pi$ results from the Coulomb forces between the charges on the plates. The charge density on a plate has an absolute value of $|\sigma| = |D|/4\pi = \varepsilon|E|/4\pi$, leading to a force per cm^2 with absolute value $\varepsilon E^2/8\pi$.

To calculate Δp_{pa}, we consider a system consisting of a capacitor with volume V_c, partially filled with a sample with volume V, where the boundary of the sample lies parallel with the capacitor plates. For the calculation of the differential of the free energy in this case we split up the volume integral $\frac{1}{4\pi}\iiint_{V_c} \boldsymbol{E}\cdot \boldsymbol{dD}\,\mathrm{d}v$ into two parts: an integration over the volume V of the sample and an integration over the volume $(V_c - V)$ of the empty part of the capacitor. Further, we use the fact that the component of $\boldsymbol{D}$ in the direction perpendicular to the boundary is continuous and that for the empty part of the capacitor $\boldsymbol{E} = \boldsymbol{D}$. Thus, we have:

$$\begin{aligned} \mathrm{d}F &= -S\,\mathrm{d}T - p_{\text{pa}}\,\mathrm{d}V + \frac{V}{4\pi}E\,\mathrm{d}D + \frac{V_c - V}{4\pi}D\,\mathrm{d}D \\ &= -S\,\mathrm{d}T - p_{\text{pa}}\,\mathrm{d}V + \frac{V_c}{4\pi}D\,\mathrm{d}D + \frac{V}{4\pi}(E - D)\,\mathrm{d}D. \end{aligned} \tag{7.139}$$

From this, we find:

$$\begin{aligned} \left(\frac{\partial p_{\text{pa}}}{\partial D}\right)_{T,V} &= -\left(\frac{\partial}{\partial V}\left(\frac{V_c}{4\pi}D + \frac{V}{4\pi}(E - D)\right)\right)_{T,D} \\ &= -\frac{1}{4\pi}(E - D) - \frac{V}{4\pi}\left(\frac{\partial E}{\partial V}\right)_{T,D} \\ &= -\frac{1}{4\pi}(E - D) - \frac{VD}{4\pi}\left(\frac{\partial}{\partial V}\frac{1}{\varepsilon}\right)_{T,D} \\ &= -\frac{1}{4\pi}(E - D) - \frac{Ed}{4\pi\varepsilon}\left(\frac{\partial \varepsilon}{\partial d}\right)_{T,D}. \end{aligned} \tag{7.140}$$

Hence:

$$\begin{aligned} \left(\frac{\partial p_{\text{pa}}}{\partial E}\right)_{T,V} &= \left(\frac{\partial D}{\partial E}\right)_{T,V}\left(\frac{\partial p_{\text{pa}}}{\partial D}\right)_{T,V} \\ &= \varepsilon\left\{-\frac{1}{4\pi}(E - D) - \frac{Ed}{4\pi\varepsilon}\left(\frac{\partial \varepsilon}{\partial d}\right)_{T,D}\right\} \\ &= -\frac{Ed}{4\pi}\left(\frac{\partial \varepsilon}{\partial d}\right)_T + \frac{\varepsilon(\varepsilon - 1)}{4\pi}E, \end{aligned} \tag{7.141}$$

if terms in higher powers of D are neglected. We find for Δp_{pa}:

$$\Delta p_{pa} = -\frac{E^2}{8\pi} d\left(\frac{\partial \varepsilon}{\partial d}\right)_T + \frac{\varepsilon(\varepsilon - 1)}{8\pi} E^2, \tag{7.142}$$

and using (7.131):

$$\Delta p_{pa} = \Delta p + \frac{\varepsilon(\varepsilon - 1)}{8\pi} E^2. \tag{7.143}$$

To calculate Δp_{tr}, we consider a system of a partially filled capacitor where the boundary of the sample is perpendicular to the capacitor plates. In this case the differential of the free energy:

$$dF = -S\,dT - p_{tr}\,dV + dW_{el}, \tag{7.144}$$

cannot be used, because the dielectric displacement is not the same in both parts of the capacitor. Since the electric field is continuous at the boundary, E can be used as the independent variable. To accomplish this, we apply the general definition of pressure for the case of p_{tr}:

$$p_{tr} = -\left(\frac{\partial F}{\partial V}\right)_{T,E} \tag{7.145}$$

The decrease of p_{tr} on application of the electric field is now given by:

$$-\Delta p_{tr} = -\left(\frac{\partial(F - F_0)}{\partial V}\right)_{T,E} = -\left(\frac{\partial W_{el}}{\partial V}\right)_{T,E}. \tag{7.146}$$

Here, F_0 is the free energy in the absence of the electric field. The electrical work W_{el} can be found by applying eqn. (3.42) to the case under consideration, which leads to:

$$W_{el} = \frac{1}{8\pi} VDE + \frac{1}{8\pi}(V_c - V)E^2. \tag{7.147}$$

Substitution of (7.147) into (7.146) gives:

$$\begin{aligned}\Delta p_{tr} &= \frac{1}{8\pi} DE + \frac{1}{8\pi} VE^2\left(\frac{\partial \varepsilon}{\partial V}\right)_T - \frac{1}{8\pi} E^2 \\ &= -\frac{1}{8\pi} E^2 d\left(\frac{\partial \varepsilon}{\partial d}\right)_T + \frac{\varepsilon - 1}{8\pi} E^2,\end{aligned} \tag{7.148}$$

and with (7.131):

$$\Delta p_{tr} = \Delta p + \frac{\varepsilon - 1}{8\pi} E^2. \tag{7.149}$$

Both Δp_{pa} and Δp_{tr} consist of two parts. The first part Δp is the same in both cases and is connected with the change in the energy of the dielectric on application of the electric field, which favours higher densities. The second part is connected with the energy change due to the decrease of the filled part of the capacitor, which in these cases is necessary to reach a higher density. Thus, the second part is related to the energy of a dielectric in an external field as given by eqn. (3.69). For the case of p_{pa} the external field in (3.69) must be taken equal to $\boldsymbol{D}$, the field in a disc-shaped cavity, whereas for p_{tr}, $\boldsymbol{E}$, the field in a needle-shaped cavity, should be used.

The pressures p_{pa} and p_{tr} are not equal. Since $\varepsilon > 1$, it follows from eqns. (7.143) and (7.149) that $p_{pa} > p_{tr}$. This implies that if a capacitor is partially filled with a substance with zero compressibility, the electric field will cause a deformation of the dielectric. Thus a freely floating drop will be elongated in the direction of the field.

The three pressures p_{pl}, p_{pa}, and p_{tr} can only be used in the experimental situations 1, 2, and 3, respectively, described above. To derive the behaviour of a dielectric in an inhomogeneous field, *e.g.* in the calculation of the equations of mass transport, the pressure p at a plane not in the field must be used.

References

1. J. M. Thiébaut, *Thesis* Nancy 1968.
2. J. M. Thiébaut, A. Weisbecker, and C. Ginet, *C.R.* **267** (1968) 661.
3. J. H. van Vleck, *J. Chem. Phys.* **5** (1937) 556.
4. A. Piekara and B. Piekara, *C.R.* **203** (1936) 852.
5. A. Chelkowski, *J. Chem. Phys.* **28** (1958) 1249.
6. A. Piekara and A. Chelkowski, *J. Chem. Phys.* **25** (1956) 794.
7. A. Piekara, A. Chelkowski, and S. Kielich, *Z. Phys. Chem.* **206** (1957) 375.
8. I. Danielewicz-Ferchmin, *Bull Ac. Pol. Sci., Sér. sci. math., astr., phys.* **14** (1966) 51.
9. I. Danielewicz-Ferchmin, *Bull. Ac. Pol. Sci., Sér. sci. math., astr., phys.* **16** (1968) 965.
10. J. Malecki, *J. Chem. Phys.* **36** (1962) 2144.
11. J. Malecki, *Acta Phys. Pol.* **21** (1962) 13.
12. A. Piekara, *C.R.* **204** (1937) 1106.
13. J. A. Schellman, *J. Chem. Phys.* **26** (1957) 1225.
14. J. Barriol and J. L. Greffe, *J. Chim. Phys.* **66** (1969) 575.
15. E. L. Eliel, N. L. Allinger, S. J. Angyal, and G. A. Morrison, *Conformational Analysis*, Interscience, New York 1965.
16. C. Altona and H. J. Hageman, *Rec. Trav. Chim.* **87** (1968) 279.
17. A. Piekara, *Acta Phys. Pol.* **10** (1950) 37.
18. S. Kielich, *Acta Phys. Pol.* **17** (1958) 239.
19. A. Piekara, *J. Chem. Phys.* **36** (1962) 2145.
20. K. Bergmann, M. Eigen, and L. de Maeyer, *Ber. Bunsenges.* **67** (1963) 819.
21. E. A. Guggenheim, *Thermodynamics*, North Holland, Amsterdam 1949, Ch. III.
22. A. D. Buckingham, *J. Chem. Phys.* **25** (1956) 428.
23. A. D. Buckingham and J. A. Pople, *Proc. Phys. Soc.* **A68** (1955) 905.
24. F. O. Koenig, *J. Phys. Chem.* **41** (1937) 597.
25. H. S. Frank, *J. Chem. Phys.* **23** (1955) 2023.

APPENDIX I

SOME APPLICATIONS OF VECTOR AND TENSOR CALCULUS

In this appendix some vector-analytic rules and the elements of tensor calculus will be summarized. The concept of a vector and the combination rules are assumed to be known.[1,2]

§1. Differentiation of scalar and vector fields

A rule that produces a scalar or a vector for each point of a 3-dimensional space, generates a scalar or a vector field, respectively. Physical examples of a scalar field are the temperature distribution and the electric potential. Vector fields are met with, for instance, in hydrodynamics (fields of flow) and electromagnetics (electric and magnetic fields).

Scalar fields and the components of vector fields are functions of the space coordinates and can be differentiated with respect to them. A useful way of combining the results of the differentiating operations is symbolized by the vector operator $\boldsymbol{\nabla}$ (nabla), which may be treated formally as a vector. The definition of the nabla operator in a Cartesian coordinate system is:

$$\boldsymbol{\nabla} = \boldsymbol{i}\frac{\partial}{\partial x} + \boldsymbol{j}\frac{\partial}{\partial y} + \boldsymbol{k}\frac{\partial}{\partial z}, \tag{A1.1}$$

where $\boldsymbol{i}$, $\boldsymbol{j}$, and $\boldsymbol{k}$ are the unit vectors along the x-, y-, and z-axes, respectively.

Limiting ourselves to scalar and vector fields, we have three possible differentiations using this nabla operator:

$$\boldsymbol{\nabla}\,\Phi = \operatorname{grad}\Phi, \tag{A1.2}$$

$$\boldsymbol{\nabla}\cdot\boldsymbol{A} = \operatorname{div}\boldsymbol{A}, \tag{A1.3}$$

$$\boldsymbol{\nabla}\times\boldsymbol{A} = \operatorname{curl}\boldsymbol{A}, \tag{A1.4}$$

where Φ is a scalar field and $\boldsymbol{A}$ a vector field. The gradient and the curl or rotation are vector fields, the divergence is a scalar field. Combining (A1.1) with (A1.2), A(1.3), and (A1.4) we can write explicitly:

$$\text{grad}\,\Phi = \boldsymbol{i}\frac{\partial\Phi}{\partial x} + \boldsymbol{j}\frac{\partial\Phi}{\partial y} + \boldsymbol{k}\frac{\partial\Phi}{\partial z}, \tag{A1.5}$$

$$\begin{aligned}\text{div}\,\boldsymbol{A} &= \frac{\partial}{\partial x}(\boldsymbol{A}\cdot\boldsymbol{i}) + \frac{\partial}{\partial y}(\boldsymbol{A}\cdot\boldsymbol{j}) + \frac{\partial}{\partial z}(\boldsymbol{A}\cdot\boldsymbol{k}) \\ &= \frac{\partial}{\partial x}A_x + \frac{\partial}{\partial y}A_y + \frac{\partial}{\partial z}A_z,\end{aligned} \tag{A1.6}$$

$$\text{curl}\,\boldsymbol{A} = \boldsymbol{i}\left(\frac{\partial}{\partial y}A_z - \frac{\partial}{\partial z}A_y\right) + \boldsymbol{j}\left(\frac{\partial}{\partial z}A_x - \frac{\partial}{\partial x}A_z\right) + \boldsymbol{k}\left(\frac{\partial}{\partial x}A_y - \frac{\partial}{\partial y}A_x\right). \tag{A1.7}$$

The nabla operator cannot be applied repeatedly in an arbitrary manner because of the restrictions on the operands. The five possible combinations are:

$$\text{div grad}\,\Phi = \nabla\cdot\nabla\Phi = \nabla^2\Phi = \Delta\Phi, \tag{A1.8}$$

$$\text{curl grad}\,\Phi = \nabla\times\nabla\Phi, \tag{A1.9}$$

$$\text{div curl}\,\boldsymbol{A} = \nabla\cdot\nabla\times\boldsymbol{A}, \tag{A1.10}$$

$$\text{curl curl}\,\boldsymbol{A} = \nabla\times\nabla\times\boldsymbol{A}, \tag{A1.11}$$

$$\text{grad div}\,\boldsymbol{A} = \nabla(\nabla\cdot\boldsymbol{A}). \tag{A1.12}$$

The operator div grad is often called the del operator or the Laplace operator. In a Cartesian coordinate system it is given by:

$$\Delta\Phi = \left(\frac{\partial^2}{\partial x^2} + \frac{\partial^2}{\partial y^2} + \frac{\partial^2}{\partial z^2}\right)\Phi, \tag{A1.13}$$

as is easily seen by substitution of (A1.5) and (A1.6) into (A1.8). Upon substitution, the following identities are found to hold for (A1.9) and (A1.10):

$$\text{curl grad}\,\Phi \equiv \mathbf{0}, \tag{A1.14}$$

$$\text{div curl}\,\boldsymbol{A} \equiv 0 \tag{A1.15}$$

(compare eqns. (1.12) and (1.14)).

In a Cartesian coordinate system (A1.11) may be rewritten as:

$$\text{curl curl}\,\boldsymbol{A} = \text{grad div}\,\boldsymbol{A} - \Delta\boldsymbol{A}. \tag{A1.16}$$

§2. Some vector-analytic equations

(a) In Chapter I the gradient operation is applied to a function of the distance of two points or of the radius vector $\boldsymbol{r}$. It is then important to know at which end-point of the vector the differentiation is performed. This is easily shown by taking $\boldsymbol{r}$ as the distance between two points P and Q with radius vectors $\boldsymbol{r}_P$ and $\boldsymbol{r}_Q$, where $\boldsymbol{r}$ is directed from P to Q. We then have:

$$\boldsymbol{r} = \boldsymbol{r}_Q - \boldsymbol{r}_P. \tag{A1.17}$$

The operation of taking the gradient of $f(r)$, where $f(r)$ is any differentiable function of r, will then lead to:

$$\text{grad}\, f(r) = \text{grad}\, f(|\boldsymbol{r}|) = \text{grad}\, f(|\boldsymbol{r}_Q - \boldsymbol{r}_P|), \tag{A1.18}$$

or:

$$\text{grad}\, f(r) = \text{grad}_Q\, f(r), \tag{A1.19}$$

where the subscript Q indicates that the differentiation is performed at the end-point Q of the vector $\boldsymbol{r}$.

In some problems, however, the differentiation has to be performed at the point P. Since we may write:

$$\frac{\partial}{\partial x_Q}|\boldsymbol{r}_Q - \boldsymbol{r}_P| = -\frac{\partial}{\partial x_P}|\boldsymbol{r}_Q - \boldsymbol{r}_P|,$$

and analogous equations for the derivatives with respect to y and z, we have with (A1.18):

$$\text{grad}_Q\, f(r) = -\text{grad}_P\, f(r). \tag{A1.20}$$

This formula has to be used when the solution of Poisson's equation, which gives the potential in the origin, has to be transformed to the potential at an arbitrary point (compare eqn. (1.108)).

(b) Applying (A1.5) to the function $\Phi = r^n$, with $r^2 = x^2 + y^2 + z^2$, we find:

$$\nabla r^n = nr^{n-1}\nabla r = nr^{n-2}\boldsymbol{r}. \tag{A1.21}$$

An application of this general equation for $n = -1$ is:

$$\text{grad}\, \frac{1}{r} = -\frac{1}{r^3}\boldsymbol{r}. \tag{A1.22}$$

This is used to derive eqn. (1.23). Compare also the derivation of eqn. (1.29).

(c) When the nabla operator is applied to a product $\Phi \boldsymbol{A} \cdot \boldsymbol{B}$, we find, using Leibniz' rule for the differentiation of products:

$$\operatorname{grad} \Phi \boldsymbol{A} \cdot \boldsymbol{B} = (\boldsymbol{A} \cdot \boldsymbol{B}) \operatorname{grad} \Phi + \Phi \operatorname{grad} (\boldsymbol{A} \cdot \boldsymbol{B}). \tag{A1.23}$$

For the special case $\boldsymbol{B} = \boldsymbol{r}$, the radius vector of each point, and $\boldsymbol{A} = \boldsymbol{m}$, a constant vector, the last term of (A1.23) may be reduced further:

$$\operatorname{grad} \boldsymbol{m} \cdot \boldsymbol{r} = \boldsymbol{i}\frac{\partial}{\partial x}(m_x x) + \boldsymbol{j}\frac{\partial}{\partial y}(m_y y) + \boldsymbol{k}\frac{\partial}{\partial z}(m_z z),$$

where m_x, m_y, and m_z are the three components of $\boldsymbol{m}$ in the x, y, and z directions. Since the constant factors are not affected by the differentiation, we have:

$$\operatorname{grad} \boldsymbol{m} \cdot \boldsymbol{r} = \boldsymbol{i}m_x + \boldsymbol{j}m_y + \boldsymbol{k}m_z = \boldsymbol{m}. \tag{A1.24}$$

This identity is used to derive eqn. (1.29).

Another differentiation of a product is used in eqns. (3.37) and (3.63):

$$\operatorname{div} \Phi \boldsymbol{A} = \frac{\partial}{\partial x}(\Phi A_x) + \frac{\partial}{\partial y}(\Phi A_y) + \frac{\partial}{\partial z}(\Phi A_z),$$

and thus:

$$\operatorname{div} \Phi \boldsymbol{A} = \Phi\left\{\frac{\partial}{\partial x}A_x + \frac{\partial}{\partial y}A_y + \frac{\partial}{\partial z}A_z\right\} + A_x\frac{\partial \Phi}{\partial x} + A_y\frac{\partial \Phi}{\partial y} + A_z\frac{\partial \Phi}{\partial z},$$

or:

$$\operatorname{div} \Phi \boldsymbol{A} = \Phi \operatorname{div} \boldsymbol{A} + \boldsymbol{A} \cdot \operatorname{grad} \Phi. \tag{A1.25}$$

(d) From the nabla operator other operators can be derived, as for instance the scalar operator $\boldsymbol{n} \cdot \nabla$, where $\boldsymbol{n}$ is a unit vector. By using (A1.1), this may be written explicitly as:

$$\boldsymbol{n} \cdot \nabla = (\boldsymbol{n} \cdot \boldsymbol{i})\frac{\partial}{\partial x} + (\boldsymbol{n} \cdot \boldsymbol{j})\frac{\partial}{\partial y} + (\boldsymbol{n} \cdot \boldsymbol{k})\frac{\partial}{\partial z}. \tag{A1.26}$$

Since $\boldsymbol{n} \cdot \boldsymbol{i}$ is the projection of a unit vector on the x-axis, it may also be written as the direction cosine α of $\boldsymbol{n}$ with respect to the x-axis of the co-ordinate system. In this way we obtain, denoting the direction cosines of $\boldsymbol{n}$ with respect to the y- and z-axes by β and γ:

$$\boldsymbol{n} \cdot \nabla = \alpha \frac{\partial}{\partial x} + \beta \frac{\partial}{\partial y} + \gamma \frac{\partial}{\partial z}. \tag{A1.27}$$

This operator is used in section 5.

§3. Integration of scalar and vector fields

Fields, being functions of the space coordinates, can also be integrated with respect to them. If the integration is performed with respect to the three independent coordinates in a three-dimensional space, it is called a volume integration and written as:

$$\iiint_V \Phi \, \mathrm{d}x \, \mathrm{d}y \, \mathrm{d}z \quad \text{or} \quad \iiint_V \boldsymbol{A} \, \mathrm{d}x \, \mathrm{d}y \, \mathrm{d}z,$$

where $\mathrm{d}x\,\mathrm{d}y\,\mathrm{d}z$ is often denoted by $\mathrm{d}v$. In this example we have used Cartesian coordinates; by a transformation of variables we can express the fields as functions of other variables (*e.g.* spherical coordinates) and derive the corresponding expression for the volume element $\mathrm{d}v$.

By imposing a constraint on the independent variables we limit the integration to an integration over a surface. If we take $x^2 + y^2 + z^2 = a^2$ as a constraint, the integration will be over the values of the field on the surface of a sphere with radius a:

$$\oiint_{\text{sphere}} \Phi \, \mathrm{d}S \quad \text{or} \quad \oiint_{\text{sphere}} \boldsymbol{A} \, \mathrm{d}S.$$

The circle across the integration sign is used to express the fact that the surface integration is over a closed surface.

When two constraints are imposed, *e.g.* $x^2 + y^2 + z^2 = a^2$ and $z = b$, we have a line integration:

$$\oint_C \Phi \, \mathrm{d}s \quad \text{or} \quad \oint_C \boldsymbol{A} \, \mathrm{d}s,$$

where C denotes the line, or contour, along which the integration is to be performed (in our example the circle in the plane $z = b$ with radius $\sqrt{a^2 - b^2}$ and centre on the z-axis).

In most cases we do not want to integrate the complete vector field, but only the component normal to the surface of integration, or the component tangent to the line of integration:

$$\oiint_{\text{sphere}} A_{\text{n}}\, \mathrm{d}S \quad \text{and} \quad \oint_{C} A_{\text{t}}\, \mathrm{d}s.$$

Introducing the convention that $\boldsymbol{dS}$ stands for a vector with length $\mathrm{d}S$ and direction along the outward normal of the (closed) surface, and that $\boldsymbol{ds}$ is a vector with length $\mathrm{d}s$ and direction along the tangent to the line, this may be written as:

$$\oiint_{\text{sphere}} \boldsymbol{A} \cdot \boldsymbol{dS} \quad \text{and} \quad \oint_{C} \boldsymbol{A} \cdot \boldsymbol{ds}.$$

With this convention we can also express the differentials in terms of each other:

$$\mathrm{d}v = \boldsymbol{dS} \cdot \boldsymbol{ds}, \tag{A1.28}$$

$$\boldsymbol{dS} = \boldsymbol{ds}_2 \times \boldsymbol{ds}_3, \tag{A1.29}$$

$$\mathrm{d}v = \boldsymbol{ds}_1 \cdot \boldsymbol{ds}_2 \times \boldsymbol{ds}_3. \tag{A1.30}$$

Examples of these integrals are to be found in eqns. (1.6), (1.13), and (3.3).

When the integrand of the integration is a first derivative of a scalar or vector field, one integration can sometimes be performed without taking into account the functional dependence on the space coordinates. In this respect the volume integration, the surface integration, and the line integration have to be considered separately. In the case of the volume integration of the divergence of a vector field $\boldsymbol{A}$ we have:*

$$\iiint_{V} \operatorname{div} \boldsymbol{A}\, \mathrm{d}v = \oiint_{S} \boldsymbol{A} \cdot \boldsymbol{dS}, \tag{A1.31}$$

where S is the surface of the volume of integration V. Eqn. (A1.31) expresses Green's theorem, also called the divergence theorem or Gauss' theorem. It is used in eqn. (1.9) to change a surface integration into a volume integration.

For the analogous case of a surface integration we have Stokes' theorem:

* See for the proof of this theorem and related theorems in the following ref. 1 and 2.

$$\iint_S (\text{curl } \boldsymbol{A}) \cdot \boldsymbol{dS} = \oint_C \boldsymbol{A} \cdot \boldsymbol{ds}, \tag{A1.32}$$

where C is the contour of the surface of integration S, and where the contour C is followed in the clock-wise sense when looking in the direction of $\boldsymbol{dS}$. This theorem is used to pass from eqn. (1.12) to eqn. (1.13).

The line integration leads to the equation:

$$\int_C (\text{grad } \Phi) \cdot \boldsymbol{ds} = \Phi(P) - \Phi(Q), \tag{A1.33}$$

where P and Q are the end-points of the line C, which is traversed in the direction from Q to P.

From these theorems a number of related theorems can be derived. We shall only give three theorems connected with the divergence theorem.

When $\boldsymbol{A} = \boldsymbol{M}\Phi$, where $\boldsymbol{M}$ is a constant vector, is substituted into eqn. (A1.31), we find, using (A1.25) and dividing out the constant vector $\boldsymbol{M}$:

$$\iiint_V \text{grad } \Phi \, \mathrm{d}v = \oiint_S \Phi \boldsymbol{dS}. \tag{A1.34}$$

This theorem is used in section 3b.

By substitution of $\boldsymbol{A} = \Psi \text{ grad } \Phi$ into the divergence theorem, another theorem is derived, called Green's first or asymmetric theorem. Performing the differentiation div (Ψ grad Φ) one obtains:

$$\iiint_V (\Psi\Delta\Phi + \boldsymbol{\nabla}\Phi \cdot \boldsymbol{\nabla}\Phi) \, \mathrm{d}v = \oiint_S \Psi\boldsymbol{\nabla}\Phi \cdot \boldsymbol{dS}. \tag{A1.35}$$

Interchanging Ψ and Φ in this equation, and subtraction of the result leads to:

$$\iiint_V (\Psi\Delta\Phi - \Phi\Delta\Psi) \, \mathrm{d}v = \oiint_S (\Psi\boldsymbol{\nabla}\Phi - \Phi\boldsymbol{\nabla}\Psi) \cdot \boldsymbol{dS}. \tag{A1.36}$$

This expresses the second, or symmetric theorem of Green.

§4. Spherical coordinates

In many problems of electrostatics it is advantageous to use spherical coordinates r, θ, φ instead of the Cartesian coordinates x, y, z used so far.

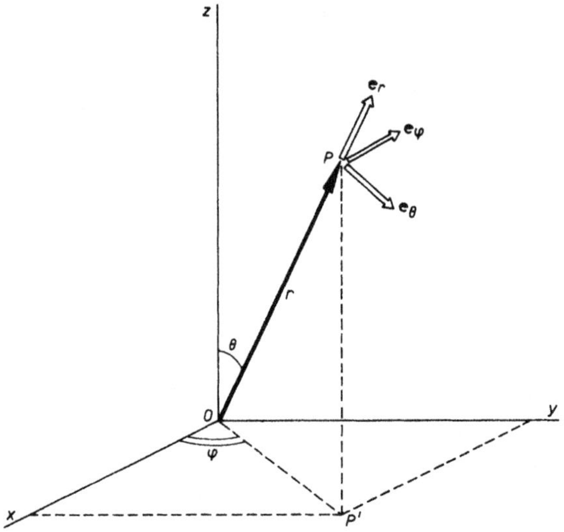

Fig. A1
Spherical coordinates.

The transformation rules from spherical to Cartesian coordinates are derived easily from geometrical considerations (see Fig. A1 with $OP' = r\sin\theta$):

$$\left.\begin{aligned} x &= r\sin\theta\cos\varphi \\ y &= r\sin\theta\sin\varphi \\ z &= r\cos\theta \end{aligned}\right\}, \tag{A1.37}$$

with $0 \leqslant r < \infty, 0 \leqslant \theta \leqslant \pi, 0 \leqslant \varphi < 2\pi$.
The inverse transformation is:

$$\left.\begin{aligned} r^2 &= x^2 + y^2 + z^2 \\ tg\,\theta &= \frac{1}{z}\sqrt{x^2 + y^2} \\ tg\,\varphi &= y/x \end{aligned}\right\}. \tag{A1.38}$$

The unit vectors $\boldsymbol{e}_r$, $\boldsymbol{e}_\theta$, and $\boldsymbol{e}_\varphi$ in the spherical coordinate system are found by considering the planes of constant r, of constant θ, and of constant φ. As for the case of Cartesian coordinates, the unit vectors are the normals to these planes, pointing in the direction of increasing values of the variable (see Fig. A1).

The transformation rules for the unit vectors can also be found from

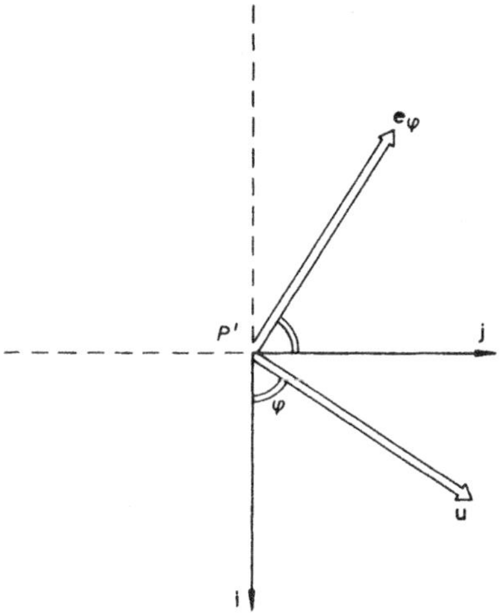

Fig. A2
The xy-plane.

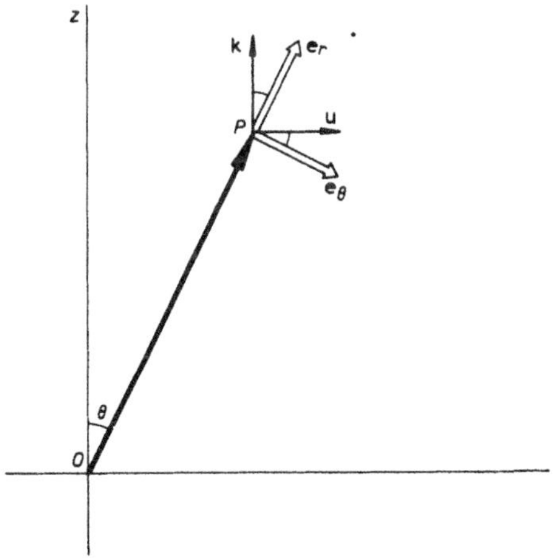

Fig. A3
The plane φ = constant.

geometrical considerations (see Fig. A2 and A3). First, we introduce an auxiliary unit vector $\boldsymbol{u}$, pointing in the direction of OP'. We can now transform the unit vectors $\boldsymbol{i}, \boldsymbol{j}, \boldsymbol{k}$ into $\boldsymbol{u}, \boldsymbol{e}_\varphi, \boldsymbol{k}$ by a rotation through φ around $\boldsymbol{k}$, and then transform $\boldsymbol{u}, \boldsymbol{e}_\varphi, \boldsymbol{k}$ into $\boldsymbol{e}_r, \boldsymbol{e}_\theta, \boldsymbol{e}_\varphi$ by a rotation through θ around $\boldsymbol{e}_\varphi$:

$$\left.\begin{aligned} \boldsymbol{u} &= \cos\varphi\boldsymbol{i} + \sin\varphi\boldsymbol{j} \\ \boldsymbol{e}_\varphi &= -\sin\varphi\boldsymbol{i} + \cos\varphi\boldsymbol{j} \\ \boldsymbol{k} &= \boldsymbol{k} \end{aligned}\right\}, \tag{A1.39}$$

and:

$$\left.\begin{aligned} \boldsymbol{e}_r &= \sin\theta\boldsymbol{u} + \cos\theta\boldsymbol{k} \\ \boldsymbol{e}_\theta &= \cos\theta\boldsymbol{u} - \sin\theta\boldsymbol{k} \\ \boldsymbol{e}_\varphi &= \boldsymbol{e}_\varphi \end{aligned}\right\}. \tag{A1.40}$$

From this we find:

$$\left.\begin{aligned} \boldsymbol{e}_r &= \sin\theta\cos\varphi\boldsymbol{i} + \sin\theta\sin\varphi\boldsymbol{j} + \cos\theta\boldsymbol{k} \\ \boldsymbol{e}_\theta &= \cos\theta\cos\varphi\boldsymbol{i} + \cos\theta\sin\varphi\boldsymbol{j} - \sin\theta\boldsymbol{k} \\ \boldsymbol{e}_\varphi &= -\sin\varphi\boldsymbol{i} + \cos\varphi\boldsymbol{j} \end{aligned}\right\}, \tag{A1.41}$$

and for the inverse transformation:

$$\left.\begin{aligned} \boldsymbol{i} &= \sin\theta\cos\varphi\boldsymbol{e}_r + \cos\theta\cos\varphi\boldsymbol{e}_\theta - \sin\varphi\boldsymbol{e}_\varphi \\ \boldsymbol{j} &= \sin\theta\sin\varphi\boldsymbol{e}_r + \cos\theta\sin\varphi\boldsymbol{e}_\theta + \cos\varphi\boldsymbol{e}_\varphi \\ \boldsymbol{k} &= \cos\theta\boldsymbol{e}_r - \sin\theta\boldsymbol{e}_\theta \end{aligned}\right\}. \tag{A1.42}$$

To obtain the expression for the differential volume element in spherical coordinates we use (A1.30) and find, by inspection of Fig. A1, that the differential line elements are given by:

$$\left.\begin{aligned} d\boldsymbol{s}_1 &= \mathrm{d}r\,\boldsymbol{e}_r \\ d\boldsymbol{s}_2 &= r\,\mathrm{d}\theta\,\boldsymbol{e}_\theta \\ d\boldsymbol{s}_3 &= r\sin\theta\,\mathrm{d}\varphi\,\boldsymbol{e}_\varphi \end{aligned}\right\}, \tag{A1.43}$$

where the fact that a line element of a circle is given by the product of radius and supporting angle in radians is used. From (A1.43) and $\boldsymbol{e}_r \cdot \boldsymbol{e}_\theta \times \boldsymbol{e}_\varphi = 1$ we infer immediately:

$$\mathrm{d}v = r^2 \sin\theta\,\mathrm{d}r\,\mathrm{d}\theta\,\mathrm{d}\varphi. \tag{A1.44}$$

As an example of the use of spherical coordinates we perform the calculation of eqn. (1.6). Given that:

$$E = \frac{e}{r^3}\boldsymbol{r},$$

we have to calculate $\oint\!\!\oint \boldsymbol{E}\cdot d\boldsymbol{S}$ for an arbitrary surface enclosing the charge e. Taking the surface of a sphere with radius R, we have:

$$\oint\!\!\oint \boldsymbol{E}\cdot d\boldsymbol{S} = \int_0^{2\pi}\!\!\int_0^{\pi} \frac{e}{R^3}\boldsymbol{R}\cdot(R\,\mathrm{d}\theta\,\boldsymbol{e}_\theta)\times(R\sin\theta\,\mathrm{d}\varphi\,\boldsymbol{e}_\varphi) = \int_0^{2\pi}\!\!\int_0^{\pi} e\sin\theta\,\boldsymbol{e}_r\cdot\boldsymbol{e}_\theta\times\boldsymbol{e}_\varphi\,\mathrm{d}\theta\,\mathrm{d}\varphi,$$

or:

$$\oint\!\!\oint \boldsymbol{E}\cdot d\boldsymbol{S} = e\int_0^{2\pi}\!\!\int_0^{\pi}\sin\theta\,\mathrm{d}\theta\,\mathrm{d}\varphi = 2\pi e\int_0^{\pi}\sin\theta\,\mathrm{d}\theta = 4\pi e.$$

§5. The nabla operator in spherical coordinates

The transformation rules for the nabla operator have to take into account the fact that it is an operator, *i.e.*, the results of the operations on scalar and vector fields should remain the same. Thus, it is not possible simply to substitute (A1.37) and (A1.42) into the definition (A1.1). Instead, we use eqns. (A1.31), (A1.32), and (A1.33), since the integrations are independent of the coordinate system used. In this way we derive the following alternative definitions for the three operations with the nabla operator:

$$\operatorname{div}\boldsymbol{A} = \lim_{\Delta V\to 0}\frac{1}{\Delta V}\oint\!\!\oint_S \boldsymbol{A}\cdot d\boldsymbol{S}, \tag{A1.45}$$

$$(\operatorname{curl}\boldsymbol{A})_{\mathrm{n}} = \lim_{\Delta S\to 0}\frac{1}{\Delta S}\oint \boldsymbol{A}\cdot d\boldsymbol{s}, \tag{A1.46}$$

$$(\operatorname{grad}\Phi)_s = \lim_{\Delta s\to 0}\frac{1}{\Delta s}\{\Phi(P)-\Phi(Q)\}, \tag{A1.47}$$

where ΔV, ΔS, and Δs are respectively small volumes, surfaces, and line intervals of integration, and where the subscript n denotes the component normal to ΔS and the subscript s the component in the direction of Δs.

From (A1.47) we infer:

$$(\text{grad}\,\Phi)_s = \frac{\partial \Phi}{\partial s}. \tag{A1.48}$$

For the case of three line elements in the direction of $\boldsymbol{e}_r$, $\boldsymbol{e}_\theta$, and $\boldsymbol{e}_\varphi$, respectively, as given by (A1.43), we derive:

$$\left.\begin{aligned}(\text{grad}\,\Phi)_r &= (\text{grad}\,\Phi)\cdot\boldsymbol{e}_r = \frac{\partial\Phi}{\partial r}\\ (\text{grad}\,\Phi)_\theta &= (\text{grad}\,\Phi)\cdot\boldsymbol{e}_\theta = \frac{1}{r}\frac{\partial\Phi}{\partial\theta}\\ (\text{grad}\,\Phi)_\varphi &= (\text{grad}\,\Phi)\cdot\boldsymbol{e}_\varphi = \frac{1}{r\sin\theta}\frac{\partial\Phi}{\partial\varphi}\end{aligned}\right\}. \tag{A1.49}$$

Choosing suitable surfaces and volumes of integration, one can derive in an analogous way the following expressions:

$$\text{div}\,\boldsymbol{A} = \left(\frac{\partial}{\partial r}+\frac{2}{r}\right)A_r + \frac{1}{r}\left(\frac{\partial}{\partial\theta}+\text{cotg}\,\theta\right)A_\theta + \frac{1}{r\sin\theta}\frac{\partial}{\partial\varphi}A_\varphi, \tag{A1.50}$$

and:

$$\begin{aligned}\text{curl}\,\boldsymbol{A} = &\left[\frac{1}{r}\left(\frac{\partial}{\partial\theta}+\text{cotg}\,\theta\right)A_\varphi - \frac{1}{r\sin\theta}\frac{\partial}{\partial\varphi}A_\theta\right]\boldsymbol{e}_r +\\ &+\left[\frac{1}{r\sin\theta}\frac{\partial}{\partial\varphi}A_r - \left(\frac{\partial}{\partial r}+\frac{1}{r}\right)A_\varphi\right]\boldsymbol{e}_\theta +\\ &+\left[\left(\frac{\partial}{\partial r}+\frac{1}{r}\right)A_\theta - \frac{1}{r}\frac{\partial}{\partial\theta}A_r\right]\boldsymbol{e}_\varphi.\end{aligned} \tag{A1.51}$$

Combining (A1.49) with (A1.50) we finally derive:

$$\Delta\Phi = \left(\frac{\partial}{\partial r}+\frac{2}{r}\right)\frac{\partial\Phi}{\partial r} + \frac{1}{r^2}\left(\frac{\partial}{\partial\theta}+\text{cotg}\,\theta\right)\frac{\partial\Phi}{\partial\theta} + \frac{1}{r^2\sin^2\theta}\frac{\partial^2\Phi}{\partial\varphi^2}. \tag{A1.52}$$

The first term of this expression for the Laplace operator may also be written as:

$$\left(\frac{\partial}{\partial r}+\frac{2}{r}\right)\frac{\partial\Phi}{\partial r} = \frac{1}{r^2}\frac{\partial}{\partial r}\left(r^2\frac{\partial\Phi}{\partial r}\right) = \frac{1}{r}\frac{\partial^2}{\partial r^2}(r\Phi). \tag{A1.53}$$

This equation is often useful to obtain the solutions of the Laplace equation, $\Delta\Phi = 0$ (see Appendix II).

An analogous form of the second term is:

$$\frac{1}{r^2}\left(\frac{\partial}{\partial\theta} + \operatorname{cotg}\theta\right)\frac{\partial\Phi}{\partial\theta} = \frac{1}{r^2 \sin\theta}\frac{\partial}{\partial\theta}\left(\sin\theta\frac{\partial\Phi}{\partial\theta}\right). \tag{A1.54}$$

§6. Tensors and matrix representations of tensors

An example of a linear relation between two vectors is the result of the multiplication of a vector $\boldsymbol{v}$ with a scalar λ leading to another vector $\boldsymbol{w}$, having the same direction as $\boldsymbol{v}$ and with a length λ times as great. In many cases, however, we find a linear relation between two vectors that do not have the same direction (compare the relation between dipole field and dipole moment, eqn. (1.29), and the relation between the dielectric displacement and electric field intensity in the case of an anisotropic substance, eqn. (2.49)).

To be able to describe these relations we introduce the concept of a tensor:

$$\boldsymbol{w} = \boldsymbol{T}\cdot\boldsymbol{v}, \tag{A1.55}$$

defined as the quantity which produces $\boldsymbol{w}$ when post-multiplied by $\boldsymbol{v}$. Numbering the components of $\boldsymbol{v}$ and $\boldsymbol{w}$ in some coordinate system with subscripts 1, 2, and 3, and splitting up eqn. (A1.55) into three equations for the three components of $\boldsymbol{w}$, we see that we need nine elements for $\boldsymbol{T}$ to write the required linear relationships:

$$\left.\begin{aligned} w_1 &= t_{11}v_1 + t_{12}v_2 + t_{13}v_3 \\ w_2 &= t_{21}v_1 + t_{22}v_2 + t_{23}v_3 \\ w_3 &= t_{31}v_1 + t_{32}v_2 + t_{33}v_3 \end{aligned}\right\}. \tag{A1.56}$$

Since the components of $\boldsymbol{v}$ and $\boldsymbol{w}$ differ in different coordinate systems, this will generally be the case for the components t_{ij} of $\boldsymbol{T}$ as well. The tensor equation, eqn. (A1.55), however, remains valid in all coordinate systems. When some coordinate system has been chosen, it is possible to write down the components of $\boldsymbol{v}$, $\boldsymbol{w}$, and $\boldsymbol{T}$, and to form three matrices with them:

$$v = \begin{pmatrix} v_1 \\ v_2 \\ v_3 \end{pmatrix}, \quad w = \begin{pmatrix} w_1 \\ w_2 \\ w_3 \end{pmatrix}, \quad T = \begin{pmatrix} t_{11} & t_{12} & t_{13} \\ t_{21} & t_{22} & t_{23} \\ t_{31} & t_{32} & t_{33} \end{pmatrix}.$$

The 3×1 matrices v and w are often called column-vectors. From the definition of matrix multiplication:[3]

$$T = AB \text{ is equivalent to}: t_{ij} = \sum_k a_{ik} b_{kj}, \tag{A1.57}$$

we see that eqn. (A1.56) can now also be written as a matrix equation:

$$w = Tv. \tag{A1.58}$$

Eqn. (A1.58) is the matrix representation of the tensor equation (A1.55) and is different in different coordinate systems. On the other hand, it is quite easy to obtain the transformation rules in matrix notation. For the special case of the transformation from Cartesian coordinates to spherical coordinates, we use the identity:

$$v_x \boldsymbol{i} + v_y \boldsymbol{j} + v_z \boldsymbol{k} = \boldsymbol{v} = v_r \boldsymbol{e}_r + v_\theta \boldsymbol{e}_\theta + v_\varphi \boldsymbol{e}_\varphi. \tag{A1.59}$$

Substituting (A1.41) in the righthand side and collecting the coefficients of $\boldsymbol{i}$, $\boldsymbol{j}$, and $\boldsymbol{k}$, we find:

$$\left.\begin{aligned} v_x &= \sin\theta\cos\varphi v_r + \cos\theta\cos\varphi v_\theta - \sin\varphi v_\varphi \\ v_y &= \sin\theta\sin\varphi v_r + \cos\theta\sin\varphi v_\theta + \cos\varphi v_\varphi \\ v_z &= \cos\theta v_r - \sin\theta v_\theta \end{aligned}\right\}. \tag{A1.60}$$

This can be written as a matrix equation:

$$v_{\text{Cartesian}} = C v_{\text{spherical}}, \tag{A1.61}$$

where the matrix C is given by:

$$C = \begin{pmatrix} \sin\theta\cos\varphi & \cos\theta\cos\varphi & -\sin\varphi \\ \sin\theta\sin\varphi & \cos\theta\sin\varphi & \cos\varphi \\ \cos\theta & -\sin\theta & 0 \end{pmatrix}. \tag{A1.62}$$

It is of course also possible to substitute (A1.42) in the lefthand side of eqn. (A1.59) to find the inverse transformation from $v_{\text{spherical}}$ to $v_{\text{Cartesian}}$. The result of this procedure can be written as:

$$v_{\text{spherical}} = C^{-1} v_{\text{Cartesian}}, \tag{A1.63}$$

where C^{-1} is the inverse matrix of C, defined by:

$$CC^{-1} = C^{-1}C = I. \tag{A1.64}$$

Here, I is the 3×3 unit matrix, consisting of ones on the principal diagonal and zeros elsewhere. Thus:

$$(I)_{ij} = \delta_{ij}, \tag{A1.65}$$

where δ_{ij} is the Kronecker delta, defined by:

$$\left.\begin{array}{lll} \delta_{ij} = 1 & \text{for} & i = j \\ \delta_{ij} = 0 & \text{for} & i \neq j \end{array}\right\} . \tag{A1.66}$$

With the help of the defined properties of the inverse matrix it is easy to see that eqn. (A1.63) follows from eqn. (A1.61) by premultiplication with the matrix C^{-1}.

From the transformation rules for vectors, as given in (A1.61) and (A1.63), we can now obtain the transformation rules for matrices. Eqn. (A1.55), being a tensor equation, will be valid in a Cartesian, as well as in a spherical coordinate system. Thus, we have:

$$w_{\text{Cartesian}} = T_{\text{Cartesian}} v_{\text{Cartesian}}, \tag{A1.67}$$

and:

$$w_{\text{spherical}} = T_{\text{spherical}} v_{\text{spherical}}. \tag{A1.68}$$

If we now premultiply eqn. (A1.67) with C^{-1}, and substitute the form analogous to (A1.63) for w in the lefthand side and (A1.61) in the righthand side, we obtain:

$$w_{\text{spherical}} = C^{-1} T_{\text{Cartesian}} C v_{\text{spherical}}. \tag{A1.69}$$

Comparing (A1.69) with (A1.68) we infer:

$$T_{\text{spherical}} = C^{-1} T_{\text{Cartesian}} C. \tag{A1.70}$$

This is the required transformation rule for matrices. From the method of the proof we may also infer that (A1.70) forms the necessary and sufficient condition for a matrix T to be the representation of a tensor $\boldsymbol{T}$.

Calculation of C^{-1} with the help of the procedure outlined for eqn. (A1.63) leads to the expression:

$$C^{-1} = \begin{pmatrix} \sin\theta\cos\varphi & \sin\theta\sin\varphi & \cos\theta \\ \cos\theta\cos\varphi & \cos\theta\sin\varphi & -\sin\theta \\ -\sin\varphi & \cos\varphi & 0 \end{pmatrix}. \tag{A1.71}$$

Inspection shows that this expression is equal to the expression for C in (A1.62) with rows and columns interchanged. This operation, the transposition of the matrix C, generally leads to a new matrix C', the transpose of C. Thus, we have in this case:

$$C' = C^{-1}. \tag{A1.72}$$

A matrix with this property is called an orthogonal matrix, since all matrices connecting orthogonal coordinate systems have this property.

The operation of transposition may also be applied to column vectors; we then obtain 1×3 matrices called row vectors:

$$v' = (v_1, v_2, v_3). \tag{A1.73}$$

With the help of row vectors, the multiplication of vectors can be given a matrix representation:

$$v'w = v_1 w_1 + v_2 w_2 + v_3 w_3, \tag{A1.74}$$

according to (A1.57), and thus:

$$v'w = \boldsymbol{v} \cdot \boldsymbol{w}. \tag{A1.75}$$

There is a second possibility for the product of a row vector and a column vector, leading to a 3×3 matrix denoted by S:

$$S = wv' \tag{A1.76}$$

or:

$$s_{ij} = w_i v_j. \tag{A1.77}$$

This product is sometimes called the dyadic product of two vectors. The matrix equation (A1.76) is the representation of a tensor equation:

$$\boldsymbol{S} = \boldsymbol{wv}. \tag{A1.78}$$

This can be proved easily by checking the result of a transformation of eqn. (A1.76) to spherical coordinates:

$$S_{\text{spherical}} = w_{\text{spherical}} v'_{\text{spherical}} = (C^{-1} w_{\text{Cartesian}})(C^{-1} v_{\text{Cartesian}})',$$

or:

$$S_{\text{spherical}} = C^{-1} S_{\text{Cartesian}} (C^{-1})'. \tag{A1.79}$$

Here, the following property of the transpose of a product is used:

$$(AB)' = B'A'. \tag{A1.80}$$

This relation can be checked easily by applying the definitions of matrix multiplication and transposition. By substitution of (A1.72) we see that eqn. (A1.79) is indeed the formula for the transformation of a tensor, eqn. (A1.70).

The tensor $\boldsymbol{S}$ in eqn. (A1.78) is called the tensor product of the two vectors $\boldsymbol{w}$ and $\boldsymbol{v}$. According to (A1.77), it is given in some coordinate system by:

$$S = \begin{pmatrix} w_1 v_1 & w_1 v_2 & w_1 v_3 \\ w_2 v_1 & w_2 v_2 & w_2 v_3 \\ w_3 v_1 & w_3 v_2 & w_3 v_3 \end{pmatrix}. \tag{A1.81}$$

The sum of the elements on the principal diagonal forms the inner product of the two vectors $\boldsymbol{w}$ and $\boldsymbol{v}$ (compare eqns. (A1.74) and (A1.75)). It is also called the contraction of the tensor product. In general, contraction is denoted by a dot between the symbols in question; its result is a summation over the neighbouring indices. Thus, we have for two vectors (compare (A1.75)):

$$\boldsymbol{w} \cdot \boldsymbol{v} = \sum_i w_i v_i,$$

and for a tensor and a vector (compare (A1.56)):

$$(\boldsymbol{T} \cdot \boldsymbol{v})_i = \sum_j t_{ij} v_j.$$

In the same way we have for two tensors $\boldsymbol{T}$ and $\boldsymbol{S}$:

$$(\boldsymbol{T} \cdot \boldsymbol{S})_{ik} = \sum_j t_{ij} s_{jk},$$

and:

$$\boldsymbol{T} : \boldsymbol{S} = \sum_{i,j} t_{ij} s_{ji},$$

since after the first contraction i and k become neighbouring indices.

§7. Transformations and invariants of symmetric matrices

Equations formulated in terms of vectors and tensors have the advantage that they are valid in any coordinate system. As soon as the components are written explicitly, the equations change from coordinate system to coordinate system (compare, for example, (A1.5) with (A1.49), (A1.6) with (A1.50), etc.) However, it is possible to distinguish on the one hand between expressions that change their form but not their value, and on the other hand expressions that take another value as well as another form in a different coordinate system. An expression of which the value is not changed by a transformation is called invariant under the transformation.

An example is the length of a vector:

$$l(\boldsymbol{v}) = \sqrt{\boldsymbol{v} \cdot \boldsymbol{v}}. \tag{A1.82}$$

Reverting to the matrix representation, we have:

$$l^2(v) = v'v. \tag{A1.83}$$

A transformation with orthogonal transformation matrix C now leads to a new column vector:

$$w = Cv \tag{A1.84}$$

(compare (A1.61)) with length:

$$l^2(w) = v'C'Cv = v'v = l^2(v). \tag{A1.85}$$

For the matrices that represent tensors, one of the invariants is the determinant $|T|$ of the matrix T. This is easily seen from the product rule for determinants:[3]

$$|C^{-1}TC| = |C^{-1}||T||C| = |C^{-1}||C||T| = |T|. \tag{A1.86}$$

Another invariant is the trace or spur (the German word for "trace"), which is defined as:

$$\mathrm{Tr}\,(T) = \sum_i t_{ii}, \tag{A1.87}$$

the sum of the elements on the principal diagonal.

In general, there are three invariants that can be given by the three eigenvalues of the matrix, or any three independent combinations of them. An eigenvalue of a matrix, λ_i, is defined as a solution of the matrix equation:

$$Tv = \lambda v. \tag{A1.88}$$

This corresponds with the tensor equation:

$$\boldsymbol{w} = \boldsymbol{T} \cdot \boldsymbol{v} = \lambda \boldsymbol{v}. \tag{A1.89}$$

From this equation it is seen that in these particular cases the multiplication with a tensor reduces to scalar multiplication, *i.e.* the vector does not change its direction, but only its length by a factor λ.

In general, for a given tensor in three-dimensional space, vectors in only three, mutually orthogonal directions show this peculiarity. These directions are called the principal axes of the tensor. When the coordinate system is changed to one that has the principal axes as coordinate axes, one speaks of a transformation to principal axes.

When such a transformation to principal axes is applied to the matrix representation of the tensor, the transformed matrix is diagonal, *i.e.* only its elements on the principal diagonal differ from zero. The values of the diagonal elements are equal to the eigenvalues. Thus, we have:

$$\begin{pmatrix} t_{11} & t_{12} & t_{13} \\ t_{21} & t_{22} & t_{23} \\ t_{31} & t_{32} & t_{33} \end{pmatrix} \xrightarrow[\text{transformation}]{\text{principal axes}} \begin{pmatrix} \lambda_1 & 0 & 0 \\ 0 & \lambda_2 & 0 \\ 0 & 0 & \lambda_3 \end{pmatrix}$$

Although it is often a quite intricate calculation to obtain the transformation matrix for a transformation to principal axes, the resulting matrices have such a simple form that all subsequent calculations are simplified correspondingly.

It is not always possible to transform a matrix to its principal axes. The condition for this to be possible is that the matrix is symmetric, *i.e.*:

$$T = T' \quad \text{or} \quad t_{ij} = t_{ji}. \tag{A1.90}$$

If the matrix is not symmetric, two of the eigenvalues will be complex.

The symmetry of a tensor is defined in the same way as for a matrix. The tensor $\boldsymbol{T}$ is symmetric when its components t_{ij} are related as:

$$t_{ij} = t_{ji}. \tag{A1.91}$$

All tensors $\boldsymbol{T}$ contain a symmetric part, called the symmetrant of $\boldsymbol{T}$ and denoted as Sym ($\boldsymbol{T}$) or $\boldsymbol{T}^S$. The components of the symmetrant are given by:

$$[\mathrm{Sym}\,(\boldsymbol{T})]_{ij} = (\boldsymbol{T}^S)_{ij} = \tfrac{1}{2}(t_{ij} + t_{ji}). \tag{A1.92}$$

Up to now only tensors of the second order have been considered. It is possible to introduce tensors of a higher order; then the number of indices of the components is extended, and appropriate transformation rules are given. For example, a tensor $\boldsymbol{U}^{(n)}$ of the n-th order will have components $u_{ij\cdots k}$ with n indices. Expression (A1.78) for a tensor product of two vectors can now be generalized to a tensor product of tensors of arbitrary order. For instance, when $\boldsymbol{T}$ and $\boldsymbol{S}$ are second-order tensors and $\boldsymbol{U}^{(4)}$ is a fourth-order tensor we can write:

$$\boldsymbol{U}^{(4)} = \boldsymbol{TS}, \tag{A1.93}$$

or in components:

$$u_{ijkl} = t_{ij}s_{kl}. \tag{A1.94}$$

In three-dimensional space the indices i, j, k, and l can have the values 1, 2, or 3. Another example is the tensor product of n unit vectors $\boldsymbol{n}_1, \boldsymbol{n}_2, \cdots, \boldsymbol{n}_n$, which has been used in section 5; denoting it by $\boldsymbol{U}^{(n)}$ we have:

$$\boldsymbol{U}^{(n)} = \boldsymbol{n}_1\boldsymbol{n}_2 \cdots \boldsymbol{n}_n, \tag{A1.95}$$

or:

$$u_{ij\cdots k} = (n_1)_i(n_2)_j \cdots (n_n)_k. \tag{A1.96}$$

The definition of the symmetrant or completely symmetric part of a tensor can be extended to tensors of higher order:

$$\text{Sym}\,(\boldsymbol{U}^{(n)}) = \frac{1}{n!} \sum_{\text{permutations}} u_{ij\cdots k}, \tag{A1.97}$$

where the summation is over all $n!$ permutations of the indices of $u_{ij\cdots k}$. This definition is used in section 5.

When second-order tensors are used to describe experimentally observable quantities, it is necessary that they be symmetric to avoid the occurrence of complex eigenvalues. This requirement on a tensor $\boldsymbol{T}$ can be derived by another method when $\boldsymbol{T}$ is related to the energy of the system under consideration in the following way.

A vector $\boldsymbol{w}$ is given by $\boldsymbol{w} = \boldsymbol{T} \cdot \boldsymbol{v}$, or, equivalently:

$$\boldsymbol{v} = \boldsymbol{T}^{-1} \cdot \boldsymbol{w}. \tag{A1.98}$$

A variation of the energy δV is given either by $\boldsymbol{w} \cdot \delta\boldsymbol{v}$ or $\boldsymbol{v} \cdot \delta\boldsymbol{w}$. To be explicit, we shall take:

$$\delta V = \boldsymbol{v} \cdot \boldsymbol{\delta w}. \tag{A1.99}$$

When the variation of the energy is a reversible process, thermodynamics require δV to be a total differential. Hence:

$$\delta V = \frac{\partial V}{\partial w_1}\delta w_1 + \frac{\partial V}{\partial w_2}\delta w_2 + \frac{\partial V}{\partial w_3}\delta w_3. \tag{A1.100}$$

Combining (A1.100) with (A1.99), we have:

$$\frac{\partial V}{\partial w_i} = v_i \qquad i = 1, 2, 3. \tag{A1.101}$$

If we now differentiate for instance $\partial V/\partial w_i = v_i$ with respect to w_j, we have with the help of (A1.98):

$$\frac{\partial^2 V}{\partial w_j \, \partial w_i} = \frac{\partial v_i}{\partial w_j} = t_{ij}^{-1}. \tag{A1.102}$$

If we now change the dummy index in (A1.101) to j and differentiate with respect to w_i, we find:

$$\frac{\partial^2 V}{\partial w_i \, \partial w_j} = \frac{\partial v_j}{\partial w_i} = t_{ji}^{-1}.$$

When V has continuous second derivatives, the order of differentiation may be interchanged. Thus, we obtain:

$$t_{ij}^{-1} = t_{ji}^{-1}, \tag{A1.103}$$

or T^{-1} and hence T also is symmetric.

References

1. H. Margenau and G. M. Murphy, *The Mathematics of Physics and Chemistry*, van Nostrand, Princeton 1956, Ch. 4.
2. M. R. Spiegel, *Theory and problems of vector analysis and an introduction to tensor analysis*, Schaum, New York 1959.
3. Ref. 1, Ch. 10.

APPENDIX II

THE SOLUTION OF LAPLACE'S AND POISSON'S EQUATIONS

§1. Laplace's equation

Laplace's equation for the potential ϕ, $\Delta\phi = 0$, is valid in regions where no true or apparent charge is present. Since in general the potential inside a charge-free region is determined by the values of the potential (or its normal derivative) at the boundary of the region,[1] the solution of Laplace's equation for a particular region can be determined from the general solution and the values for the potential at the boundary of that region, which follow from the conditions of the problem. If the region extends to infinity, the asymptotic behaviour for large distances should be prescribed.

The equation $\Delta\phi = 0$ has a variety of general solutions, depending on the particular coordinate system chosen. It is advantageous to choose the coordinate system such that the coordinates of the boundary of the region take a simple form. For most problems in the theory of electric polarization, a spherical coordinate system (r, θ, φ) appears to be suitable. It is especially useful when the problem has axial symmetry (also called cylindrical symmetry) or spherical symmetry, *i.e.*, in the first case $\partial\phi/\partial\varphi = 0$, and in the second also $\partial\phi/\partial\theta = 0$.

In this appendix we will limit ourselves to the solution for cases with at least axial symmetry. Further, we will assume in our derivation of the general solution that the potential remains finite and continuous at the boundary. If the region has the outer boundary at infinity, singularities of a prescribed form are allowed to be present. Finally, we will require the potential to be finite and continuous on the z-axis, because in this book the regions where the potential has to be calculated will generally contain points on the z-axis.

§2. The general solution of Laplace's equation in spherical coordinates for the case of axial symmetry

From eqn. (A1.52) we see that Laplace's equation in spherical coordinates for the case of at least axial symmetry reduces to:

$$\Delta\phi = \left(\frac{\partial}{\partial r} + \frac{2}{r}\right)\frac{\partial\phi}{\partial r} + \frac{1}{r^2}\left(\frac{\partial}{\partial\theta} + \operatorname{cotg}\theta\right)\frac{\partial\phi}{\partial\theta} = 0. \tag{A2.1}$$

This partial differential equation can be reduced to two ordinary differential equations by the method of separation of variables. To accomplish this, we assume ϕ to be of the form:

$$\phi(r, \theta) = U(r)V(\theta). \tag{A2.2}$$

Substitution of (A2.2) into (A2.1) and multiplication with r^2/ϕ leads to:

$$\frac{r^2}{U(r)}\left(\frac{\partial}{\partial r} + \frac{2}{r}\right)\frac{\partial U(r)}{\partial r} + \frac{1}{V(\theta)}\left(\frac{\partial}{\partial\theta} + \operatorname{cotg}\theta\right)\frac{\partial V(\theta)}{\partial\theta} = 0. \tag{A2.3}$$

The lefthand side is the sum of an r-dependent and an r-independent term and has the constant value 0. Therefore the value of the first term cannot change with r; putting this term equal to C, the separation constant, we can write:

$$r^2\frac{d^2U}{dr^2} + 2r\frac{dU}{dr} - CU = 0, \tag{A2.4}$$

and:

$$\frac{d^2V}{d\theta^2} + \operatorname{cotg}\theta\frac{dV}{d\theta} + CV = 0. \tag{A2.5}$$

Both differential equations can be solved in a straightforward manner. Assuming U to have the form:

$$U(r) = r^\alpha, \tag{A2.6}$$

we obtain upon substitution into (A2.4):

$$\alpha(\alpha - 1) + 2\alpha - C = 0, \tag{A2.7}$$

provided r does not reach the values 0 or ∞.

Eqn. (A2.7) is a quadratic equation in α with solutions α_1 and α_2. Thus, we find:

$$U(r) = ar^{\alpha_1} + br^{\alpha_2} \tag{A2.8}$$

where a and b are arbitrary constants. (A2.8) is the general solution of the ordinary differential equation, eqn. (A2.4), since it contains two adjustable constants.

Eqn. (A2.5) can be reduced to Legendre's differential equation by the substitutions $\cos\theta = x$ and $V(\theta) = y(x)$. Using the relations:

$$\frac{\mathrm{d}}{\mathrm{d}\theta} = -\sin\theta\frac{\mathrm{d}}{\mathrm{d}x} \quad \text{and} \quad \frac{\mathrm{d}^2}{\mathrm{d}\theta^2} = -\cos\theta\frac{\mathrm{d}}{\mathrm{d}x} + \sin^2\theta\frac{\mathrm{d}^2}{\mathrm{d}x^2},$$

we obtain:

$$(1 - x^2)\frac{\mathrm{d}^2y}{\mathrm{d}x^2} - 2x\frac{\mathrm{d}y}{\mathrm{d}x} + Cy = 0. \tag{A2.9}$$

The general solution of this equation is (see Appendix III):

$$y(x) = AP_\nu(x) + BQ_\nu(x), \tag{A2.10}$$

where P_ν and Q_ν are the Legendre functions of the first and second kind, respectively. The index ν is given by $C = \nu(\nu + 1)$ and A and B are arbitrary constants.

The general solution given in eqn. (A2.10) must be used in cases when the z-axis ($\cos\theta = x = \pm 1$) is excluded from the region where the potential has to be calculated. As already mentioned, this is not the case in the potential problems treated in this book; therefore our continuity requirements on the potential require $y(x)$ to be finite and continuous for $x = \pm 1$. Since both $P_\nu(x)$ with fractional index ν and $Q_\nu(x)$ for all values of ν have singularities at $x = \pm 1$, we can use as our general solution:

$$y(x) = AP_n(x), \tag{A2.11}$$

where n is an integer and P_n is a Legendre polynomial. Thus, we also have:

$$C = n(n + 1), \tag{A2.12}$$

which leads to the following solution of eqn. (A2.7):

$$\alpha_1 = n, \qquad \alpha_2 = -(n + 1). \tag{A2.13}$$

Collecting our results, we may write with (A2.8), (A2.11), and (A2.13):

$$\phi_n(r, \theta) = UV = (a_n r^n + b_n r^{-(n+1)})P_n(\cos\theta). \tag{A2.14}$$

Here, $a_n = Aa$ and $b_n = Ab$; the index n is added to emphasize the fact that this solution belongs to a particular choice for the integer n.

Since the differential equation (A2.1) is linear and homogeneous, the solutions are linearly independent and any linear combination of solutions of the form (A2.14) will again be a solution, satisfying the continuity requirements.

Therefore we can take as our general solution a linear combination of solutions of the form (A2.14) with undetermined coefficients a_n and b_n:

$$\phi(r, \theta) = \sum_{n=0}^{\infty} (a_n r^n + b_n r^{-(n+1)}) P_n(\cos \theta). \tag{A2.15}$$

The coefficients a_n and b_n are determined from the values of the potential at the boundary, or, when appropriate, from the asymptotic behaviour at infinity.

With the help of the general theory of differential equations,[2] it can be shown that it is always possible to determine the appropriate coefficients a_n and b_n for a given problem, so that (A2.15) is indeed the general solution of Laplace's equation in spherical coordinates, when $\partial\phi/\partial\varphi = 0$.

It is sometimes required to find the potential in a region where a point charge or a point dipole or, in general, an ideal multipole is present. Formally, we now have to use Poisson's instead of Laplace's equation (see section 2a). However, it is possible instead to exclude the source-point from our region by surrounding it with a closed surface and prescribing the values of the potential on this surface. An equivalent method, which accounts for the source completely, is to take the appropriate coefficients a_n or b_n corresponding to the source-strength (see section 4, p. 37). This method is used in sections 9f and 9h, and in section 18.

The case of spherical symmetry is easily obtained from the general solution (A2.15), by noting that all Legendre polynomials $P_n(\cos \theta)$ depend on θ except the first one, P_0 (see Appendix III). Thus for spherical symmetry we have:

$$\phi(r) = a_0 + b_0 r^{-1}. \tag{A2.16}$$

§3. The three-dimensional δ-function

From (A2.16) we infer:

$$\Delta \frac{1}{r} = 0 \quad (r \neq 0), \tag{A2.17}$$

as can be verified by direct calculation. The condition $r \neq 0$ has to be added, because $1/r$ becomes infinite at $r = 0$.

Sometimes, however, it is necessary to know the behaviour of $\Delta(1/r)$ when r goes to zero. Although $\Delta(1/r)$ is undefined for $r = 0$, it is possible to find the volume integral of $\Delta(1/r)$ over a region containing the origin. Using the divergence theorem for a sphere with radius a around the origin, we have:

$$\iiint\limits_{\text{sphere}} \Delta\frac{1}{r}\,\mathrm{d}v = \iiint\limits_{\text{sphere}} \operatorname{div}\operatorname{grad}\frac{1}{r}\,\mathrm{d}v = \oiint\limits_{\substack{\text{surface}\\\text{sphere}}} \left(\operatorname{grad}\frac{1}{r}\right)\cdot \boldsymbol{dS}. \tag{A2.18}$$

Introducing spherical coordinates, this reduces to:

$$\iiint\limits_{\text{sphere}} \Delta\frac{1}{r}\,\mathrm{d}v = -\int_0^{\pi}\int_0^{2\pi} \frac{1}{a^2}a^2 \sin\theta\,\mathrm{d}\theta\,\mathrm{d}\varphi = -4\pi. \tag{A2.19}$$

The non-zero value of the integral is caused by the behaviour of $\Delta(1/r)$ at the origin, since $\Delta(1/r) = 0$ for $r \neq 0$. Thus the value will be independent of the shape of the region of integration.

Strictly speaking, the divergence theorem need not be applicable, since the function grad $(1/r)$ is singular for $r = 0$. It is possible, however, to take a function $\lim_{b\to 0}(1/r)(1 - \exp(-r/b))$ which differs negligibly from $1/r$. For this function the divergence theorem may be applied.

If we now consider the function:

$$\delta(\boldsymbol{r}) = -\frac{1}{4\pi}\Delta\frac{1}{r}, \tag{A2.20}$$

we see that it is zero for $r \neq 0$, and has a singularity for $r = 0$ of such a form that the volume integral over a region containing the origin is equal to 1. This function is called the three-dimensional δ-function, because of its analogy with the one-dimensional δ-function $\delta(x)$.

The function $\delta(x)$ is defined by the properties:

$$\left.\begin{aligned} \delta(x) &= 0 \quad \text{for} \quad x \neq 0 \\ \int_{\mathrm{I}} \delta(x)\,\mathrm{d}x &= 1 \\ \int_{\mathrm{I}} \delta(x) f(x)\,\mathrm{d}x &= f(0) \end{aligned}\right\}, \tag{A2.21}$$

where I is any interval containing the origin $x = 0$.

For the three-dimensional δ-function the third property also holds, so that we have:

$$\iiint_V f(\boldsymbol{r})\delta(\boldsymbol{r})\,\mathrm{d}v = f(0), \tag{A2.22}$$

when V contains the origin.

Although δ-functions themselves are not functions in the mathematical sense but only symbols occurring in integrands, they may be treated in almost all physical problems as if they were functions.[3]

§4. Poisson's equation

The solution of Poisson's equation for the potential ϕ, $\Delta\phi = -4\pi\rho$, can be found easily with the help of the three-dimensional δ-function introduced in the last section.

If we substitute $\Phi = \phi$ and $\Psi = 1/r$ into the symmetric theorem of Green, eqn. (A1.36), we find with the help of (A2.20):

$$\iiint_V \left\{\frac{-4\pi\rho}{r} + \phi 4\pi\delta(\boldsymbol{r})\right\}\mathrm{d}v = \oiint_S \left(\frac{1}{r}\nabla\phi + \phi\frac{\boldsymbol{r}}{r^3}\right)\cdot \boldsymbol{dS}. \tag{A2.23}$$

With (A2.22) this leads to:

$$\phi(0) = \iiint_V \frac{\rho}{r}\mathrm{d}v + \frac{1}{4\pi}\oiint_S \left(\frac{1}{r}\nabla\phi + \phi\frac{\boldsymbol{r}}{r^3}\right)\cdot \boldsymbol{dS}. \tag{A2.24}$$

When the volume of integration is extended to infinity, and ϕ is assumed to go to zero as fast as $1/r$ or faster for $r \to \infty$, the surface integral vanishes, and we have:

$$\phi(0) = \iiint_\infty \frac{\rho}{r}\mathrm{d}v. \tag{A2.25}$$

This is the general solution of Poisson's equation.

Formally it should also include the general solution of the homogeneous equation, $\Delta\phi = 0$. This is still the case in eqn. (A2.24), where the surface integral represents the solution of $\Delta\phi = 0$, Laplace's equation. However, when we take the limit $V \to \infty$ and require the contribution to the potential due to the solution of the homogeneous equation to remain finite everywhere, we see from eqn. (A2.15) that $a_n = b_n = 0$, *i.e.* this contribution vanishes. Therefore, if we take all sources of the potential into account by integrating over the whole of space, we find the correct expression for the potential.

When the only source is a point charge e at a point $\boldsymbol{r}_e$, the charge distribution is given by:

$$\rho = e\delta(\boldsymbol{r} - \boldsymbol{r}_e). \tag{A2.26}$$

Substitution into (A2.25) immediately leads to the well-known equation:

$$\phi(0) = \frac{e}{r_e} \tag{A2.27}$$

(Compare eqn. (1.17)).

If the potential at an arbitrary point $\boldsymbol{r}_0$ has to be calculated, eqn. (A2.25) must be transformed to:

$$\phi(\boldsymbol{r}_0) = \iiint_\infty \frac{\rho(\boldsymbol{r}')}{|\boldsymbol{r}' - \boldsymbol{r}_0|} \mathrm{d}v', \tag{A2.28}$$

where $r = |\boldsymbol{r}' - \boldsymbol{r}_0|$. Dropping the index 0, we have for the general solution of Poisson's equation:

$$\phi(\boldsymbol{r}) = \iiint_\infty \frac{\rho(\boldsymbol{r}')}{|\boldsymbol{r}' - \boldsymbol{r}|} \mathrm{d}v'. \tag{A2.29}$$

Here, the unprimed variables denote field points and the primed variables source points.

References

1. W. K. H. Panovsky and M. Phillips, *Classical Electricity and Magnetism*, Addison-Wesley, Reading 1962, p. 42.
2. R. Courant and D. Hilbert, *Methods of Mathematical Physics*, Vol. II, Interscience, New York 1962, pp. 290 and 306.
3. Ref. 2, p. 766.

APPENDIX III

SOME PROPERTIES OF THE LEGENDRE POLYNOMIALS

§1. The generating function and recursion formulas

The Legendre polynomials $P_n(\cos\theta)$ were defined in section 3 by the equation:

$$F(x, z) = \frac{1}{\sqrt{1 - 2xz + z^2}} = \sum_{n=0}^{\infty} P_n(x)z^n, \qquad \text{(A3.1)}$$

where $x = \cos\theta$ and $0 < z < 1$. The function $F(x, z)$ is called the generating function.

With the help of the generating function three well-known recursion formulas for the Legendre polynomials can be derived. First we differentiate (A3.1) with respect to z:

$$\frac{\partial F}{\partial z} = -\frac{z - x}{(1 - 2xz + z^2)^{3/2}} = \sum_{n=0}^{\infty} nP_n(x)z^{n-1}. \qquad \text{(A3.2)}$$

Substituting (A3.1), we obtain:

$$-(z - x)\sum_{n=0}^{\infty} P_n(x)z^n = (1 - 2xz + z^2)\sum_{n=0}^{\infty} nP_n(x)z^{n-1}. \qquad \text{(A3.3)}$$

For each value of n, the coefficients of z^n on both sides must be equal. Thus, we find for the coefficient of z^n:

$$-P_{n-1}(x) + xP_n(x) = (n + 1)P_{n+1}(x) - 2nxP_n(x) + (n - 1)P_{n-1}(x),$$

or:

$$(n + 1)P_{n+1}(x) - (2n + 1)xP_n(x) + nP_{n-1}(x) = 0. \qquad \text{(A3.4)}$$

With the help of this recursion formula all Legendre polynomials can be expressed in terms of the first two, *e.g.*:

$$P_2(x) = \tfrac{1}{2}[3xP_1(x) - P_0(x)],$$

$$P_3(x) = \tfrac{1}{3}[5x\{\tfrac{3}{2}xP_1(x) - \tfrac{1}{2}P_0(x)\} - 2P_1(x)]$$

$$= \tfrac{1}{3}[\tfrac{1}{2}(15x^2 - 4)P_1(x) - \tfrac{5}{2}xP_0(x)].$$

We can also differentiate the generating function with respect to x, obtaining:

$$\frac{\partial F}{\partial x} = \frac{z}{(1 - 2xz + z^2)^{3/2}} = \sum_{n=0}^{\infty} \frac{\mathrm{d}P_n(x)}{\mathrm{d}x} z^n. \tag{A3.5}$$

Substitution of (A3.1) now leads to:

$$z \sum_{n=0}^{\infty} P_n(x) z^n = (1 - 2xz + z^2) \sum_{n=0}^{\infty} \frac{\mathrm{d}P_n(x)}{\mathrm{d}z} z^n. \tag{A3.6}$$

Comparing coefficients of z^{n+1}, we find:

$$P_n(x) = \frac{\mathrm{d}P_{n+1}(x)}{\mathrm{d}x} - 2x\frac{\mathrm{d}P_n(x)}{\mathrm{d}x} + \frac{\mathrm{d}P_{n-1}(x)}{\mathrm{d}x}. \tag{A3.7}$$

We can also compare (A3.2) and (A3.5), and equate:

$$z \sum_{n=0}^{\infty} n P_n(x) z^{n-1} = (x - z) \sum_{n=0}^{\infty} \frac{\mathrm{d}P_n(x)}{\mathrm{d}x} z^n. \tag{A3.8}$$

Equating coefficients of z^n we obtain:

$$nP_n(x) = x\frac{\mathrm{d}P_n(x)}{\mathrm{d}x} - \frac{\mathrm{d}P_{n-1}(x)}{\mathrm{d}x}. \tag{A3.9}$$

Eliminating $x\dfrac{\mathrm{d}P_n(x)}{\mathrm{d}x}$ from (A3.7) and (A3.9), we find:

$$(2n + 1)P_n(x) = \frac{\mathrm{d}P_{n+1}(x)}{\mathrm{d}x} - \frac{\mathrm{d}P_{n-1}(x)}{\mathrm{d}x}. \tag{A3.10}$$

§2. Legendre's differential equation

The Legendre polynomials, defined by (A3.1), satisfy a differential equation called Legendre's differential equation. This can be shown with the help of the recursion formulas (A3.9) and (A3.10). Differentiating (A3.9) with respect to x, we find:

$$n\frac{\mathrm{d}P_n(x)}{\mathrm{d}x} = x\frac{\mathrm{d}^2P_n(x)}{\mathrm{d}x^2} + \frac{\mathrm{d}P_n(x)}{\mathrm{d}x} - \frac{\mathrm{d}^2P_{n-1}(x)}{\mathrm{d}x^2}. \tag{A3.11}$$

Raising in eqn. (A3.11) the index n to $(n + 1)$ we also have:

$$(n+1)\frac{dP_{n+1}(x)}{dx} = x\frac{d^2P_{n+1}(x)}{dx^2} + \frac{dP_{n+1}(x)}{dx} - \frac{d^2P_n(x)}{dx^2}. \qquad \text{(A3.12)}$$

If we now multiply (A3.11) through with x, and add the result to (A3.12), we have:

$$(n-1)x\frac{dP_n(x)}{dx} + n\frac{dP_{n+1}(x)}{dx}$$

$$= (x^2-1)\frac{d^2P_n(x)}{dx^2} + x\left\{\frac{d^2P_{n+1}(x)}{dx^2} - \frac{d^2P_{n-1}(x)}{dx^2}\right\}. \qquad \text{(A3.13)}$$

The last term can be simplified, since we note that it is equal to x times the derivative of the righthand side of (A3.10). Substituting this and rearranging, we obtain:

$$(1-x^2)\frac{d^2P_n(x)}{dx^2} - (n+2)x\frac{dP_n(x)}{dx} + n\frac{dP_{n+1}(x)}{dx} = 0. \qquad \text{(A3.14)}$$

$\frac{dP_{n+1}(x)}{dx}$ can be expressed in terms of $P_n(x)$ and $\frac{dP_n(x)}{dx}$ by combining (A3.10) and (A3.9), which leads to:

$$\frac{dP_{n+1}(x)}{dx} = (n+1)P_n(x) + x\frac{dP_n(x)}{dx}. \qquad \text{(A3.15)}$$

Substituting this into (A3.14), we finally arrive at:

$$(1-x^2)\frac{d^2P_n(x)}{dx^2} - 2x\frac{dP_n(x)}{dx} + n(n+1)P_n(x) = 0. \qquad \text{(A3.16)}$$

This is Legendre's differential equation:

$$(1-x^2)\frac{d^2y}{dx^2} - 2x\frac{dy}{dx} + n(n+1)y = 0,$$

or:

$$\frac{d}{dx}\left[(1-x^2)\frac{dy}{dx}\right] + n(n+1)y = 0. \qquad \text{(A3.17)}$$

From (A3.16) it follows that the Legendre polynomials P_n are particular solutions of the Legendre equation when n is an integer.

§3. The solution of Legendre's equation

The general solution of Legendre's equation on the interval $-1 \leqslant x \leqslant +1$ can be found by assuming the solution $y(x)$ to be expressed as a power series in x:

$$y(x) = \sum_{j=0}^{\infty} c_j x^j. \tag{A3.18}$$

Provided that this series converges, we can substitute it into Legendre's equation (A3.17) to find a relation for the coefficients c_j. We shall write ν instead of n in (A3.17) so as not to prejudge the issue of whether or not ν is an integer. When the differentiations have been performed on the power series, we obtain:

$$\sum_{j=0}^{\infty} j(j-1)c_j x^{j-2} - \sum_{j=0}^{\infty} \{j(j-1) + 2j - \nu(\nu+1)\} c_j x^j = 0. \tag{A3.19}$$

Equating the coefficients of the same power in x we obtain the recurrence relation:

$$c_{j+2} = \frac{j(j+1) - \nu(\nu+1)}{(j+1)(j+2)} c_j. \tag{A3.20}$$

Thus, all coefficients with even indices may be expressed in terms of c_0, and all coefficients with odd indices in terms of c_1:

$$y(x) = c_0\left[1 + \frac{-\nu(\nu+1)}{2}x^2 + \frac{6-\nu(\nu+1)}{12}\,\frac{-\nu(\nu+1)}{2}x^4 + \cdots\right] +$$

$$+ c_1\left[x + \frac{2-\nu(\nu+1)}{6}x^3 + \frac{12-\nu(\nu+1)}{20}\,\frac{2-\nu(\nu+1)}{6}x^5 + \cdots\right]. \tag{A3.21}$$

In this way the general solution of Legendre's equation is obtained as a linear combination of two power series in x with two arbitrary constants c_0 and c_1. Each series converges when $|x| < 1$; when $x = \pm 1$, however, both series will generally diverge, and the solution will have a singularity for these values of x.

From the recurrence relation (A3.20) it follows that c_{j+2} will become zero for some j when $j = \nu$, *i.e.* when ν is an integer n. All higher coefficients c_{j+4}, c_{j+6}, etc. will then also be zero, so that the series terminates at the n-th term and there will be no question of divergence whatever value x may have. When n is even, the power series with coefficients c_0 will reduce to a polynomial in x and the power series with coefficient c_1 will continue to diverge for $x = \pm 1$. When n is odd, it will be the other way around.

Thus if a solution of Legendre's equation is required that has no singularity for $x = \pm 1$, one has to take $v = n$ and either c_1 or c_0 equal to zero, depending on whether n is even or odd. The solutions obtained in this way are the Legendre polynomials defined in eqn. (A3.1) with the help of a generating function. From this it immediately follows that $P_n(x)$ is even or odd in x when n is even or odd, respectively.

When instead c_1 or c_0 are taken equal to zero for n odd or even, respectively, a power series in x is obtained that diverges for $x = \pm 1$. This power series is called the Legendre function of the second kind[1] and denoted by $Q_n(x)$ (see also eqns. (A2.10) and (4.49)).

§4. A table of the Legendre polynomials

With the help of (A3.20), taking c_0 or c_1 equal to 0 as required, all Legendre polynomials can be calculated explicitly. It is also possible to use the generating function defined in eqn. (A3.1), or, when the first two polynomials have been calculated, the recursion formula (A3.4). When a polynomial with a high value of n is required, it is easier to use the equation of Rodrigues:

$$P_n(x) = \frac{1}{2^n n!} \frac{\mathrm{d}^n}{\mathrm{d}x^n}(x^2 - 1)^n. \tag{A3.22}$$

The result of the calculations for the first eight polynomials is the following table:

$$P_0(x) = 1$$

$$P_1(x) = x$$

$$P_2(x) = \tfrac{1}{2}(3x^2 - 1)$$

$$P_3(x) = \tfrac{1}{2}(5x^3 - 3x)$$

$$P_4(x) = \tfrac{1}{8}(35x^4 - 30x^2 + 3)$$

$$P_5(x) = \tfrac{1}{8}(63x^5 - 70x^3 + 15x)$$

$$P_6(x) = \tfrac{1}{16}(231x^6 - 315x^4 + 105x^2 - 5)$$

$$P_7(x) = \tfrac{1}{16}(429x^7 - 693x^5 + 315x^3 - 35x)$$

§5. Orthogonality of the Legendre polynomials

Two real functions $\Phi(x)$ and $\Psi(x)$ are called orthogonal on an interval I when they satisfy the relation:

$$\int_{\mathrm{I}} \Phi(x)\Psi(x)\,\mathrm{d}x = 0. \tag{A3.23}$$

The Legendre polynomials are orthogonal to each other on the interval $-1 \leqslant x \leqslant +1$. This can be shown with the help of the generating function $F(x, z)$ (eqn. (A3.1)):

$$F(x, z) = \frac{1}{\sqrt{1 - 2xz + z^2}} = \sum_{n=0}^{\infty} P_n(x)z^n.$$

Using a different variable v and a different index m, we may also write:

$$F(x, v) = \frac{1}{\sqrt{1 - 2xv + v^2}} = \sum_{m=0}^{\infty} P_m(x)v^m. \tag{A3.24}$$

Now

$$\int_{-1}^{+1} F(x, z)F(x, v)\,\mathrm{d}x = \int_{-1}^{+1} \frac{\mathrm{d}x}{\sqrt{1 - 2xz + z^2}\sqrt{1 - 2xv + v^2}}$$

can be solved with the help of the relation:

$$\frac{\mathrm{d}}{\mathrm{d}x} \log\left\{\sqrt{v(1 + z^2) - 2vzx} - \sqrt{z(1 + v^2) - 2vzx}\right\}$$

$$= \frac{vz}{\sqrt{v(1 + z^2) - 2vzx}\sqrt{z(1 + v^2) - 2vzx}}.$$

Thus we find:

$$\int_{-1}^{+1} F(x, z)F(x, v)\,\mathrm{d}x = \frac{1}{\sqrt{vz}} \log \frac{\sqrt{v(1 + z^2) - 2vz} - \sqrt{z(1 + v^2) - 2vz}}{\sqrt{v(1 + z^2) + 2vz} - \sqrt{z(1 + v^2) + 2vz}}$$

$$= \frac{1}{\sqrt{vz}} \log \frac{1 + \sqrt{vz}}{1 - \sqrt{vz}} = f(vz), \tag{A3.25}$$

and:

$$f(vz) = \sum_{m=0}^{\infty} \sum_{n=0}^{\infty} \left\{ \int_{-1}^{+1} P_m(x)P_n(x)\,\mathrm{d}x \right\} v^m z^n. \qquad \text{(A3.26)}$$

Since the lefthand side of (A3.26) is a function of the product vz only, development in a power series will lead to terms of the form $c_l v^l z^l$. Therefore, the coefficients on the righthand side for $m \neq n$ are zero, or:

$$\int_{-1}^{+1} P_m(x)P_n(x)\,\mathrm{d}x = 0 \quad \text{for} \quad m \neq n. \qquad \text{(A3.27)}$$

Thus, the Legendre polynomials are orthogonal to each other on the interval $-1 \leqslant x \leqslant +1$.

When $m = n$, we have, taking $v = z$ in (A3.25) and (A3.26):

$$\sum_{n=0}^{\infty} z^{2n} \int_{-1}^{+1} \{P_n(x)\}^2\,\mathrm{d}x = \frac{1}{z} \log \frac{1+z}{1-z}. \qquad \text{(A3.28)}$$

The series development of the righthand side with the help of the well-known development for $\log(1 \pm z)$ with $|z| < 1$ now leads to:

$$\sum_{n=0}^{\infty} z^{2n} \int_{-1}^{+1} \{P_n(x)\}^2\,\mathrm{d}x = \sum_{n=0}^{\infty} \frac{2z^{2n}}{2n+1}. \qquad \text{(A3.29)}$$

Comparison of coefficients gives:

$$\int_{-1}^{+1} \{P_n(x)\}^2\,\mathrm{d}x = \frac{2}{2n+1}. \qquad \text{(A3.30)}$$

Eqns. (A3.27) and (A3.30) can now be expressed by one relation:

$$\int_{-1}^{+1} P_m(x)P_n(x)\,\mathrm{d}x = \frac{2}{2n+1}\delta_{mn}, \qquad \text{(A3.31)}$$

where δ_{mn}, the Kronecker delta, is equal to zero for $m \neq n$ and equal to one for $m = n$.

Reference

1. E. W. Hobson, *The Theory of Spherical and Ellipsoidal Harmonics*, University Press, Cambridge 1955.

AUTHOR INDEX

SUBJECT INDEX

CHEMICAL NAME INDEX